Parametric Resonance in Dynamical Systems

Thor I. Fossen • Henk Nijmeijer
Editors

Parametric Resonance in Dynamical Systems

 Springer

Editors
Thor I. Fossen
Department of Engineering Cybernetics
Norwegian University of Science
and Technology
NO-7491 Trondheim
Norway
Thor.Fossen@itk.ntnu.no

Henk Nijmeijer
Department of Mechanical Engineering
Eindhoven University of Technology
PO Box 513
5600 MB Eindhoven
The Netherlands
h.nijmeijer@tue.nl

ISBN 978-1-4614-1042-3   e-ISBN 978-1-4614-1043-0
DOI 10.1007/978-1-4614-1043-0
Springer New York Dordrecht Heidelberg London

Library of Congress Control Number: 2011943003

Printed on acid-free paper

Springer is part of Springer Science+Business Media (www.springer.com)

Preface

Parametric Resonance in Dynamical Systems contains a collection of contributions presented at an invited workshop with the same name held from June 22–26, 2011 in Longyearbyen, Svalbard, Norway.

The subject of the book is parametric resonance in marine and mechanical systems with focus on detection, mathematical modeling, and control. The book contains new results on modeling, detection, and control of parametric resonance and it is a supplement to engineers who are familiar with nonlinear systems.

What is Parametric Resonance?

Parametric resonance is a phenomenon not caused by external excitation, but by time-varying changes in the parameters. The archetypical example is the *Mathieu–Hill equation*:

$$\ddot{y} + a(t)y = 0$$

where $a(t + T) = a(t)$. If the period T or $2T$ is an integer multiple of the natural period N, a resonance occurs causing the origin to become unstable.

Not only mechanical systems, vehicles, motorcycles, aircraft, and marine craft but also micro–electro-mechanical systems are prone to parametric resonance. Sparse offshore platforms and ships also exhibit parametric resonance. For ships, parametric resonance is known to occur in roll in certain conditions. The resulting heavy roll motion, which can reach 30–40 degrees of roll angle, may bring the vessel into conditions dangerous for the ship, the cargo, and the crew. Container ships, fishing vessels, and cruise ships are also known to be prone to parametric roll and several incidents have been reported with significant damage to cargo as well as structural damages for millions of dollars. The origin of this unstable motion is the time-varying geometry of the submerged hull-more specific, time-varying changes in water area produce periodic variations of the transverse stability properties of the ship. This is seen as periodic oscillations of the ship's meta-centric height and consequently the spring stiffness. Other examples feature similar effects as described in the aforementioned ship example.

Can Parametric Resonance be Controlled?

There are a few potential methods for controlling the parametric resonance phenomenon. Detuning the resonance condition by (semi-)active control is possible if the control objective can be properly defined. Alternatively, feedback can be used to provide additional damping. In order to understand the effect of detuning and active control, it is necessary to study the *Poincare maps* and *Ince–Strutt* diagrams for bounded and unbounded solutions of the *Mathieu–Hill equation*. This gives insight in how nonlinear observers and controllers can be designed for systems exhibiting parametric oscillations. However, the subject of parametric resonance and passive and/or active control of it, is still far from being fully understood, and it forms the theme of this book. Our particular aim is to bring together contributions and insights from the different disciplines where parametric resonance occurs. It is our belief that this will be of great importance for researchers in these disciplines.

Acknowledgments

We are grateful to our sponsors, the:

- *Center for Ships and Ocean Structures (CeSOS)* at the Norwegian University of Science and Technology, Trondheim, Norway
- *Strategic University Program (SUP) on Control, Information and Communication Systems for Environmental and Safety Critical Systems*, Department of Engineering Cybernetics, Norwegian University of Science and Technology, Trondheim, Norway

who made this possible.

Svalbard, Norway   Thor I. Fossen
Eindhoven, Netherlands   Henk Nijmeijer

Contents

Contributors

Vadim Belenky David Taylor Model Basin, 9500 MacArthur Blvd., West Bethesda, MD 20817, USA, vadim.belenky@navy.mil

Mogens Blanke Department of Electrical Engineering, Technical University of Denmark, DTU Electrical Engineering, DK 2800 Kgs. Lyngby, Denmark

Center for Ships and Ocean Structures, Norwegian University of Science and Technology, NO-7491 Trondheim, Norway, mb@elektro.dtu.dk

Dominik A. Breu Centre for Ships and Ocean Structures, Norwegian University of Science and Technology, NO-7491 Trondheim, Norway, breu@itk.ntnu.no

Gabriele Bulian Department of Mechanical Engineering and Naval Architecture, University of Trieste, Via A. Valerio 10, 34127 Trieste, Italy, gbulian@units.it

Le Feng Centre for Ships and Ocean Structures, Norwegian University of Science and Technology, NO-7491 Trondheim, Norway, feng.le@ntnu.no

Rob H.B. Fey Department of Mechanical Engineering, Eindhoven University of Technology, PO Box 513, 5600 MB Eindhoven, The Netherlands, R.H.B.Fey@tue.nl

Thor I. Fossen Department of Engineering Cybernetics, Norwegian University of Science and Technology, NO-7491 Trondheim, Norway

Centre for Ships and Ocean Structures, Norwegian University of Science and Technology, NO-7491 Trondheim, Norway, fossen@ieee.org

Alberto Francescutto Department of Mechanical Engineering and Naval Architecture, University of Trieste, Via A. Valerio 10, 34127 Trieste, Italy, francesc@units.it

Roberto Galeazzi Department of Electrical Engineering, Technical University of Denmark, DTU Electrical Engineering, DK 2800 Kgs. Lyngby, Denmark

Center for Ships and Ocean Structures, Norwegian University of Science and Technology, NO-7491 Trondheim, Norway, rg@elektro.dtu.dk

Hirotada Hashimoto Department of Naval Architecture and Ocean Engineering, Osaka University, 2-1 Yamadaoka, Suita, Osaka 565-0871, Japan, h_hashi@naoe.eng.osaka-u.ac.jp

Christian Holden Centre for Ships and Ocean Structures, Norwegian University of Science and Technology, NO-7491 Trondheim, Norway, c.holden@ieee.org

Mikael Huss Wallenius Marine AB, PO Box 17086, S-104 62, Stockholm, Sweden, mikael.huss@walleniusmarine.com

Jørgen Juncher Jensen DTU Mechanical Engineering, Department of Mechanical Engineering, Technical University of Denmark, Bygning 403, Niels Koppels Alle, DK 2800 Kgs. Lyngby, Denmark, jjj@mek.dtu.dk

C. Stefan Kraaij IHC Lagersmit BV, PO Box 5, 2960 AA Kinderdijk, The Netherlands, CS.Kraaij@ihcmerwede.com

Niels J. Mallon Centre for Mechanical and Maritime Structures, TNO Built Environment and Geosciences, PO Box 49, 2600 AA Delft, The Netherlands, Niels.Mallon@tno.nl

Marcelo A.S. Neves LabOceano, COPPE, Universidade Federal do Rio de Janeiro, Rio de Janeiro, Brazil, masn@peno.coppe.ufrj.br

Henk Nijmeijer Department of Mechanical Engineering, Eindhoven University of Technology, PO Box 513, 5600 MB Eindhoven, The Netherlands, H.Nijmeijer@tue.nl

Mikael Palmquist Seaware AB, PO Box 1244, SE-131 28, Nacka Strand, Sweden, mikael.palmquist@seaware.se

Kristin Y. Pettersen Department of Engineering Cybernetics, Norwegian University of Science and Technology, NO-7491 Trondheim, Norway, Kristin.Ytterstad.Pettersen@itk.ntnu.no

Niels K. Poulsen Department of Electrical Engineering, Technical University of Denmark, DTU Informatics, 2800 Lyngby, Denmark, nkp@imm.dtu.dk

Jonatan Peña-Ramírez Department of Mechanical Engineering, Eindhoven University of Technology, PO Box 513, 5600 MB Eindhoven, The Netherlands, J.Pena@tue.nl

Claudio A. Rodríguez LabOceano, COPPE, Universidade Federal do Rio de Janeiro, Rio de Janeiro, Brazil, claudiorc@peno.coppe.ufrj.br

Anders Rosén KTH Royal Institute of Technology, Centre for Naval Architecture, Teknikringen 8, SE-10044, Stockholm, Sweden, aro@kth.se

Yasuhiro Sogawa Department of Naval Architecture and Ocean Engineering, Graduate School of Engineering, Osaka University, 2-1 Yamadaoka, Suita, Osaka 565-0871, Japan, sogawa0610@hotmail.co.jp

Izumi Tsukamoto Department of Naval Architecture and Ocean Engineering, Graduate School of Engineering, Osaka University, 2-1 Yamadaoka, Suita, Osaka 565-0871, Japan, izuko-okuzi@hotmail.co.jp

Naoya Umeda Department of Naval Architecture and Ocean Engineering, Osaka University, 2-1 Yamadaoka, Suita, Osaka 565-0871, Japan, umeda@naoe.eng.osaka-u.ac.jp

Kenneth Weems Science Application International Corporation, Ste 250, 4321 Collington Rd, Bowie, MD 20716, USA, kenneth.m.weems@saic.com

Acronyms

DOF	Degree of Freedom
DNV	Det Norske Veritas
FORM	First-Order Reliability Method
IMO	International Maritime Organization
ITTC	International Towing Tank Conference
JONSWAP	Joint North Sea Wave Project
LAMP	Large-Amplitude Motions Program
LCTC	Large Car and Truck Carriers
MCS	Monte Carlo Simulation
MPC	Model Predictive Control
ODE	Ordinary Differential Equation
PAD	Parametric Amplification Domain
PCTC	Pure Car and Truck Carriers
PDE	Partial Differential Equation
PSD	Power Spectral Density
RO–RO	Roll-On/Roll-Off
SOLAS	The International Convention for the Safety of Life at Sea
STF	The Salvesen, Tuck, and Faltinsen Method

Chapter 1
An Introduction to Parametric Resonance

Jonatan Peña-Ramírez, Rob H.B. Fey, and Henk Nijmeijer

1.1 Introduction

In many engineering, physical, electrical, chemical, and biological systems, oscillatory behavior of the dynamic system due to periodic excitation is of great interest.
Two kinds of oscillatory responses can be distinguished: forced oscillations and
parametric oscillations. Forced oscillations appear when the dynamical system is
excited by a periodic input. If the frequency of an external excitation is close to
the natural frequency of the system, then the system will experience *resonance*, i.e.
oscillations with a large amplitude. Parametric oscillations are the result of having
time-varying (periodic) parameters in the system. In this case, the system could
experience *parametric resonance*, and again the amplitude of the oscillations in the
output of the system will be large.

Systems with time-varying parameters are called parametrically excited systems
[19]. A very classical example of a parametrically excited system is a swing. To
increase the amplitude of the motion, the person must crouch in the extreme position
and sit straight up in the middle position. Consequently, the distance between the
hanging point and the center of gravity of the person varies periodically. This system
(person on a swing) can be seen as pendulum with varying length [24].

In this book, many of the chapters deal with a specific class of dynamical systems,
namely with the dynamics describing the motion of a ship sailing in the ocean.
Indeed, a ship can be seen as an autoparametric system, where the dynamics corresponding to the roll motion are parametrically excited by the other motions in the
ship and under some circumstances, it will experience parametric resonance known
as *parametric roll*, which consists of large oscillations in the roll motion, which
may become dangerous for the ship, its cargo, and its crew. The phenomenon of

J. Peña-Ramírez (✉) • R.H.B. Fey • H. Nijmeijer
Department of Mechanical Engineering, Eindhoven University of Technology,
P.O. Box 513, 5600 MB Eindhoven, The Netherlands
e-mail: J.Pena@tue.nl; R.H.B.Fey@tue.nl; H.Nijmeijer@tue.nl

T.I. Fossen and H. Nijmeijer (eds.), *Parametric Resonance in Dynamical Systems*,
DOI 10.1007/978-1-4614-1043-0_1, © Springer Science+Business Media, LLC 2012

parametric roll is related to the periodic change of stability as the ship is sailing against the waves (with length close to the length of the ship and height exceeding a critical value) at a speed such that the wave excitation frequency (called encounter frequency) is approximately twice the natural roll frequency and the roll damping of the ship is insufficient to avoid the onset of parametric roll [9, 11].

Perhaps the lead in research on parametric roll was taken by William Froude (1810–1879), who in 1861 discovered that the roll angle can increase rapidly when the period of the ship is in resonance with the period of wave encounter [8]. He also came to the conclusion that the roll motion is not produced by the waves hitting the side of the hull, but rather because of the pressure of the waves acting on the hull. For a historical note on the early days of parametric roll, the reader is referred to [6] and [10]. It is worthwhile mentioning that in 1998 there was an incident with a post-Panamax C11 class container ship, which was caught by a violent storm and experienced parametric roll with roll angles close to 40°. As a consequence, one third of the on-deck containers were lost overboard and a similar amount were severely damaged [9]. This event converted the study of parametric roll into a hot topic for research.

When studying parametric resonance in a dynamical system, at some point in the analysis, a very familiar equation will pop up. We refer to *Mathieu's equation*, which is a special case of second order differential equation with a periodic coefficient. Therefore, this introductory chapter is devoted to present the Mathieu equation and some of its applications. The idea is to motivate the reader to investigate the properties of this equation, its solutions, and the stability of the solutions and then apply it in practical problems. The last part of this chapter presents a short introduction of *autoparametric systems*, which are interconnected systems where parametric resonance appears in one of the constituting subsystems due to the vibrations in one of the other constituting subsystems, which can be externally excited, parametrically excited, or self excited.

1.2 The Mathieu Equation

In 1868, M. Émile Mathieu (1835–1890) published his celebrated work about the vibrational movement of a stretched membrane having an elliptical boundary [16]. By transforming the two dimensional wave equation

$$\frac{\partial^2 V}{\partial x^2} + \frac{\partial^2 V}{\partial y^2} + c^2 V = 0 \tag{1.1}$$

into elliptical coordinates and separating it into two ordinary differential equations, Mathieu obtained the differential equation

$$\frac{d^2 u}{dt^2} + (a - 2q\cos 2t)\, u = 0. \tag{1.2}$$

This equation is called the Mathieu equation. In the physical problem studied by Mathieu the constants a and q are real. This is also true for any practical problem. Some of the properties of this equation are [3]:

- Mathieu's equation always has one odd and one even solution.
- (Floquet's theorem) Mathieu's equation always has at least one solution $x(t)$ such that $x(t + \pi) = \sigma x(t)$, where the constant σ, called the periodicity factor, depends on the parameters a and q and may be real or complex.
- Mathieu's equation always has at least one solution of the form $e^{\mu t} \phi(t)$, where the constant μ is called the periodicity exponent and $\phi(t)$ has period π.
- In the case that the parameters a and q are real, then it holds for the periodicity exponent μ that either $\text{Re}(\mu) = 0$ or $\text{Im}(\mu)$ is an integer.

As a consequence of Floquet's theorem, the general solution of (1.2) has the form

$$u(t) = Ae^{\mu t}\phi(t) + Be^{-\mu t}\phi(-t), \tag{1.3}$$

where A and B are constants of integration and μ and $\phi(t)$ are as described above. In the case that $\text{Re}(\mu) = 0$ and $\text{Im}(\mu) = r/s$, a rational fraction with $s \geq 2$, the general solution (1.3) is *periodic* and remains bounded as $t \longrightarrow \infty$ and therefore (1.3) is called a *stable solution*. If $\text{Re}(\mu) = 0$ and $\text{Im}(\mu)$ is irrational, then solution (1.3) is *aperiodic* and *stable*. In the particular case that $\text{Re}(\mu) = 0$, $\text{Im}(\mu) = n$ with n integer, solution (1.3) is classified as *unstable*. In the same way, if $\text{Re}(\mu) \neq 0$ and $\text{Im}(\mu)$ is an integer then (1.3) is an *unstable solution*. Since the value of μ depends on a and q, it follows that the solution of Mathieu's equation is stable for certain values of a and q, whereas it is unstable for other values. For a complete explanation of solutions of the Mathieu equation and its stability regions in the plane (a, q), the reader is referred to [3, 17, 27].

In the analysis of parametric resonance, it is common to use a generalized form of Mathieu's equation given by:

$$\frac{d^2x}{dt^2} + F(t)x = 0, \tag{1.4}$$

where $F(t)$ is a periodic function of time. Equation (1.4) is called Hill's equation. Its solution is of the form (1.3) and it has similar stability properties as the Mathieu equation.

By using Floquet's theorem, it follows that if $x(t)$ is an arbitrary solution of (1.4), then

$$x(t + 2\pi) = \sigma x(t). \tag{1.5}$$

Three cases are considered for (1.5) in order to determine its stability (see [27]):

$$\begin{cases} 1 \ |\sigma| > 1 & \text{then } x(t) \text{ is unstable} \\ 2 \ |\sigma| < 1 & \text{then } x(t) \text{ is stable} \\ 3 \ |\sigma| = 1 & \text{then } x(t) \text{ is unstable.} \end{cases} \tag{1.6}$$

For a complete analysis about the properties (solutions and their stability) of (1.4), the reader is again referred to [3,17,27] and to [7] for a geometrical approach.

1.3 Applications of Mathieu's Equation

The Mathieu equation can be used in practical applications where the problem at hand is either a boundary value problem or an initial value problem. An example of a boundary value problem is the solution of the wave equation expressed in elliptical coordinates. An application illustrating an initial value problem is that one of a pendulum with periodically varying length.

For the sake of clarity, three classical examples where the Mathieu equation appears are presented. The choice of the examples comes from the fact that we want to show the broad application of Mathieu's equation in several disciplines. Examples 1 and 2 correspond to the class of initial value problems, whereas Example 3 belongs to the class of boundary value problems.

1.3.1 Example 1: A Mechanical Mechanism

Figure 1.1 shows a link–mass–spring mechanism which can be viewed as a dynamical system covered by Mathieu's equation [17]. The mass m can move in the vertical direction and the spring and the links are assumed to be massless. The spring is unstretched when the mass m is at D. Moreover, it is assumed that $y/l \ll 1$. The driving force $F_0 = lf_0 \cos 2\omega t$ applied at the pin-joint B can be resolved into components, one along AB and the other along DA. Then, it follows that the force driving the mass in the vertical direction is $F_1 = (f_0 \cos 2\omega t) y$.

Using Newton's second law, it follows that the equation of motion of the system in the vertical direction is

$$m\frac{d^2y}{dt^2} + ky = F_1. \tag{1.7}$$

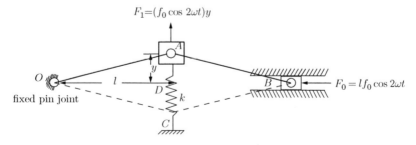

Fig. 1.1 Schematic diagram of a dynamical mechanical system covered by Mathieu's equation

Since $F_1 = (f_0 \cos 2\omega t) y$, (1.7) becomes

$$m\frac{d^2 y}{dt^2} + (k - f_0 \cos 2\omega t) y = 0. \tag{1.8}$$

Introducing a new time variable $\tau = \omega t$ leads to the standard form of the Mathieu equation

$$\frac{d^2 y}{d\tau^2} + (a - 2q\cos 2\tau) y = 0, \tag{1.9}$$

where $a = k/(\omega^2 m)$ and $q = f_0/(2\omega^2 m)$.

1.3.2 Example 2: An Oscillatory Electrical Circuit

Another interesting application of the Mathieu equation is in the analysis of oscillatory circuits having time varying parameters, which have been highly important in the development of communication systems since the introduction of the super-regenerative receiver by Edwin Howard Armstrong (1890–1954) in 1922 [2]. The example at hand is the RLC circuit depicted in Fig. 1.2, which is described in [4]. The circuit contains a coil with constant inductance L, a capacitor with constant capacitance C and a time varying resistance R which is given by the expression:

$$R = \gamma + \rho(i) + R_m \sin \omega t, \tag{1.10}$$

where γ is an ohmic resistance, $\rho(i)$ is a negative resistance (accounting for regeneration, i.e., supplying energy to the circuit to reinforce the oscillations), which is dependent on the current i and can be either less, equal, or greater than γ. Finally, $R_m \sin \omega t$ is a periodic resistance.

Using Kirchhoff's voltage law it follows that the governing differential equation of the circuit of Fig. 1.2 is

$$L\frac{di}{dt} + (\gamma + \rho(i) + R_m \sin \omega t) i + \frac{1}{C} \int idt = 0. \tag{1.11}$$

Under the assumption that $\gamma + \rho(i) \ll R_m \sin \omega t$, (1.11) verifies [4]:

$$L\frac{di}{dt} + R_m \sin \omega t\, i + \frac{1}{C} \int idt = 0. \tag{1.12}$$

After taking the time derivative of (1.12), it follows that

$$\frac{d^2 i}{dt^2} + \frac{R_m}{L} \sin \omega t \frac{di}{dt} + \left(\omega_0^2 + \frac{\omega R_m}{L} \cos \omega t\right) i = 0, \tag{1.13}$$

Fig. 1.2 Equivalent circuit of
a super-regenerative receiver

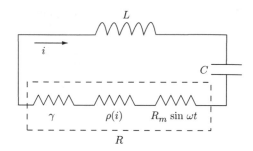

where $\omega_0 = 1/\sqrt{LC}$ is the resonance frequency of the circuit. The damping term
may be removed by the substitution $i = Iye^{(k/\omega)\cos\omega t}$ with $k = R_m/2L$. With this
substitution, (1.13) becomes

$$Ie^{(k/\omega)\cos\omega t}\left(\frac{d^2y}{dt^2} + \left(\omega_0^2 + 2k\omega\cos\omega t + \frac{k^2}{2} - \frac{k^2}{2}\cos 2\omega t\right)y\right) = 0. \qquad (1.14)$$

By considering the extreme cases $k \ll \omega$ and $k \gg \omega$, (1.14) reduces to

$$\frac{d^2y}{dt^2} + \left(\omega_0^2 + 2\omega k\cos\omega t\right)y = 0 \quad \text{and} \quad \frac{d^2y}{dt^2} + \left(\omega_0^2 + \frac{k^2}{2} - \frac{k^2}{2}\cos 2\omega t\right)y = 0, \qquad (1.15)$$

respectively. Both equations in (1.15) can be written in the standard form of Mathieu's equation by defining a new time variable. For instance, consider the second
equation in (1.15) and define $\tau = \omega t$. This leads to

$$\frac{d^2y}{d\tau^2} + (a - 2q\cos 2\tau)y = 0, \qquad (1.16)$$

where $a = (2\omega_0^2 + k^2)/2\omega^2$ and $q = k^2/4$. Clearly, (1.16) is the Mathieu equation.
   Further applications of Mathieu's equation on oscillatory electrical circuits
containing time varying parameters can be found in [5] and [17].

1.3.3 Example 3: Hydrodynamics

A considerable part of this book is dedicated to the study of parametric roll occurring
in ships sailing in the sea; therefore, in this introductory chapter, an application of
the Mathieu equation related to hydrodynamics is obvious. The example at hand
is related with the free oscillations of water in an elliptical lake (see Fig. 1.3),
whose eccentricity is close to 1. This example is described in [13]. The position
of any point referred to the major and minor axes of the lake is given by the
horizontal coordinates x and y with corresponding velocities u and v. Assume that
the coordinate system rotates at constant angular velocity ω about the z-axis (earth's

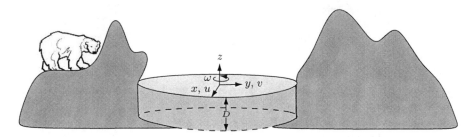

Fig. 1.3 Elliptical lake

rotation axis). It can be shown that the equations of horizontal motion (assuming infinitely small relative motion) are [14]:

$$\frac{\partial u}{\partial t} - 2\omega v = -g\frac{\partial \zeta}{\partial x} \quad \text{and} \quad \frac{\partial v}{\partial t} + 2\omega x = -g\frac{\partial \zeta}{\partial y}, \tag{1.17}$$

where g is the acceleration of gravity and ζ is the height of the free surface above its equilibrium position. Next, the equation of continuity is given by the expression

$$-\frac{\partial \zeta}{\partial t} = \frac{\partial (Du)}{\partial x} + \frac{\partial (Dv)}{\partial y}, \tag{1.18}$$

where D is the depth of the lake, which is assumed to be uniform.

Next, explicit expressions for u and v are computed by considering "perturbative" solutions for u, v, and ζ of the form $e^{i\sigma t}$ (see [12, 14]), with $\sigma = 2\pi/T$ with T being the period of tidal oscillation [25]. Hence

$$u = u_1 e^{i\sigma t} = u_1(\cos \sigma t + i\sin \sigma t) \tag{1.19}$$

$$v = v_1 e^{i\sigma t} = v_1(\cos \sigma t + i\sin \sigma t) \tag{1.20}$$

$$\zeta = \zeta_1 e^{i\sigma t} = \zeta_1(\cos \sigma t + i\sin \sigma t). \tag{1.21}$$

Substitution of (1.19)–(1.21) in (1.17)–(1.18) yields

$$i\sigma u - 2\omega v = -g\frac{\partial \zeta}{\partial x} \quad \text{and} \quad i\sigma v + 2\omega u = -g\frac{\partial \zeta}{\partial y} \tag{1.22}$$

and

$$-i\sigma \zeta = D\left(\frac{\partial u}{\partial x} + \frac{\partial v}{\partial y}\right). \tag{1.23}$$

From (1.22) follows that

$$u = \frac{g}{\sigma^2 - 4\omega^2}\left(i\sigma \frac{\partial \zeta}{\partial x} + 2\omega \frac{\partial \zeta}{\partial y}\right) \quad \text{and} \quad v = \frac{g}{\sigma^2 - 4\omega^2}\left(i\sigma \frac{\partial \zeta}{\partial y} - 2\omega \frac{\partial \zeta}{\partial x}\right).$$

$$\tag{1.24}$$

By using (1.24), (1.23) is represented as:

$$\frac{\partial^2 \zeta}{\partial x^2} + \frac{\partial^2 \zeta}{\partial y^2} + \frac{\sigma^2 - 4\omega^2}{gD} \zeta = 0. \tag{1.25}$$

Equation (1.25) is the well known two-dimensional wave equation. Since the problem at hand involves a boundary condition, which is elliptical in shape, it is convenient to express (1.25) in elliptical coordinates ξ and η, which are related to x and y in the following manner

$$x + iy = h \cosh(\xi + i\eta), \tag{1.26}$$

or

$$x = h \cosh \xi \cos \eta, \quad y = h \sinh \xi \cos \eta. \tag{1.27}$$

Expressed in elliptical coordinates, (1.25) becomes (for the transformation procedure, see [17] or [28])

$$\frac{\partial^2 \zeta}{\partial \xi^2} + \frac{\partial^2 \zeta}{\partial \eta^2} + 2k^2 (\cosh 2\xi - \cos 2\eta) \zeta = 0, \tag{1.28}$$

where $k^2 = (\sigma^2 - 4\omega^2) h^2 / 8gD$.

By substituting in (1.28) a solution of the form

$$\zeta = XY \tag{1.29}$$

with X being a function only depending on ξ and Y a function only depending on η, it follows that

$$-\frac{1}{X}\left(\frac{d^2X}{d\xi^2}\right) + 2k^2 \cosh 2\xi = \frac{1}{Y}\left(\frac{d^2Y}{d\eta^2}\right) - 2k^2 \cos 2\eta. \tag{1.30}$$

It is straightforward to see that both sides of (1.30) should be equal to the same constant, say R. In consequence, it follows that

$$\frac{d^2X}{d\xi^2} + (2k^2 \cosh 2\xi - R) X = 0 \tag{1.31}$$

$$\frac{d^2Y}{d\eta^2} + (R - 2k^2 \cos 2\eta) Y = 0. \tag{1.32}$$

Equation (1.31) is the so-called *modified* Mathieu equation [17], whereas (1.32) is the standard Mathieu equation. Then, it is clear that the problem of free oscillations of water in an elliptical lake is an example of an application of the Mathieu equation.

1.3.4   Other Modern Applications

Nowadays, the literature is very rich in applications related to the Mathieu equation. For example, this equation is considered in

- the study of the stability of structural elements such as plates and shells, which are widely used in aerospace and mechanical applications [1, 23].
- the analysis of the dynamical behavior of micro- and nano-electromechanical systems, which are very often used in actuators, sensors, and in data and communication applications [18, 21].
- the study of parametric resonance in civil structures like bridges [20].

For more applications of the Mathieu equation the reader is referred to [22] and the references there in.

As a final note, an example of parametric resonance in a biological system is briefly discussed. The squid giant axon membrane is taken as example. This axon controls part of the water jet propulsion system in squid. When considering the membrane capacitance as a periodic time varying parameter, it has been found that the membrane sensitivity to stimulation is increased due to parametric resonance [15]. However, it should be stressed that the equations describing the membrane potential response are parametrically excited first order equations, which can not be written in the form of the Mathieu equation.

1.4   Autoparametric Systems

The phenomenon of parametric resonance also occurs in a special class of non-linear dynamical systems called *autoparametric* systems. In its simplest form, an autoparametric system consists of two nonlinearly coupled subsystems. One of the subsystems (called primary system) can be externally excited, parametrically excited, or self-excited. Due to the coupling, the other subsystem (called secondary system) can be seen as a parametrically excited system. A particular feature of autoparametric systems is that for certain values of the excitation frequency and/or certain values of the parameters, the primary system will have an oscillating response, whereas the secondary system will be at rest. When this solution becomes unstable, the system will experience autoparametric resonance, i.e., the oscillations of the primary system will produce an oscillating behavior in the secondary subsystem. In some cases, the oscillations of the secondary system will grow unbounded [26].

Perhaps one of the simplest examples of an autoparametric system is given by a pendulum attached to a mass–spring–damper system, where the mass can move in the vertical direction and is driven by a harmonic force. In such case, the primary system is given by the mass–spring–damper subsystem, whereas the pendulum is considered as the secondary subsystem. For certain intervals of the excitation frequency, the behavior of the system is as follows: the mass oscillates in the vertical axis, whereas the pendulum remains at rest. However, there are intervals of the excitation frequency such that the pendulum is parametrically excited by the oscillatory mass–spring–damper system and the pendulum will no longer stay at rest but it will show an oscillating behavior.

In the study of the stability of autoparametric systems, often the Mathieu equation plays a fundamental roll. This is demonstrated by the following academic example where the primary system is assumed to be externally excited. Note that the considered system does not model any physical system; however, it is quite useful in demonstrating some of the basic properties of autoparametric systems. This example has been presented in [26].

Consider an autoparametric system (in dimensionless form) given by the set of equations:

$$\ddot{x} + \beta_1 \dot{x} + x + \alpha_1 y^2 = k\eta^2 \cos \eta t, \tag{1.33}$$

$$\ddot{y} + \beta_2 \dot{y} + q^2 y + \alpha_2 xy = 0, \tag{1.34}$$

where the primary system (1.33) is externally excited by a harmonic force, and the secondary system (1.34) is parametrically excited. The secondary system is coupled to the first system by the term $\alpha_2 xy$, whereas the primary system is coupled to the secondary system with the term $\alpha_1 y^2$. Parameters $\beta_i \in \mathbb{R}^+$ $(i = 1, 2)$ are the damping coefficients, $q = \omega_2/\omega_1$ is the ratio of the natural frequency ω_2 of the secondary system and the natural frequency ω_1 of the primary system. The amplitude of the external excitation is given by k and the driving frequency η is given by the ratio of the (dimensional) excitation frequency ω and the natural frequency of the primary system ω_1, i.e., $\eta = \omega/\omega_1$.

This system has a semi-trivial solution (i.e., a solution where $x(t)$ and $\dot{x}(t)$ have oscillatory behavior and $y(t) = \dot{y}(t) = 0$), which is determined by substituting

$$x(t) = R\cos(\eta t + \psi) \tag{1.35}$$

$$y(t) = 0 \tag{1.36}$$

into (1.33)–(1.34). This yields an expression for R, which is

$$R = \frac{k\eta^2}{\sqrt{(1 - \eta^2)^2 + \beta_1^2 \eta^2}}. \tag{1.37}$$

Then, the semi-trivial solution of (1.33)–(1.34) is given by:

$$x(t) = \frac{k\eta^2}{\sqrt{(1 - \eta^2)^2 + \beta_1^2 \eta^2}} \cos(\eta t + \psi), \quad y(t) = 0, \tag{1.38}$$

$$\dot{x}(t) = -\frac{k\eta^3}{\sqrt{(1-\eta^2)^2 + \beta_1^2 \eta^2}} \sin(\eta t + \psi), \quad \dot{y}(t) = 0. \tag{1.39}$$

The stability of the semi-trivial solution (1.38)–(1.39) is determined by inserting the expressions

$$x = R\cos(\eta t + \psi_1) + u(t), \quad y = 0 + v(t), \tag{1.40}$$

where $u(t)$ and $v(t)$ are small perturbations, into (1.33)–(1.34). This leads to the linear approximation

$$\ddot{u} + \beta_1 \dot{u} + u = 0 \tag{1.41}$$

$$\ddot{v} + \beta_2 \dot{v} + \left[q^2 + \alpha_2 R\cos(\eta t + \psi_1) \right] v = 0. \tag{1.42}$$

From (1.41) it is clear that u is asymptotically stable. In consequence, the stability of (1.38)–(1.39) is completely determined by (1.42), which indeed is the Mathieu equation (1.2). In [26], it has been found that the main instability domain of (1.42) corresponds to values of $q \approx \frac{1}{2}\eta$. Indeed, by using the averaging method, it is possible to find the boundary of the main instability region and then to determine, for which values of the amplitude of the external excitation, the semi-trivial solution (1.38)–(1.39) becomes unstable, i.e., for which values of k the response $y(t) \neq 0$.

Now, an example of an autoparametric system occurring in engineering applications is considered. When studying parametric roll resonance in ships, the dynamics describing the motions of the ship in the vertical plane (heave, pitch, and roll) can be seen as an autoparametric system, where the primary system, consisting of the dynamics of heave and pitch motion, is externally excited by the ocean waves, whereas the secondary system, corresponding to the roll dynamics, is parametrically excited by the oscillations in heave and pitch. As in the previous example, the heave–pitch–roll system also accepts a semi-trivial solution similar to (1.38)–(1.39), i.e. under certain conditions confer [11], the ship will exhibit oscillations in heave and pitch directions, whereas no oscillations in roll will appear. However, it has been found that this solution is prone to become unstable when the frequency, to which the system approaches waves, is almost twice the value of the roll natural frequency. In such case parametric roll will appear.

1.5 Conclusions

This introductory chapter has been written in order to show the reader that a solid knowledge of the Mathieu equation, its properties, and solutions will facilitate the analysis of the parametric resonance phenomenon occurring in dynamical systems with time-varying parameters. By using some examples from diverse disciplines, it has been shown that since its introduction in 1868, the Mathieu equation has been of paramount importance in the development of the theory of parametrically excited systems.

A short summary about a special class of systems called autoparametric systems has been presented. In this kind of systems, parametric resonance will occur due to the interconnection of the constituting subsystems. Therefore, the theory of autoparametric systems is a very useful framework when analyzing, for instance, the parametric roll effect, because the influence of other motions of the ship can be included into the analysis.

As a final remark, it should be clear to the reader that when analyzing a single degree of freedom dynamical system with time varying parameters, the solutions of the Mathieu equation or Hill's equation are of key importance to analyze the behavior of the system, whereas when analyzing an interconnected system, where one of the subsystems acts as a parametric excitation of the other subsystem, then the theory related to autoparametric systems should be used in order to determine the behavior of the system.

In general, the chapters contained in this book are indeed applications strongly related to the Mathieu equation or Hill's equation and/or applications, where the system under consideration belongs to the class of autoparametric systems.

References

1. Amabili, M., Paidoussis, M. P.: Review of studies on geometrically nonlinear vibrations and dynamics of circular cylindrical shells and panels, with and without fluid-structure interaction. Applied Mechanics Reviews, 56(4):349–381, (2003).
2. Armstrong, E. H.: Some recent developments of regenerative circuits. Proceedings of the Institute of Radio Engineers, 10:244–260, (1922).
3. Arscott, F.: Periodic differential equations. Pergamon Press. New York, (1964).
4. Ataka, H.: On superregeneration of an ultra-short-wave receiver. Proceedings of the Institute of Radio Engineers, 23(8):841–884, (1935).
5. Barrow, W. L., Smith, D. B., Baumann, F. W.: A further study of oscillatory circuits having periodically varying parameters. Journal of The Franklin Institute, 221(3):403–416, (1936).
6. Biran, A.: Ship hydrostatics and stability. Butterworth-Heinemann. Oxford, (2003).
7. Broer, H. W., Levi, M.: Geometrical aspects of stability theory for Hill's equations. Archive for rational mechanics and analysis, 131(3):225–240, (1995).
8. Brown, D. K.: The way of a ship in the midst of the sea: the life and work of William Froude. Periscope Publishing Ltd. ISBN-10: 1904381405, (2006).
9. France, W. N., Levadou, M., Trakle, T. W., Paulling, J. R., Michel, R. K., Moore, C.: An investigation of head-sea parametric rolling and its influence on container lashing systems, Marine Technology, 40(1):1–19, (2003).
10. Galeazzi, R.: Autonomous supervision and control of parametric roll resonance. Ph.D. Thesis, Kongens Lyngby, Denmark, (2009).
11. Holden, C., Galeazzi, R., Rodríguez, C., Perez, T., Fossen, T. I., Blanke, M., Neves, M. A. S.: Nonlinear container ship model for the study of parametric roll resonance. Modeling, identification and control, 28(4):87–103, (2007).
12. Ivanov, M. I.: Free tides in two-dimensional uniform depth basins. Fluid Dynamics, 39(5): 779–789, (2004).
13. Jeffreys, H.: The free oscillation of water in an elliptical lake. Proceedings of the London Mathematical Society (second edition), 23:455–476, (1923).

14. Lamb, H.: Hydrodynamics. Sixth edition. Cambridge University Press. Cambridge, United Kingdom (1945).
15. Markevich, N. I., Sel'kov, E. E.: Parametric resonance and amplification in excitable membranes. The Hodgkin-Huxley model. Journal of Theoretical Biology, 140(1):27–38, (1989).
16. Mathieu, É.: Mémoire sur le mouvement vibratoire d'une membrane de forme elliptique. Journal of Mathematics, 13:137–203, (1868).
17. McLachlan, N. W.: Theory and Applications of Mathieu Functions. Oxford Press, London (1951).
18. Mestrom, R. M. C., Fey, R. H. B., van Beek, J. T. M., Phan, K. L., Nijmeijer, H.: Modelling the dynamics of a MEMS resonator: simulations and experiments. Sensors and Actuators A: Physical, 142(1):306–315, (2008).
19. Nayfeh, A. H., Mook, D. T.: Nonlinear Oscillations. John Wiley & Sons. New York, (1979).
20. Piccardo, G., Tubino, F.: Parametric resonance of flexible footbridges under crowd-induced lateral excitation. Journal of Sound and Vibration, 311:353–371, (2008).
21. Rhoads, J. F., Shaw, S. W., Turner, K. L.: Nonlinear dynamics and its applications in micro- and nanoresonators. Journal of Dynamic Systems, Measurement, and Control, 132(034001):1–14, (2010).
22. Ruby, L.: Applications of the Mathieu equation. American Journal of Physics, 64(1):39–44, (1996).
23. Sahu, S. K. and Datta, P. K.: Research advances in the dynamic stability behavior of plates and shells: 1987–2005 Part I: Conservative Systems. Applied Mechanics Reviews, 60:65–75, (2007).
24. Seyranian, A. P.: The swing: parametric resonance. Journal of Applied Mathematics and Mechanics, 68(5):757–764, (2004).
25. Taylor, G. I.: Tidal oscillations in gulfs and rectangular basins. Proceedings of the London Mathematical Society, 20:148–181, (1920).
26. Tondl, A., Ruijgrok, T., Verhulst, F., Nabergoj, R.: Autoparametric resonance in mechanical systems. Cambridge University Press, Cambridge, United Kingdom (2000).
27. van der Pol B, Strutt, M. J. O.: On the stability of the solutions of Mathieu's equation. Philosophical Magazine, Series 7, 5(27):18–38, (1928).
28. Whittaker, E. T., Watson, G. N.: A Course of Modern Analysis: an introduction to the general theory of infinite processes and of analytic functions; with an account of the principal transcendental functions. Cambridge University Press, 3rd edition, Cambridge, United Kingdom (1920).

Part I
Detection and Estimation of Parametric Resonance

Chapter 2
Detection of Parametric Roll for Ships

Roberto Galeazzi, Mogens Blanke, and Niels K. Poulsen

2.1 Introduction

Observations of parametric resonance on ships were first done by Froude [9, 10] who reported that a vessel, whose frequency of oscillation in heave/pitch is twice its natural frequency in roll, shows undesirable seakeeping characteristics that can lead to the possibility of exciting large roll oscillations. Theoretical explanations appeared in the 20th century, see [26, 35] and references herein, and parametrically induced roll has been a subject in maritime research since the early 1950s [23] and [32]. The report by France et al. [8] about the *APL China* incident in October 1998 accelerated the awareness and parametric roll resonance became an issue of key concern. Døhlie [7] emphasized parametric resonance as a very concrete phenomenon, which will be able to threaten some of the giants of the sea in common passage conditions, which were previously considered to be of no danger.

Publications addressing parametric roll on container ships include [3,5,17,20,21, 24,33,34,36]. Fishing vessels were in focus in [27,28]. A main topic of this research has been to analyse the nonlinear interactions between roll and other ships' motions and develop models, which could predict vessels' susceptibility to parametric roll at the design stage. However, commercial interest is to maximize cargo capacity.

R. Galeazzi (✉) • M. Blanke
Department of Electrical Engineering, Technical University of Denmark, DTU Electrical Engineering, DK 2800 Kgs. Lyngby, Denmark

Center for Ships and Ocean Structures, Norwegian University of Science and Technology, NO-7491 Trondheim, Norway
e-mail: rg@elektro.dtu.dk; mb@elektro.dtu.dk

N.K. Poulsen
Department of Electrical Engineering, Technical University of Denmark, DTU Informatics, 2800 Lyngby, Denmark
e-mail: nkp@imm.dtu.dk

T.I. Fossen and H. Nijmeijer (eds.), *Parametric Resonance in Dynamical Systems*,
DOI 10.1007/978-1-4614-1043-0_2, © Springer Science+Business Media, LLC 2012

Hull designs have not been significantly changed and parametric resonance is left as a calculated risk. Therefore, there is a need to enhance safety against parametric roll through on-board detection and decision support systems.

First generation warning systems are based on longer horizon analysis of ship's responses and they provide polar diagrams with risk zones in speed and heading. These are found in commercial products as the SeaSense [30] and the Amarcon's OCTOPUS Resonance.[1] For detection of the resonant bifurcation mode, Holden et al. [19] proposed an observer based predictor that estimates the eigenvalues of a linear second-order oscillatory system. This algorithm issues a warning when eigenvalues have positive real parts. The method works convincingly but was designed to cope with excitation by narrow band regular waves. Irregular sea conditions were studied by McCue and Bulian [25] who used finite time Lyapunov exponents to detect the onset of parametric roll, but this method was not found to possess sufficiently robustness when validated against experimental data.

Starting from early results outlined in [12, 13] this chapter re-visits the core of the theory of parametric resonance and proposes signal-based methods for detection of parametric roll [11]. Development of a robust warning system for detecting the onset of parametric roll is discussed, and it is shown possible to obtain based solely on signals. The core of the method is shown to consist of two detection schemes: one in the frequency domain, a second in the time domain. The frequency-based detector uses an indicator of spectral correlation between pitch or heave and the roll. A time-based detector exploits the phase synchronization between the square of the roll and of pitch. A generalized likelihood ratio test (GLRT) is derived for a Weibull distribution that is observed from data and adaptation is employed to obtain robustness in reality with time-varying weather conditions. Robustness to forced roll motion is also discussed and the detection system's performance is evaluated on two data sets: model test data from towing tank experiments and data from a container vessel experiencing an Atlantic storm.

2.2 Parametric Roll – Conditions and Underlying Physics

This section presents empirical experience and introduces a mathematical treatment of parametric roll resonance.

2.2.1 Empirical Conditions

Empirical conditions have been identified that may trigger parametric roll resonance:

---

[1]http://www.amarcon.com

1. the period of the encounter wave is approximately equal to half the roll natural period ($T_e \approx \frac{1}{2}T_\phi$)
2. the wave length is approximately equal to the ship's length ($\lambda_w \approx L_{PP}$)
3. the wave height is greater than a ship-dependent threshold ($h_w > \bar{h}_s$)

When these conditions are met, and the ship sails in moderate to heavy longitudinal or oblique seas, then the wave passage along the hull and the wave excited vertical motions result in variations of the intercepted water-plane area and in turn, they change the roll restoring characteristics. The onset of parametric resonance causes a quick rise of roll oscillation, which can reach amplitudes larger than $\pm 40°$ ([6, 8]), and it may bring the vessel into conditions dangerous for cargo, crew, and hull integrity. Damages produced by parametric roll to the post-Panamax container ship *APL China* had a price tag of USD 50 millions in 1998 [15].

2.2.2   Mathematical Background

Consider a vessel sailing in moderate head regular seas and let the wave elevation be modeled as a single frequency sinusoid

$$\zeta(t) = A_w \cos(kx\cos\chi - ky\sin\chi - \omega_e t),$$

where A_w is the wave amplitude, ω_e the wave encounter frequency, k the wave number, and χ the wave encounter angle. In head seas the wave encounter angle is $\chi = 180°$, and

$$\zeta(t) = A_w \cos(kx + \omega_e t).$$

The incident wave gives rise to forces and moments acting on the hull. In head seas, conventional forced roll cannot occur since forces and moments from wave pressure on the hull have no components perpendicular to the ship, but motions in the vertical plane are clearly excited. Heaving and pitching cause periodic variations of the submerged hull geometry. In particular, during a wave passage, the intercepted water-plane area S_w changes from the still water case S_{W_0}, causing a variation of the position of the center of buoyancy [29]. This in turn gives rise to a modification of the transverse metacentric height GM and also to a new position of the metacenter M. The center of gravity G depends upon the ship's loading condition and is fixed. Consequently the periodic fluctuation of GM, which can be considered sinusoidal,

$$\mathrm{GM}(t) = \overline{\mathrm{GM}} + \mathrm{GM}_a \cos(\omega_e t)$$

influences the stability properties of the vessel through the roll restoring moment that is approximated by:

$$\tau(t) \approx \rho g \nabla \mathrm{GM}(t) \sin\phi,$$

where $\overline{GM}$ is the mean value of the metacentric height, GM_a is the amplitude of the variations of the metacentric height in waves, ρ is the water density, g is the acceleration constant of gravity, and ∇ is the displaced volume.

The following situations alternate in a periodic manner:

- a wave trough is amidships: in this case $S_W > S_{W_0}$ causing a larger restoring moment ($\tau > \tau_0$) and increased stability
- a wave crest is amidships: in this case $S_W < S_{W_0}$ inducing a smaller restoring moment ($\tau < \tau_0$) and reduced stability.

If a disturbance occurs in roll when the ship is between the wave crest and trough at amidships position, then its response will be greater than in calm water since it is approaching a situation of instantaneous increased stability. Therefore the vessel will roll back to a larger angle than it would have done in calm water. After the first quarter of the roll period T_ϕ the vessel has rolled back to the zero degree attitude but it continues towards port side due to the inertia. However now the ship encounters a wave crest amidships, which determines a reduced restoring moment with respect to that in calm water; therefore the ship rolls to a larger angle than it would have done in calm water. As a result the roll angle is increased again over the second quarter of the roll period, reaching a higher value than at the end of the first quarter. This alternate sequence of instantaneous increased and reduced restoring moment causes the roll angle to keep increasing unless some other factors start counteracting it.

Formally, this can be described as the interaction between coupled modes of an autoparametric system, where the primary system is externally forced by a sinusoidal excitation. In particular, let θ be the pitch angle, and ϕ be the roll angle. Then the system reads

$$\left(I_y - M_{\ddot\theta}\right)\ddot\theta + M_{\dot\theta}\dot\theta + M_\theta\theta + M_{\phi^2}\phi^2 = M_w\cos(\omega_w t), \tag{2.1}$$

$$\left(I_x - K_{\ddot\phi}\right)\ddot\phi + K_{\dot\phi}\dot\phi + K_\phi\phi + K_{\phi^3}\phi^3 + K_{\phi\theta}\phi\theta = 0, \tag{2.2}$$

where I_x, I_y are the rigid body inertia in roll and pitch; $K_{\ddot\phi}$, $M_{\ddot\theta}$ are the added inertia; $K_{\dot\phi}$, $M_{\dot\theta}$ are the linear damping due to viscous effects; K_ϕ, K_{ϕ^3}, $K_{\phi\theta}$, M_θ, and M_{ϕ^2} are the linear and the nonlinear coefficients of the restoring moments due to hydrostatic actions; M_w, ω_w are the amplitude and frequency of the wave induced pitch moment. The model introduced above is not meant to precisely describe the hull–wave interactions that determine the onset and development of parametric roll on ships, but it simply tries to cast the roll–pitch dynamics within the autoparametric resonance framework along the lines of [31].

System (2.1)–(2.2) can be rewritten as:

$$\ddot\theta + \mu_1\dot\theta + \omega_1^2\theta + \alpha_1\phi^2 = \kappa\cos(\omega_w t) \tag{2.3}$$

$$\ddot\phi + \mu_2\dot\phi + \omega_2^2\phi + \varepsilon\phi^3 + \alpha_2\phi\theta = 0 \tag{2.4}$$

with coefficients

$$\mu_1 = \frac{M_{\dot{\theta}}}{I_y - M_{\ddot{\theta}}} \ , \quad \omega_1 = \sqrt{\frac{M_{\theta}}{I_y - M_{\ddot{\theta}}}} \ , \quad \alpha_1 = \frac{M_{\phi^2}}{I_y - M_{\ddot{\theta}}} \ , \quad \kappa = \frac{M_w}{I_y - M_{\ddot{\theta}}}$$

$$\mu_2 = \frac{K_{\dot{\phi}}}{I_x - K_{\dot{\phi}}} \ , \quad \omega_2 = \sqrt{\frac{K_{\phi}}{I_x - K_{\ddot{\phi}}}} \ , \quad \alpha_2 = \frac{K_{\phi\theta}}{I_x - K_{\ddot{\phi}}} \ , \quad \varepsilon = \frac{K_{\phi^3}}{I_x - K_{\ddot{\phi}}} \ .$$

The so-called *semi-trivial* solution of the system (2.3)–(2.4) can be determined by posing

$$\theta(t) = \theta_0 \cos(\omega_w t + \varsigma) \tag{2.5}$$

$$\phi(t) = 0 \tag{2.6}$$

and by substituting $\theta(t)$ and $\phi(t)$ into (2.3) and (2.4) it yields

$$\theta_0 = \frac{\kappa}{\sqrt{\left(\omega_1^2 - \omega_w^2\right)^2 + \mu_1^2 \omega_w^2}}. \tag{2.7}$$

The stability of the semi-trivial solution is investigated by looking at its behavior in a neighborhood defined as:

$$\theta(t) = \theta_0 \cos(\omega_w t + \varsigma) + \delta_\theta(t), \tag{2.8}$$

$$\phi(t) = 0 + \delta_\phi(t), \tag{2.9}$$

where δ_θ and δ_ϕ are small perturbations. Substituting (2.8) and (2.9) into the system (2.3)–(2.4), and linearizing about the semi-trivial solution the following system is obtained

$$\ddot{\delta}_\theta + \mu_1 \dot{\delta}_\theta + \omega_1^2 \delta_\theta = 0 \tag{2.10}$$

$$\ddot{\delta}_\phi + \mu_2 \dot{\delta}_\phi + \left(\omega_2^2 + \alpha_2 \theta_0 \cos\left(\omega_w t + \varsigma\right)\right) \delta_\phi = 0. \tag{2.11}$$

Equation (2.10) has the solution $\delta_\theta = 0$, which is exponentially stable since $\mu_1 > 0$. Therefore, the stability of the semi-trivial solution is fully determined by (2.11), which is referred to as the damped Mathieu equation. By applying Floquet theory [16] it is possible to show that (2.11) has its principal instability region for $\omega_2 \approx \frac{1}{2}\omega_w$, and its boundary is given by:

$$\frac{1}{4}\frac{\mu_2^2}{\omega_w^2} + \left(\frac{\omega_2^2}{\omega_w^2} - \frac{1}{4}\right)^2 = \frac{1}{4}\frac{\alpha_2^2 \theta_0^2}{\omega_w^4}. \tag{2.12}$$

This boundary condition can be used to determine the critical value κ_c of the external excitation, which triggers the parametric resonance in the secondary system. In particular substituting (2.7) into (2.12) we obtain

$$\kappa_c = 2 \frac{\omega_w^2 \sqrt{\left(\omega_1^2 - \omega_w^2\right)^2 + \mu_1^2 \omega_w^2}}{\alpha_2} \sqrt{\frac{1}{4} \frac{\mu_2^2}{\omega_w^2} + \left(\frac{\omega_2^2}{\omega_w^2} - \frac{1}{4}\right)^2}. \qquad (2.13)$$

For $\kappa > \kappa_c$ the semi-trivial solution becomes unstable, and a nontrivial solution appears which is given by:

$$\theta(t) = \theta_1 \cos(\omega_w t + \varsigma_1), \qquad (2.14)$$

$$\phi(t) = \phi_0 \cos\left(\frac{1}{2}\omega_w t + \varsigma_2\right), \qquad (2.15)$$

where

$$\theta_1 = \frac{2\omega_w^2}{\alpha_2} \sqrt{\frac{1}{4} \frac{\mu_2^2}{\omega_w^2} + \left(\frac{\omega_2^2}{\omega_w^2} - \frac{1}{4}\right)^2} \qquad (2.16)$$

and ϕ_0 grows over time.

The system (2.3)–(2.4) shows a saturation phenomenon both in pitch and in roll. For values of the excitation amplitudes between 0 and κ_c the semi-trivial solution is hence stable with an amplitude that grows linearly with κ, as shown in (2.7). When the amplitude of the external excitation becomes larger than κ_c then the semi-trivial solution loses stability and a nontrivial solution appears. In particular (2.16) shows that the amplitude of the solution of the primary system stays constant, whereas the amplitude of the secondary system grows with increasing κ. Therefore, when the excitation amplitude increases, the amount of energy stored in the primary system stays constant and the entire energy rise flows to the secondary system. The rate at which energy is pumped into the secondary system is not constant but varies according to the change of the phase ς_2, which is connected to the variation of the amplitude ϕ_0 through the nonlinearity in the restoring moment. When the rate at which energy being dissipated by viscous effects has matched the rate at which energy is transferred to the roll subsystem, the system reaches a steady state motion characterized by a constant amplitude ϕ_0 and a phase shift $\varsigma_2 = \pi$. Figure 2.1 shows the development of parametric roll resonance while the amplitude of the excitation $\bar{\kappa} = \kappa/\kappa_c$ increases: the stability chart clearly illustrates in the parameter space how the stability properties of the secondary system changes in response to a variation of the amplitude of the external excitation.

Concluding, parametric roll is a resonance phenomenon triggered by existence of the frequency coupling $\omega_w \approx 2\omega_2$, and whose response shows a phase synchronization of 180° with the parametric excitation.

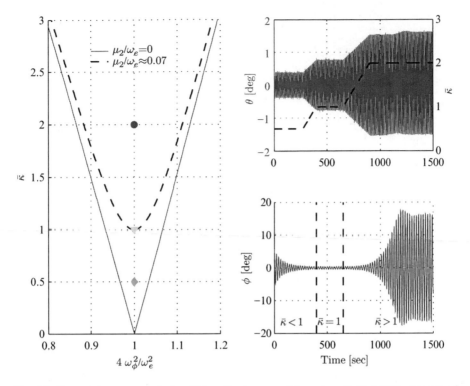

Fig. 2.1 Parametric resonance in the pitch–roll auto-parametric system: (*left*) stability diagram of the secondary system for different levels of damping, and different amplitude of the excitation; (*right*) pitch and roll time series evolution

2.3   Detection Methods

Change detection is often based on a statistical test between a hypothesis $\mathscr{H}_0$ and an alternative $\mathscr{H}_1$. The hypothesis $\mathscr{H}_0$ is related to the normal situation whereas the alternative is related to a deviation from normal.

Assume the data available for the test are $\mathbf{Y} = [y(1), \dots, y(N)]$, and that it is possible to assign a distribution of the data for the normal (fault free) case $p(\mathbf{Y}; \mathscr{H}_0)$ and for the not-normal (faulty) case $p(\mathbf{Y}; \mathscr{H}_1)$, as shown in Fig. 2.2.

Applying the Neyman–Pearson strategy (see, e.g., [2, 22] or [4]) $\mathscr{H}_1$ will be decided if

$$L(\mathbf{Y}) = \frac{p(\mathbf{Y}; \mathscr{H}_1)}{p(\mathbf{Y}; \mathscr{H}_0)} > \gamma, \tag{2.17}$$

where γ is a design parameter. The function $L(\mathbf{Y})$ is referred to as the likelihood ratio. Then the detection process can be seen as a mapping from a data manifold

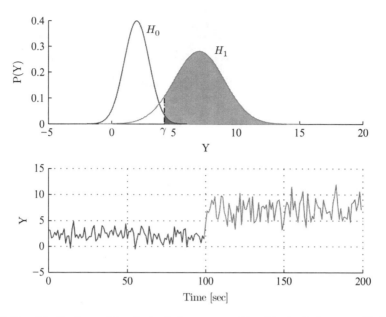

Fig. 2.2 *Top*: Distributions of data in the fault free case $\mathcal{H}_0$ and in the faulty case $\mathcal{H}_1$. *Bottom*: Data when a fault occur at $t = 100$

into a decision manifold. When testing between two simple hypothesis the decision manifold is divided into two regions defined as:

$$R_0 = \{\mathbf{Y} : \text{ decide } \mathcal{H}_0 \text{ or reject } \mathcal{H}_1\}$$
$$R_1 = \{\mathbf{Y} : \text{ decide } \mathcal{H}_1 \text{ or reject } \mathcal{H}_0\},$$

where R_1 is called the critical region.

Performing a statistical test, two types of erroneous decisions can be made. A false alarm if deciding $\mathcal{H}_1$ while $\mathcal{H}_0$ is true, or a missed detection if deciding $\mathcal{H}_0$ while $\mathcal{H}_1$ is true.

While $\mathcal{H}_1$ true, the probability of false alarm is $P_{FA} = P(\mathcal{H}_1; \mathcal{H}_0)$ and the probability of correct detection is $P_D = P(\mathcal{H}_1; \mathcal{H}_1)$. Both depend on the threshold γ chosen, as illustrated in Fig. 2.2. In a simple test situation (the distribution of data is known in both the normal and in the faulty situation) the Neyman–Pearson test in (2.17) maximizes P_D for a given P_{FA}.

In the composite situation, where the two distributions are not precisely known, but they depend on some unknown parameters, the generalized likelihood ratio test results in deciding $\mathcal{H}_1$ if

$$L(\mathbf{Y}) = \frac{p(\mathbf{Y}; \hat{\theta}_1, \mathcal{H}_1)}{p(\mathbf{Y}; \hat{\theta}_0, \mathcal{H}_0)} > \gamma, \qquad (2.18)$$

where $\hat{\theta}_i$ is the maximum likelihood estimate (MLE) of θ_i, i.e.,

$$\hat{\theta}_i = \max_{\theta} L(\mathbf{Y}|\theta_i; \mathcal{H}_i) . \tag{2.19}$$

The specific detection methods that follow below are patent pending [11].

2.3.1 Detection in the Frequency Domain

In Sect. 2.2 it was shown that the onset and development of parametric roll is attributable to the transfer of energy from the pitch mode (but also heave motion can contribute), directly excited by the wave motion, to the roll mode, at a frequency about twice the natural roll frequency. Therefore, an increase of power of roll square close to the frequencies where pitch is pumping energy into roll may be exploited as an indicator of parametric resonance.

Given two signals, e.g., $x(t)$ and $y(t)$, the cross-correlation provides a measure of similarity of the two waveforms as a function of time lag. If the two signals are discrete sequences then the cross-correlation and cross-spectrum are defined as:

$$r_{xy}[m] \triangleq \sum_{m=-\infty}^{\infty} x^*[m]y[n+m],$$

$$P_{xy}(\omega) \triangleq \sum_{m=-\infty}^{\infty} r_{xy}[m]e^{-j\omega m}, \tag{2.20}$$

where m is the time lag, and * denotes complex conjugate. The functions carry information about which components are held in common between the two signals and since it is the roll sub-harmonic regime addressing the onset of parametric roll resonance, the detection problem can be set up as monitoring the cross-spectrum of $\phi^2[n]$ and $\theta[n]$.

The parametric roll detection problem is then formulated as:

$$\mathcal{H}_0 : P_{\phi^2\theta}(\omega) \leq \bar{P}, \tag{2.21}$$

$$\mathcal{H}_1 : P_{\phi^2\theta}(\omega) > \bar{P},$$

where $\bar{P}$ is a power threshold. Instead of using directly the cross-spectrum, a spectral correlation coefficient could be exploited, defined as:

$$\mathcal{S}_{\phi^2\theta} \triangleq \frac{\sigma_{\phi^2\theta}^2}{\sqrt{\sigma_{\phi^2}^2 \sigma_{\theta}^2}}. \tag{2.22}$$

where $\sigma_{\phi^2\theta}^2$ is the average power of the cross-spectrum of ϕ^2 and θ, $\sigma_{\phi^2}^2$ is the average power of the square of the roll angle, and σ_{θ}^2 is the average power of the pitch angle.

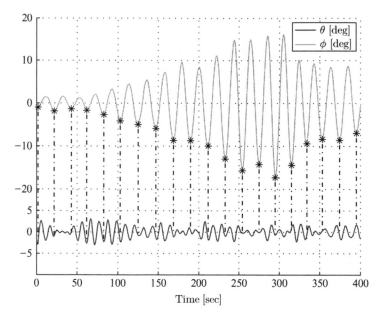

Fig. 2.3 Model tank experiment 1195: alignment of peaks between pitch θ and roll ϕ during the onset and development of parametric roll

The detection problem can then be rewritten as:

$$\mathcal{H}_0 : \mathscr{S}_{\phi^2\theta} \leq \bar{\mathscr{S}},$$
$$\mathcal{H}_1 : \mathscr{S}_{\phi^2\theta} > \bar{\mathscr{S}}, \tag{2.23}$$

where $\bar{\mathscr{S}}$ is a measure of the level of spectral correlation.

2.3.2   Detection in the Time Domain

2.3.2.1   Statistics of the Driving Signal

After onset, parametric roll resonance is characterized by nonlinear synchronization between motions. Døhlie [7] pointed out that when parametric roll develops there is a lining up of peaks between the pitch motion and the roll motion, that is, every second peak of pitch is in-phase with the peak in roll, as shown in Fig. 2.3. Figure 2.3 also shows that when this alignment is partially lost, the roll oscillations start decaying, as e.g., between 150 s and 250 s, or after 300 s. Therefore, a signal which carries the phase information of pitch and roll could be exploited for solving the detection problem.

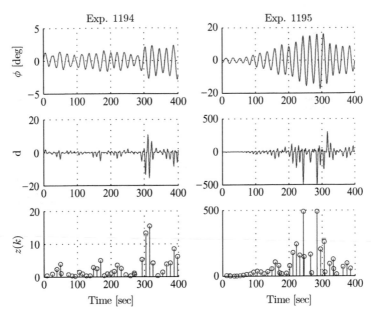

Fig. 2.4 Negative and positive peaks in d address how the amplitude of the roll oscillations increases and decreases. Data from model basin test

Following [13], given the roll angle ϕ and the pitch angle θ, the signal indicating the parametric resonance in roll is defined as:

$$d(t) \triangleq \lambda(t)\phi^2(t)\theta(t), \qquad (2.24)$$

where the time-varying scaling factor $0 < \lambda(t) \leq 1$ is introduced to reduce the sensitivity to variations in sea state. Consider Fig. 2.4, where $\phi(t)$ and $d(t)$ are plotted for one experiment without parametric roll (Exp. 1194) and another with parametric roll (Exp. 1195). The *driving* signal $d(t)$ characterizes quite well the way the amplitude grows or decays inside the signal ϕ. When the amplitude of ϕ abruptly grows, a sequence of negative spikes shows up in the driving signal. In contrast, when the amplitude of ϕ decreases, positive spikes reflect this in $d(t)$.

Moreover, when parametric roll is developing, the magnitudes of the negative spikes in the driving signal are much larger than that seen when the roll mode is not in a resonant condition (Fig. 2.4, middle plots). Therefore, a significant variation in the variance of the driving signal $d(t)$ can be expected when parametric roll is developing. An alternative to directly using the $d(t)$ signal would be to use the amplitude of local minima between up-crossings,

$$z(k) \triangleq -\min(d(t)), \quad t \in\,]T(k-1), T(k)], \qquad (2.25)$$

where $T(k)$ are the time-tags of up-crossings in $d(t)$.

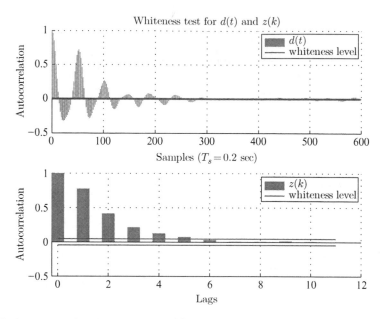

Fig. 2.5 Auto-correlation of driving signal $d(t)$ and local minima $z(k)$. Data from Atlantic passage during a storm

An important condition for subsequent statistical testing and correct selection of thresholds, is that data are independent and identically distributed (IID). A plot of the autocorrelation functions of $d(t)$ and $z(k)$ under the hypothesis $\mathcal{H}_0$ is shown in Fig. 2.5. The driving signal $d(t)$ is heavily correlated due to the narrow-band process that creates this signal. The autocorrelation of the local minima $z(k)$ have a smooth roll-off with a forgetting factor of 0.3. If whitening of $z(k)$ should subsequently be needed, a simple discrete filter could be employed for this purpose. The autocorrelation behavior makes $z(k)$ a natural choice for subsequent statistical analysis in the time-domain.

As to the distribution of $z(k)$, a scrutiny showed that a Weibull distribution characterizes $z(k)$ quite well. The Weibull distribution, which is defined only for $z > 0$, has cumulative density function, CDF:

$$P(z) = 1 - \exp\left(-\left(\frac{z}{v}\right)^{\beta}\right) \tag{2.26}$$

and probability density function, PDF:

$$p(z) = \frac{\beta}{2v^{\beta}} (z)^{\beta-1} \exp\left(-\left(\frac{z}{v}\right)^{\beta}\right), \tag{2.27}$$

where v and β are scale and shape parameters, respectively.

According to the observations from model test data, a good way to discriminate between resonant and nonresonant cases is to look for a variation in signal power. In particular, the bottom plots of Fig. 2.4 show that the onset of parametric resonance in roll is preceded by an abrupt variation of the amplitude of $z(k)$; therefore, a detector which looks for large changes in signal power is aimed at. For the Weibull distribution the variance is given by:

$$\sigma^2 = v^2 \left[\Gamma \left(1 + \frac{2}{\beta} \right) - \Gamma^2 \left(1 + \frac{1}{\beta} \right) \right].$$

Hence the detection scheme must trail variations in scale and shape parameters.

2.3.2.2 GLRT for Weibull Processes ($\mathscr{W}$-GLRT)

Assume that the local minima $z(t)$ of the driving signal is a realization of a Weibull random process. Then the distribution of N independent and identically distributed samples of z is characterized by the probability density function:

$$\mathscr{W}(\mathbf{z}; \theta) = \left(\frac{\beta}{2v^\beta} \right)^N \prod_{k=0}^{N-1} \left[z_k^{\beta-1} \exp \left(- \left(\frac{z_k}{v} \right)^\beta \right) \right], \qquad (2.28)$$

where $\theta = [v, \beta]^T$ is the parameter vector fully describing the Weibull PDF.

The detection of parametric roll can be formulated as a parameter test of the probability density function:

$$\mathscr{H}_0 : \theta = \theta_0, \qquad (2.29)$$

$$\mathscr{H}_1 : \theta = \theta_1,$$

where θ_0 is known and it represents $\mathscr{W}$ in the nonresonant case, whereas θ_1 is unknown and it describes the parametric resonant case. By applying the generalized likelihood ratio test, the detector decides $\mathscr{H}_1$ if

$$L_G(\mathbf{z}) = \frac{p\left(\mathbf{z}; \hat{\theta}_1, \mathscr{H}_1 \right)}{p(\mathbf{z}; \theta_0, \mathscr{H}_0)} > \gamma, \qquad (2.30)$$

where the unknown parameter vector θ_1 is replaced with its maximum likelihood estimate $\hat{\theta}_1$, and γ is the threshold given by the desired probability of false alarms.

The first step in computing L_G is to determine $\hat{\theta}_1 = \left[\hat{v}_1, \hat{\beta}_1 \right]^T$, therefore we need to maximize $p\left(\mathbf{z}; \hat{\theta}_1, \mathscr{H}_1 \right)$. Given $p\left(\mathbf{z}; \hat{\theta}_1 \right)$ the estimates of the parameters v_1 and β_1 are computed as:

$$\frac{\partial \ln p\left(\mathbf{z}; \hat{\theta}_1 \right)}{\partial \theta_j} = 0$$

which results in

$$\hat{\upsilon}_1 = \left(\frac{1}{N} \sum_{k=0}^{N-1} z_k^{\hat{\beta}_1} \right)^{\frac{1}{\hat{\beta}_1}} \tag{2.31}$$

$$\frac{1}{\hat{\beta}_1} = \frac{\sum_{k=0}^{N-1} z_k^{\hat{\beta}_1} \ln z_k}{\sum_{k=0}^{N-1} z_k^{\hat{\beta}_1}} - \frac{1}{N} \sum_{k=0}^{N-1} \ln z_k \ . \tag{2.32}$$

Balakrishnan and Kateri [1] have shown that $\hat{\beta}_1$ exists, it is unique, and its value is given by the intersection of the curve $1/\hat{\beta}_1$ with the right-hand side of (2.32).

Having determined the MLEs $\hat{\upsilon}_1$ and $\hat{\beta}_1$ it is then possible to derive an explicit form for the detector. By taking the natural logarithm of both sides of (2.30),

$$\ln \frac{\left(\frac{\beta_1}{2\upsilon_1^{\beta_1}} \right)^N \prod_{k=0}^{N-1} \left[z_k^{\beta_1 - 1} \exp\left(-\left(\frac{z_k}{\upsilon_1} \right)^{\beta_1} \right) \right]}{\left(\frac{\beta_0}{2\upsilon_0^{\beta_0}} \right)^N \prod_{k=0}^{N-1} \left[z_k^{\beta_0 - 1} \exp\left(-\left(\frac{z_k}{\upsilon_0} \right)^{\beta_0} \right) \right]} > \ln \gamma \Rightarrow$$

$$N \ln \left(\frac{\beta_1}{\beta_0} \frac{\upsilon_0^{\beta_0}}{\upsilon_1^{\beta_1}} \right) + (\beta_1 - \beta_0) \sum_{k=0}^{N-1} \ln z_k - \sum_{k=0}^{N-1} \left(\frac{z_k}{\upsilon_1} \right)^{\beta_1} + \sum_{k=0}^{N-1} \left(\frac{z_k}{\upsilon_0} \right)^{\beta_0} > \ln \gamma, \tag{2.33}$$

where the parameters $[\beta_1, \upsilon_1]$ are replaced by their estimates.

Data show that the shape parameter is approximately the same under both hypothesis, $\beta_1 = \beta_0 = \beta$, then the GLRT reads

$$N\beta \ln \left(\frac{\upsilon_0}{\hat{\upsilon}_1} \right) + \frac{\hat{\upsilon}_1^\beta - \upsilon_0^\beta}{(\upsilon_0 \hat{\upsilon}_1)^\beta} \sum_{k=0}^{N-1} z_k^\beta > \ln \gamma \Rightarrow$$

$$N\beta \ln \left(\frac{\upsilon_0}{\hat{\upsilon}_1} \right) + N \frac{\hat{\upsilon}_1^\beta - \upsilon_0^\beta}{\upsilon_0^\beta} > \ln \gamma. \tag{2.34}$$

Therefore the test quantity $g(k)$ is

$$g(k) = \left(\frac{\hat{\upsilon}_1(k)}{\upsilon_0} \right)^\beta - 1 - \beta \ln \left(\frac{\hat{\upsilon}_1(k)}{\upsilon_0} \right) \tag{2.35}$$

and the threshold where $\mathcal{H}_1$ is decided is

$$g(k) > \frac{\ln \gamma}{N} \equiv \gamma_g. \tag{2.36}$$

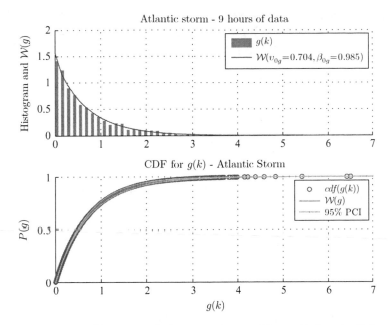

Fig. 2.6 Test statistics $g(k)$ observed in heavy weather conditions with forced roll. Data from Atlantic passage during a storm. A Weibull distribution fits the data well

Asymptotically, as $N \to \infty$ a theoretical value exists for γ, independent of distribution of $z(k)$. However, since only few peaks are used here, the distribution of $g(k)$ need be investigated and the value of the threshold γ_g need be determined from this distribution.

2.3.2.3 Selection of Threshold

Selection of the threshold γ_g to obtain a sufficiently low false alarm rate, depends on the statistics of $g(k)$ in (2.35) under assumption $\mathcal{H}_0$. Given the test signal $g(k)$, which behaves according to the PDF $p(g; \mathcal{H}_0)$ under the hypothesis $\mathcal{H}_0$, the threshold γ_g, which obtains a given false alarm probability, follows from

$$P_{FA} = \int_{\{g:L_G(g) > \gamma_g\}} p(g; \mathcal{H}_0) \, dg. \qquad (2.37)$$

Since the GLRT runs over only few peaks to obtain rapid detection, asymptotic results for the distribution of $g(k)$ are not applicable. Instead, the distribution $p(g; \mathcal{H}_0)$ can be reliably estimated from data. A plot of the histogram of the test statistics $g(k)$ is shown in Fig. 2.6 together with the estimated Weibull distribution. The data used are recordings from a container vessel during 9 h of navigation through an Atlantic storm.

Having obtained the parameters υ_{0g} and β_{0g} of the Weibull distribution for $g(k)$ under $\mathcal{H}_0$, $\mathcal{W}(g(k); \mathcal{H}_0)$, the threshold for a desired false alarm probability is obtained from

$$1 - P_{FA} = 1 - \exp\left(-\left(\frac{\gamma_g}{\upsilon_{0g}}\right)^{\beta_{0g}}\right) \tag{2.38}$$

or

$$\gamma_g = \upsilon_{0g}\left(-\ln P_{FA}\right)^{\frac{1}{\beta_{0g}}}. \tag{2.39}$$

For the hypothesis $\mathcal{H}_0$ shown in Fig. 2.6 the Weibull fit is characterized by $\upsilon_0 = 0.70 \pm 0.04$ and $\beta_0 = 0.99 \pm 0.04$; hence to obtain a probability of false alarms $P_{FA} = 0.0001$, the threshold must be set to $\gamma_g = 5.0$.

2.3.2.4   Robustness Against Forced Roll

For a real ship sailing in oblique short-crested seaways some forced roll with frequency equal to the encounter frequency will always occur. This does not obscure the proposed detection schemes since both the spectral correlation coefficient and the GLRT for nonGaussian processes are insensitive to forced roll.

Consider pitch and roll as narrow-band signals:

$$\theta(t) \text{ s.t. } \Theta(\omega) = 0 \text{ for } |\omega - \omega_\theta| \geq \Omega_\theta,$$

$$\phi(t) \text{ s.t. } \Phi(\omega) = 0 \text{ for } |\omega - \omega_\phi| \geq \Omega_\phi,$$

where $\Theta(\omega)$ and $\Phi(\omega)$ are the spectra of pitch and roll centered at the center frequency ω_θ and ω_ϕ respectively. The bands of the spectra are given by $B_\theta = \{\omega \text{ s.t. } |\omega - \omega_\theta| < \Omega_\theta\}$ and $B_\phi = \{\omega \text{ s.t. } |\omega - \omega_\phi| < \Omega_\phi\}$, where $W_\theta = 2\Omega_\theta$ and $W_\phi = 2\Omega_\phi$ are the bandwidths.

If $\omega_\theta = \omega_e = 2\omega_\phi$, as in parametric resonance, then $B_\theta = \{\omega \text{ s.t. } |\omega - 2\omega_\phi| < \Omega_\theta\}$, hence the spectrum of the square of the roll angle overlaps in large part or completely the pitch spectrum. With pitch and the roll signals:

$$\theta(t) = \theta_0(t)\cos(2\omega_\phi t + \psi_\theta(t)) \tag{2.40}$$

$$\phi(t) = \phi_0(t)\cos(\omega_\phi t + \psi_\phi(t)) \tag{2.41}$$

the spectra of pitch and of the square of roll are

$$\Theta(\omega) = \frac{1}{2}\big[\Theta_i(\omega - 2\omega_\phi) + \Theta_i(\omega + 2\omega_\phi)$$
$$- \Theta_q(\omega - 2\omega_\phi) - \Theta_q(\omega + 2\omega_\phi)\big] \qquad (2.42)$$

$$\Phi_2(\omega) = \frac{1}{2}\big[\Phi_0(\omega) + \Phi_i(\omega - 2\omega_\phi) + \Phi_i(\omega + 2\omega_\phi)$$
$$- \Phi_q(\omega - 2\omega_\phi) - \Phi_q(\omega + 2\omega_\phi)\big]. \qquad (2.43)$$

Here $\Theta_i = \mathscr{F}(\theta_0(t)\cos(\psi_\theta(t)))$, and $\Theta_q = \mathscr{F}(\theta_0(t)\sin(\psi_\theta(t)))$ are the Fourier transforms of the in-phase and quadrature components of the pitch angle; whereas $\Phi_0 = \mathscr{F}(\phi_0^2(t))$, $\Phi_i = \mathscr{F}(\phi_0^2(t)\cos(2\psi_\phi(t)))$, and $\Phi_q = \mathscr{F}(\phi_0^2(t)\sin(2\psi_\phi(t)))$ are the Fourier transform of the DC, in-phase and quadrature components of the second power of roll. Therefore, by applying the cross-correlation theorem to the signals at hand,

$$P_{\phi^2\theta} = \frac{1}{4}\big[\Phi_i\Theta_i(\omega - 2\omega_\phi) + \Phi_i\Theta_i(\omega + 2\omega_\phi) + \Phi_q\Theta_q(\omega - 2\omega_\phi)$$
$$+ \Phi_q\Theta_q(\omega + 2\omega_\phi) - \Phi_i\Theta_q(\omega - 2\omega_\phi) - \Phi_i\Theta_q(\omega + 2\omega_\phi)$$
$$- \Phi_q\Theta_i(\omega - 2\omega_\phi) - \Phi_q\Theta_i(\omega + 2\omega_\phi)\big]. \qquad (2.44)$$

The cross-spectrum is different from zero since $\phi^2(t)$ and $\theta(t)$ are centered at the same frequency; hence the spectral correlation coefficient is different from zero and it can be used for detecting parametric roll.

Consider now a ship sailing in near head seas condition. The lateral component of wave force excites roll motion directly, hence pitch and roll both respond at the same frequency ($\omega_\phi = \omega_\theta = \omega_e$). The cross-spectrum in this case is equal to zero,

$$P_{\phi^2\theta} = \frac{1}{4}\big[\Theta_i(\omega - \omega_e) + \Theta_i(\omega + \omega_e) - \Theta_q(\omega - \omega_e) - \Theta_q(\omega + \omega_e)\big]$$
$$\times \big[\Phi_0(\omega) + \Phi_i(\omega - 2\omega_e) + \Phi_i(\omega + 2\omega_e)$$
$$- \Phi_q(\omega - 2\omega_e) - \Phi_q(\omega + 2\omega_e)\big] = 0, \qquad (2.45)$$

since the spectra are different from zero only around $\omega = \omega_e$ or $\omega = 2\omega_e$. Therefore the spectral correlation coefficient is zero, showing that the proposed detection method is insensitive to forced roll.

The GLRT for the Weibull distribution of the local minima $z(k)$ is also proven to be insensitive to forced roll. Consider pitch and roll as sinusoidal signals:

$$\theta(t) = \theta_0 \cos(\omega_\theta t + \varsigma) \qquad (2.46)$$
$$\phi(t) = \phi_0 \cos(\omega_\phi t). \qquad (2.47)$$

In forced roll condition, roll and pitch are sinusoids of the same frequency ($\omega_\theta = \omega_\phi = \omega$), which yields the following driving signal:

$$d(t) = \phi^2(t)\theta(t)$$

$$= \phi_0^2\theta_0\cos^2(\omega t)\cos(\omega t + \varsigma). \tag{2.48}$$

According to (2.25) we have that

$$z(k) = \alpha(\varsigma)\phi_0^2\theta_0 \le \phi_0^2\theta_0, \quad 0 < \alpha(\varsigma) \le 1, \tag{2.49}$$

where $\alpha(\varsigma)$ is the amplitude reduction factor due to the phase shift between the two wave forms.

To prove that the GLRT detector is not sensitive to forced roll, we need to demonstrate that there exists a constant Γ such that for any $\gamma > \Gamma$ the detector does not trigger an alarm. In general Γ is function of the phase shift ς and of the time interval ΔT over which the estimates of the scaling and shape factors are performed. In particular the time interval ΔT determines how many local minima are taken into account for the detection.

To find Γ we need to prove that

$$(\beta_1 - \beta_0)\sum_{k=1}^{N}\log z(k) - \frac{1}{\upsilon_1^{\beta_1}}\sum_{k=1}^{N}z(k)^{\beta_1} + \frac{1}{\upsilon_0^{\beta_0}}\sum_{k=1}^{N}z(k)^{\beta_0} \tag{2.50}$$

is upper bounded. For any $\Delta T \in [0, T]$, where $T = 2\pi/\omega$ is the natural roll period, the GLRT detector is not sensitive to forced roll if the threshold γ is set larger than

$$\Gamma \triangleq N_{max}(\beta_1 - \beta_0)\log(\phi_0^2\theta_0) - \frac{N_{max}}{\upsilon_1^{\beta_1}}(\phi_0^2\theta_0)^{\beta_1} + \frac{N_{max}}{\upsilon_0^{\beta_0}}(\phi_0^2\theta_0)^{\beta_0}, \tag{2.51}$$

where N_{max} is the maximum number of local minima, which fall within one roll period.

2.4 Detection System Robustification

The proposed detection schemes rely on assumptions, which in general may not be completely fulfilled during real navigation operations. The spectral correlation performs best when the signals at hand have a narrow band power spectrum because in that case the Fourier transform of the convolution between the second power of roll with pitch will be zero most of time except when parametric roll is developing. However, in real sailing conditions the wave spectrum exciting the ship motions can be rather large, and it induces ship responses whose frequency content spans over a wide range of frequencies as well. Figure 2.7 compares the power spectra of time series recorded during an experiment in a towing tank, and during a container vessel voyage through an Atlantic storm.

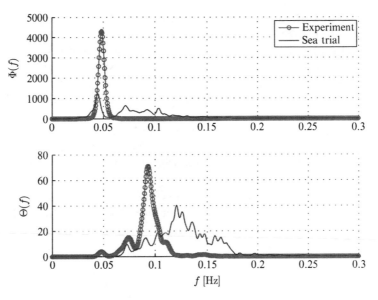

Fig. 2.7 Comparison of power spectra for model basin and Atlantic storm

During real navigation, the roll and pitch motions have an energy content different from zero over a wide range of frequencies, and this will contribute to determine a nonzero spectral correlation also in these regions of frequencies where parametric roll is not likely. Consequently robustification of the spectral correlation is needed.

This is obtained by bandpass filters that narrow in the roll and pitch signals frequency ranges of interest. The pass-bands regions are centered about ω_ϕ and $\omega_e = 2\omega_\phi$, to focus on frequency ranges where parametric roll resonance can develop. The spectral correlation hence takes the form

$$^f\mathcal{S}_{\phi^2\theta} = \frac{^f\sigma^2_{\phi^2\theta}}{\sqrt{\sigma^2_{\phi^2}\sigma^2_\theta}}, \tag{2.52}$$

where the superscript f addresses that the computation involves the filtered signals. The normalization factor in (2.52) is still calculated from the raw roll and pitch signals.

For the phase condition (GLRT) detector, a time-varying scaling factor $\lambda(t)$ is applied to the driving signal in (2.24) to adapt to weather conditions. Furthermore, the $\mathcal{H}_0$ parameters are estimated on-line. These together served to obtain desired false alarm rates and make the GLRT detector insensitive to changes in sea state. Assume that we are at time $t = T$ and the GLRT is fed with data logged within the window $[T - M + 1, T]$. The scaling factor λ is computed taking into account all the data from the time window $[T - n * M + 1, T - M]$, where $n \in \mathbb{N}$ is a design parameter.

Utilizing the $\mathscr{W}$-GLRT detector we should assume that the Weibull PDFs for the nonresonant and resonant case differ both for the shape β and scale υ parameters. In Sect. 2.3.2.2 it was shown that the MLE of the shape parameter is found as a solution of a nonlinear equation, which in practice must be solved at each iteration of the algorithm. However, data show that the shape factor remains approximately unchanged, hence, $\beta_0 = \beta_1 = \beta$. Therefore, the $\mathscr{W}$-GLRT detector only looks for variations in the scale parameter υ.

Finally, it is important to point out how the thresholds were chosen. For the cross-correlation, the spectral correlation coefficient $^f\mathscr{S}_{\phi_2\theta}$ varies between zero and one, hence the threshold $\mathscr{S}$ can be set to any value higher than 0.4 according to how conservative the detector should be.

For the GLRT-based detector it was shown that an empirical threshold can be computed based on the estimated $\mathscr{H}_0$ distribution of the test quantity $g(k)$.

2.5 Detection Scheme Validation

This section presents the validation of the detection schemes on both model scale and full scale data sets. After introducing the data sets, the performance of the Weibull GLRT detector is evaluated in both scenarios. Next, the overall robust performance of the monitoring system given by the integration of the spectral correlation detector with the $\mathscr{W}$-GLRT detector is tested. For the performance assessment of the spectral correlation detector the reader may refer to [14].

2.5.1 Experimental and Full Scale Data Sets

To assess the performance of the proposed detection schemes for parametric roll the detectors have been validated against two data sets. The first data set consists of eight experiments run in irregular waves scenario.[2] The vessel used for the experiments was a 1:45 scale model of a container ship with length overall of 294 m. The principal dimensions and hydrodynamic coefficients can be found in [18]. The time-history of roll is shown in Fig. 2.9 (top plot). Although the vessel experienced parametric roll only once, all the experiments were made to trigger the resonant phenomenon, but in the irregular wave scenario it is somewhat difficult to obtain a fully developed parametric roll resonance, because consecutive wave trains may not fulfill all conditions for its development.

---

[2]The terminology irregular wave scenario means that the wave motion used to excite the vessel is generated by the interference of multiple sinusoidal waves centered at different frequencies, and it is described by a given power spectrum. This terminology is used in opposition to the regular wave scenario where instead the vessel is excited by a single sinusoidal wave.

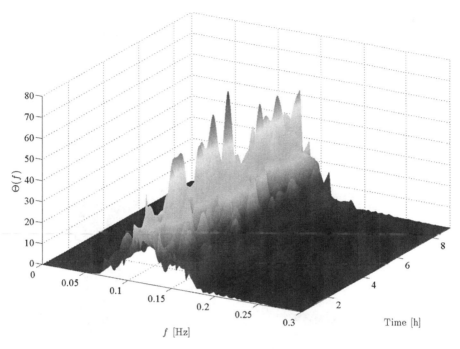

Fig. 2.8 The evolution of the pitch power spectrum provides an idea of the frequency content of the wave spectrum during the navigation

The second data set is full scale data recorded on board Clara Maersk, a 33000 dwt container ship crossing the North Atlantic. Nine hours of navigation were analyzed. Conditions were significant wave height judged by navigators to develop from 5–6 m to 7–10 m. Relative direction of waves were 150°–170° where 180° is head sea. The pitch power spectrum shown in Fig. 2.8 provides an idea of the broad frequency content of the wave spectrum exciting the container ship during the storm.

For this data set it is essential to point out that there was no prior awareness about the onset of parametric roll resonance; hence the assessment of detections and/or false alarms was done by visual inspection of the time series around the alarm time.

The model test experimental data set is used to evaluate the capability of the detectors to timely catch the onset of parametric roll; whereas the real navigation data set is used to ensure the insensitivity to usual forced roll.

In order to simulate a continuous navigation the single records of the two data set have been stitched together. A smoothing filter was applied around the stitching points to avoid that sudden fictitious variations within the signals at hand could trigger an alarm. Hence the roll time series scrutinized are those shown in Fig. 2.9.

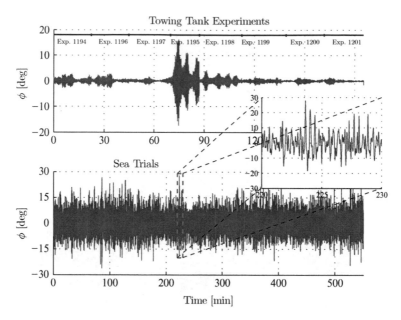

Fig. 2.9 *Top*: Roll motion time series recorded from experimental runs. Experiment 1195 is the only one where parametric roll clearly developed. *Bottom*: Roll motion time series recorded during navigation across the North Atlantic Ocean

2.5.2 Validation of Weibull GLRT Detector

Figures 2.10 and 2.11 show the results of the Weibull GLRT detector after processing the model and full scale data sets.

On the experimental data set the Weibull GLRT detector performs well. Figure 2.10 illustrates that the parametric roll event that occurs between $t = 70$ min and $t = 90$ min is timely detected when the roll angle is about $3°$. The lost of phase synchronization is also detected by the Weibull GLRT, which withdraws the alarm at $t = 90$ min when the roll motion decays and it seems that the parametric resonance is over. However, a new alarm is suddenly raised when the resonant oscillations take place again.

On the sea trial data set the Weibull GLRT detector raises five alarms, which all last for exactly one window length M, as shown in Fig. 2.11. Since no prior awareness about the presence/absence of parametric roll events was available for this data set, the alarms have been classified by visual inspection and it was concluded that all five cases are likely to be false alarms.

It is not surprising that a single detector cannot provide full information about the resonance condition since both the phase synchronization and the frequency coupling must be satisfied simultaneously. Robust detection performance therefore needs simultaneous detection of the presence of both conditions.

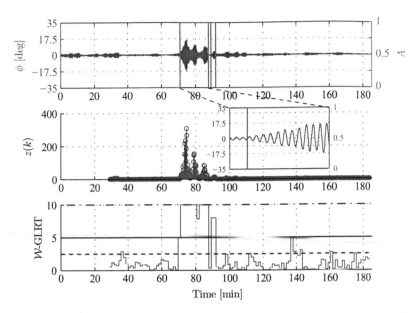

Fig. 2.10 Weibull GLRT detector on the experimental data set: the onset of the parametric roll event is timely detected when the roll angle is only about 3°

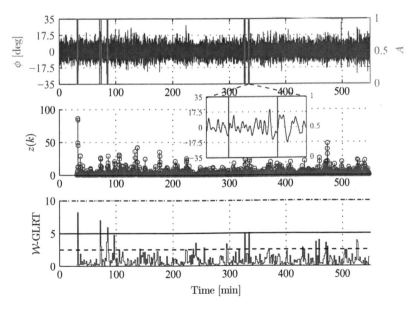

Fig. 2.11 Weibull GLRT detector on the sea trial data set: the detector raises five alarms. Visual inspection of the alarm cases suggests that those are false alarms

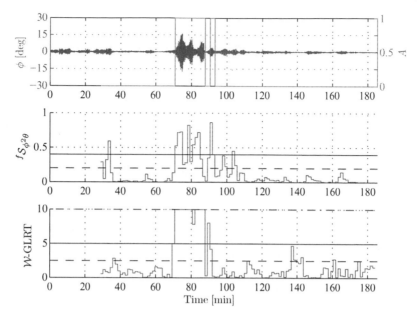

Fig. 2.12 Performance of the monitoring system on the model basin experimental data set

2.5.3 Robust Performance

The spectral correlation detector and the Weibull GLRT detector have shown fairly good performance providing a timely detection but they both give false alarms. To obtain the full picture, the two detectors are combined within a monitoring system, which issues alarms of parametric roll occurrence based upon the tests made by both detectors together. Furthermore, robustness is obtained by making the adaptation to prevailing conditions only when none of the thresholds are exceeded. This means the $\mathcal{H}_0$ statistics and the normalization of the spectral correlation in practice are calculated from data that are older than the data windows used – a few roll periods – and with appropriate forgetting to be able to track changes in weather. The performance of the monitoring system with robustified algorithms is shown in Fig. 2.12 for the model basin experimental data set, and in Fig. 2.13 for the real navigation data. The general quality of detection performance is apparent.

The performance improvement of the Weibull GLRT detector when combined with the spectral correlation detector is shown in Fig. 2.13. The reduction of false alarms is determined by the fact that in this case the update of the scaling factor $\lambda(t)$ is related to the alarms raised by the monitoring system and not to those issued by the Weibull GLRT detector alone.

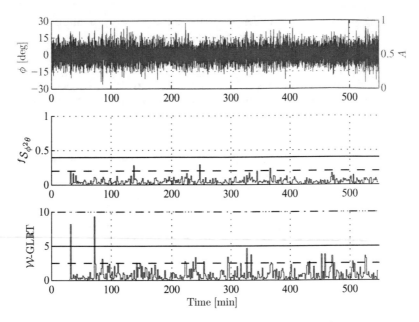

Fig. 2.13 Performance of the monitoring system on navigation data from Clara Maersk

2.5.4 Discussion

While the results have shown very convincing from model tank testing where parametric resonance was present and Atlantic passage in a storm, where it was believed, but not known with certainty that no parametric resonance was present, an independent test with full-scale data would be needed for final proof of the concepts and algorithms presented in this chapter. Such data have recently been made available from trials where also wave radar data were logged. The results with these data were convincing but this validation is outside the scope and space allocated to this chapter. The monitoring system methodology and implementation are patent pending [11], and is expected to find its way to the Seven Seas under the trade mark *PAROLL*®.

2.6 Conclusions

Detection methods were investigated for the diagnosis of parametric roll resonance and were validated against data from model basin tests and from a full-scale Atlantic crossing with a container vessel.

In the spectral domain, spectral analysis provided an indicator for energy flowing from the pitch motion, directly excited by the waves, into roll motion causing resonance. In the time domain, a Weibull GLRT detector monitored the behavior of a driving signal carrying information about the phase correlation between the square of the roll angle and the pitch angle. Robustness against usual forced roll motion was shown for both detectors.

The detectors showed to be very capable of timely detecting the onset of parametric roll, while achieving a very low false alarm rate. A necessary part of achieving excellent overall detection performance was obtained by combining the hypotheses from the two detectors.

Acknowledgements The authors are grateful for financial support from DTU for the *Proof of Concept* project *PAROLL*. The invaluable collaboration and comments from Dr. G. Storhaug and colleagues from Det Norske Veritas are gratefully acknowledged, and so is the inspiring collaboration and discussions with colleagues at the CeSOS center of excellence at NTNU. Data from model tests were kindly provided by Dr. Perez (formerly CeSOS, NTNU now Univ. of Newcastle, AUS) and Dr. Storhaug, DNV. Data from the Atlantic passage were logged by M. Blanke in the 1980s and are presented here with the permission from the owner, Maersk Lines. The help received from Mr. T. Krarup Sørensen from Maersk and the officers and crew of Clara Maersk in organizing and conducting the sea tests are much appreciated.

References

1. Balakrishnan, N., Kateri, M.: On the maximum likelihood estimation of parameters of weibull distribution based on complete and censored data. *Statistics and Probability Letters*, 78: 2971–2975 (2008).
2. Basseville, M., Nikiforov, I. V.: *Detection of Abrupt Changes: Theory and Applications.* Prentice Hall (1993).
3. Belenky, V., Yu H., Weems, K.: Numerical procedures and practical experience of assessment of parametric roll of container carriers. In *Proceedings of the 9th International Conference on Stability of Ships and Ocean Vehicles*. Brazil (2006).
4. Blanke, M., Kinnaert, M., Lunze, J., Staroswiecki, M.: *Diagnosis and fault-tolerant control* 2nd edition, Springer, Germany (2006).
5. Bulian, G., Francescutto, A., Umeda, N., Hashimoto, H.: Qualitative and quantitative characteristics of parametric ship rolling in random waves. *Ocean Engineering*, 35:1661–1675 (2008).
6. Carmel, S. M.: Study of parametric rolling event on a panamax container vessel. *Journal of the Transportation Research Board*, 1963:56–63 (2006).
7. Døhlie, K. A.: Parametric rolling – a problem solved? *DNV Container Ship Update*, 1:12–15 (2006).
8. France, W. N., Levadou, M., Treakle, T. W., Paulling, J. R., Michel, R. K., Moore, C.: An investigation of head-sea parametric rolling and its influence on container lashing systems. In: *SNAME Annual Meeting*. SNAME Marine Technology 40(1):1–19 (2003).
9. Froude, W.: On the rolling of ships. *Transactions of the Institution of Naval Architects*, 2: 180–227 (1861).
10. Froude, W.: Remarks on Mr. Scott Russel's paper on rolling. *Transactions of the Institution of Naval Architects*, 4:232–275 (1863).
11. Galeazzi, R., Blanke, M., Poulsen, N. K.: Prediction of resonant oscillation, EP09157857, filed April 2009 (2009).
12. Galeazzi, R., Blanke, M., Poulsen, N. K.: Detection of parametric roll resonance on ships from indication of nonlinear energy flow. In: *Proceedings 7th IFAC Symp. on Fault Detection, Supervision and Safety of Technical Processes*. IFAC, Spain (2009).
13. Galeazzi, R., Blanke, M., Poulsen, N. K.: Parametric roll resonance detection using phase correlation and log-likelihood testing techniques. In: *Proceedings of the 8th IFAC International Conference on Manoeuvring and Control of Marine Craft*. IFAC, Brazil (2009).

14. Galeazzi, R., Blanke, M., Poulsen, N. K.: Early detection of parametric roll resonance on container ships. *Technical report, DTU Electrical Engineering, Technical University of Denmark* (2011).
15. Ginsberg, S.: Lawsuits rock APL's boat – Cargo goes overboard; insurance lawyers surface (1998).
16. Grimshaw, R.: *Nonlinear ordinary differential equations*. CRC Press, United States of America (1993).
17. Hashimoto, H., Umeda, N.: Nonlinear analysis of parametric rolling in longitudinal and quartering seas with realistic modeling of roll-restoring moment. *Journal of Marine Science and Technology*, 9:117–126 (2004).
18. Holden, C., Galeazzi, R., Rodríguez, C., Perez, T., Fossen, T. I., Blanke, M., Neves, M. A. S.: Nonlinaer container ship model for the study of parametric roll resonance. *Modeling, Identification and Control*, 28:87–113 (2007).
19. Holden, C., Perez, T., Fossen, T. I.: Frequency-motivated observer design for the prediction of parametric roll resonance. In: *Proceedings of the 7th IFAC Conference on Control Applications in Marine Systems*. IFAC, Croatia (2007).
20. Jensen, J. J.: Efficient estimation of extreme non-linear roll motions using the first-order reliability method (FORM). *Journal of Marine Science and Technology*, 12:191–202 (2007).
21. Jensen, J. J., Pedersen, P. T., Vidic-Perunovic, J.: Estimation of parametric roll in stochastic seaway. In: *Proceedings of IUTAM Symposium on Fluid-Structure Interaction in Ocean Engineering*. Springer, Germany (2008).
22. Kay, S. M.: *Fundamental of statistical signal processing vol II: detection theory*. Prentice Hall, United States of America (1998).
23. Kerwin, J. E.: Notes on rolling in longitudinal waves. *International ShipBuilding Progress*, 2(16):597–614 (1955).
24. Levadou, M., Palazzi, L.: Assessment of operational risks of parametric roll. In: *SNAME Annual Meeting*. SNAME, United States of America (2003).
25. McCue, L. S., Bulian, G.: A numerical feasibility study of a parametric roll advance warning system. *Journal of Offshore Mechanics and Arctic Engineering*, 129:165–175 (2007).
26. Nayfeh, A. H., Mook, D. T.: *Nonlinear Oscillations*. WILEY-VCH, Germany (2004).
27. Neves, M. A. S., Rodriguez, C. A.: A coupled third order model of roll parametric resonance. In C. Guedes Soares, Y. Garbatov, and N. Fonseca, editors, *Maritime Transportation and Exploitation of Ocean and Coastal Resources*, pp. 243–253. Taylor & Francis (2005).
28. Neves, M. A.S., Rodriguez, C. A.: On unstable ship motions resulting from strong non-linear coupling. *Ocean Engineering*, 33:1853–1883 (2006).
29. Newman, J. N.: *Marine Hydrodynamics*. The MIT Press, United States of America (1977).
30. Nielsen, J. K., Pedersen, N. H., Michelsen, J., Nielsen, U. D., Baatrup, J., Jensen J. J., Petersen, E. S.: SeaSense - real-time onboard decision support. In: *World Maritime Technology Conference* (2006).
31. Oh, I. G., Nayfeh, A. H., Mook, D. T.: A theoretical and experimental investigations of indirectly excited roll motion in ships. *Philosophical Transactions: Mathematical, Physical and Engineering Sciences*, 358:1853–1881 (2000).
32. Paulling, J. R., Rosenberg, R. M.: On unstable ship motions resulting from nonlinear coupling. *Journal of Ship Research*, 3(1):36–46 (1959).
33. Shin, Y. S., Belenky, V. L., Paulling, J. R., Weems, K. M., Lin, W. M.: Criteria for parametric roll of large containeships in longitudinal seas. *Transactions of SNAME*, 112 (2004).
34. Spyrou, K. J., Tigkas, I., Scanferla, G., Pallikaropoulos, N., Themelis, N.: Prediction potential of the parametric rolling behaviour of a post-panamax container ship. *Ocean Engineering*, 35:1235–1244 (2008).
35. Tondl, A., Ruijgrok, T., Verhulst, F., Nabergoj, R.: *Autoparametric Resonance in Mechanical Systems*. Cambridge University Press, United States of America (2000).
36. Umeda, N., Hashimoto, H., Minegaki, S., Matsuda, A.: An investigation of different methods for the prevention of parametric rolling. *Journal of Marine Science and Technology*, 13:16–23 (2008).

Chapter 3
Estimation of Parametric Roll in Random Seaways

Naoya Umeda, Hirotada Hashimoto, Izumi Tsukamoto, and Yasuhiro Sogawa

Nomenclature

A_{ij}	Added mass/moment of inertia in the ith direction due to the jth motion ($j = 3$: heave, $j = 4$: roll and $j = 5$: pitch)
B_{ij}	Damping coefficient in the ith direction due to the jth motion
F_i^{DF}	Diffraction force in the ith direction
F_i^{FK+B}	Froude–Krylov force and buoyancy in the ith direction
F_n	Froude number
H	Wave height
I_{xx}	Moment of inertia in roll
m	Ship mass
N	Roll damping coefficient
ϕ	Roll angle
ϕ_a	Roll amplitude
λ	Wave length
θ	Pitch angle
ω_e	Encounter frequency
ζ	Heave displacement

N. Umeda (✉) • H. Hashimoto
Department of Naval Architecture and Ocean Engineering, Osaka University,
2-1 Yamadaoka, Suita, Osaka 565-0871, Japan
e-mail: umeda@naoe.eng.osaka-u.ac.jp; h_hashi@naoe.eng.osaka-u.ac.jp

I. Tsukamoto • Y. Sogawa
Department of Naval Architecture and Ocean Engineering, Graduate School of Engineering,
Osaka University, 2-1 Yamadaoka, Suita, Osaka 565-0871, Japan
e-mail: izuko-okuzi@hotmail.co.jp; sogawa0610@hotmail.co.jp

T.I. Fossen and H. Nijmeijer (eds.), *Parametric Resonance in Dynamical Systems*,
DOI 10.1007/978-1-4614-1043-0_3, © Springer Science+Business Media, LLC 2012

3.1 Introduction

Parametric rolling is a dangerous phenomenon for safety of cargo onboard. In regular waves, it can be evaluated with model experiments or numerical simulation in the time domain within practical accuracy [1–9]. However, this does not simply result in the ability to predict parametric roll in irregular waves because of its practical nonergodicity. Belenky et al. [10], using numerical simulation in the time domain, points out that the roll motion of a parametric rolling ship in irregular waves is nonergodic and it does not have a normal distribution. Bulian et al. [11], discusses the evaluation of parametric rolling in random waves, in the light of model experiments. These studies are executed with container ships, which are vulnerable to parametric rolling because of their exaggerated flare and transom stern [12]. In the year of 2003, however, a car carrier also suffered parametric rolling, of which the amplitude was 50° or over in the North Atlantic [13]. Using the measured data of the car carrier, a numerical study was published with the conclusion that parametric roll could not occur in its numerical experiment if the ocean waves are fully irregular [14]. A reasonably good agreement between physical and numerical experiments with a car carrier in irregular waves [15] was reported but further consideration as a random event is required.

ITTC [16] has published the "Recommended Procedure on Model Tests on Intact Stability". It states that, due to the possibility of nonergodicity in parametric rolling, several realizations of shorter durations are more desirable than one realization of long duration. However, a quantitative guidance of number and length of realizations have not been provided yet. This chapter attempts to develop a guideline for physical and/or numerical experiments in irregular waves based on the data of a car carrier. To avoid the difficulty due to practical nonergodicity, more sophisticated approaches for focusing wave groups [17, 18] or the worst wave scenario [19] are also proposed. Even so, to validate them it is essential to establish a guideline of direct experiments as mentioned above. Furthermore, physical and numerical experiments are executed for a container ship and the comparisons of these results are used to explain the applicability of the guidelines proposed in this chapter.

3.2 Model Experiment

The authors conducted model experiments with a car carrier; see Table 3.1 and Fig. 3.1. This is a typical car carrier nowadays and it is roughly similar to the ship suffering parametric rolling in the North Atlantic as reported by [13]. Its exaggerated bow flare and distinct transom stern are causing the parametric rolling.

The model was moored with elastic ropes at the bow point for realizing a head wave condition at a point located 40 m away from the wave maker. The surge, sway, and yaw motions were softly restrained and the roll, pitch, and yaw motions were measured by a gyroscope.

Table 3.1 Principal particulars of the car carrier

Items	Ship	Model
Length between perpendiculars: L_{pp}	192.0 m	3.0 m
Breadth: B	32.26 m	0.506 m
Depth: D	37.0 m	0.578 m
Mean draught: d	8.18 m	0.128 m
Displacement: W	27,908 ton	103.95 kg
Block coefficient: C_b	0.537	0.537
Metacentric height: $\overline{GM}$	1.25 m	0.020 m
Natural roll period: T_φ	22.0 s	2.75 s
Pitch radius of gyration: κ_{yy}/L_{pp}	0.25	0.25

Fig. 3.1 Body plan of the car carrier

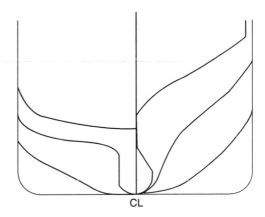

CL

Table 3.2 Experimental conditions for the car carrier

	Significant wave height ($H_{1/3}$ [m])		Mean wave period (T_{01} [s])		
	Ship scale	Model scale	Ship scale	Model scale	Number of trials
Case A	5.31	0.0830	9.76	1.22	30
Case B	2.655	0.0415	9.76	1.22	25

Long-crested irregular waves were generated using the ITTC spectrum and their conditions are shown in Table 3.2. Case A corresponds to the severe parametric rolling condition that the car carrier faced in 2003 [13]. The model experiment was conducted with more than 25 different realization and each time its duration was more than 750 s.

Prior to these systematic experiments, some trials were executed with different significant wave heights using only four realizations for each wave height. As shown in Fig. 3.2, a threshold for parametric rolling exists near the significant wave height of 2.66 m for this mean wave period. Thus Case B is chosen to be close to the threshold for parametric rolling in random waves.

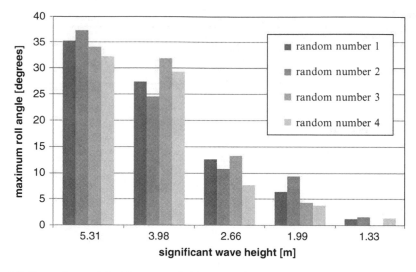

Fig. 3.2 Dependence of significant wave height on maximum roll angle

3.3 Evaluation Methodology for Parametric Rolling

It is difficult to evaluate parametric rolling in irregular waves due to its practical nonergodicity. This means that the temporal average of one long realization is not equal to the ensemble average of many different realizations.

Therefore, the standard deviations of each realization is used. First, the ensemble average of the standard deviation of the random process, $X_i(t)$, is defined as follows:

$$\sigma_{\text{erg}}^2 := \frac{\sum_{i=1}^{N} \sum_{j=1}^{n} \frac{\left(X_i(t_j) - \bar{X}_i\right)^2}{n-1}}{N}, \tag{3.1}$$

where n is the sample point number of each realization, N is the number of realizations and $\bar{X}_i$ is the mean of each realization. Second, the running standard deviation (running STD) is defined as a function of time as follows:

$$\sigma^2(t) := \frac{\sum_{j=1}^{n_s} \left(X_j(t_j) - \bar{X}(t)\right)^2}{n-1}, \tag{3.2}$$

where n_s is the sample point number of each realization up to the specified time, t and $\bar{X}(t)$ is the mean of each realization up to t. In the calculations, records during the first 180 s are excluded as transient states [11].

3.4 Experimental Results

The results of pitch motion and wave elevation in Case A are shown in Figs. 3.3 and 3.4, respectively. Since the running STDs tend to converge to the ensemble averages of the standard deviations, it is suggested that the wave elevation and pitch motion in irregular waves can be regarded as ergodic process within practical accuracy, that is, 3% or less. This assumption is well established in conventional seakeeping studies.

In contrast, the running STDs of the roll motions in irregular waves as shown in Figs. 3.5 and 3.6, do not converge to the ensemble averages of the standard deviations. In Case A the dispersion is approximately 20% and in Case B more than 40%. This is because the condition of Case B is close to the threshold of parametric rolling. In irregular waves, parametric roll occurs only when a ship meets a wave group satisfying the condition of parametric rolling. Thus if the condition is sufficiently above the threshold, parametric roll frequently occurs. As a result, the dispersion of the standard deviation could be smaller as shown in Case A. These results clearly indicate that the temporal average is different from the ensemble average within long but limited time duration of the experiments, while the ergodicity itself should be discussed with the average of the infinite time duration. Consequently, it can be concluded that parametric roll in irregular waves can be viewed as a "practically nonergodic" process [11].

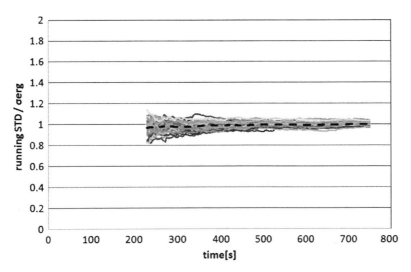

Fig. 3.3 Running STD of wave elevation normalized with ensemble average in Case A. The dotted line indicates the ensemble average of running STDs

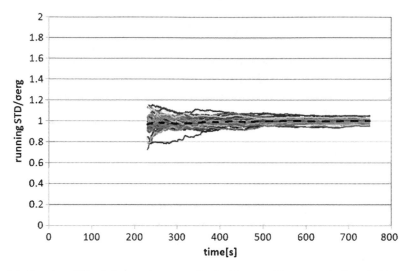

Fig. 3.4 Running STD of pitch motion normalized with ensemble average in Case A. The dotted line indicates the ensemble average of running STDs

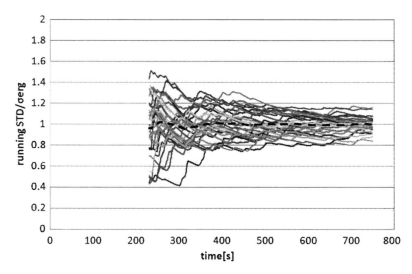

Fig. 3.5 Running STD of roll motion normalized with ensemble average in Case A. The dotted line indicates the ensemble average of running STDs

3.5 Quantitative Guidelines from Model Experiments

To obtain quantitative guidelines from the experimental results, we attempt to compare the *set average* of the realizations. The set average is defined as the average of realizations randomly selected from the total set of experimental

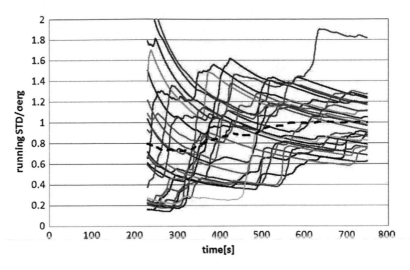

Fig. 3.6 Running STD of roll motion normalized with ensemble average in Case B. The dotted line indicates the ensemble average of running STDs

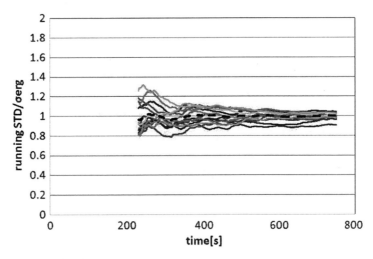

Fig. 3.7 Four realizations set average of roll STD in Case A. The dotted line indicates the ensemble average of running STDs

realizations. The numbers of selected realizations are chosen as follows: 4, 8, 12, and 16. The results for Case A are shown in Figs. 3.7–3.10 and for Case B in Figs. 3.11–3.14.

When the realization number for the set average increases, the fluctuation decreases in Case A. Here the 12-realization set average shows only 5% errors at the

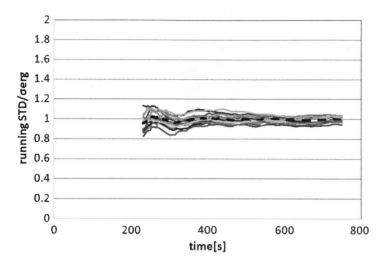

Fig. 3.8 Eight realizations set average of roll STD in Case A. The dotted line indicates the ensemble average of running STDs

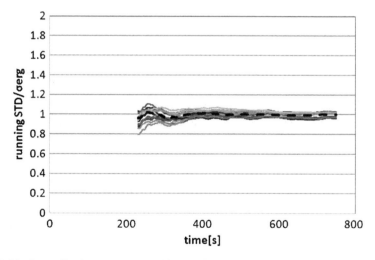

Fig. 3.9 Twelve realizations set average of STD in Case A. The dotted line indicates the ensemble average of running STDs

point of 500 s. This is comparable to the error of the pitch motion and wave elevation so this will be a permissible range. The estimation error of parametric rolling could be approximately 5% of the real value. Therefore, for this case it is recommended to use more than 12-realizations and each realization should be more than 500 s, which corresponds to 180 cycles of the natural roll period.

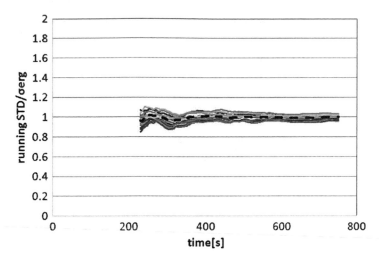

Fig. 3.10 Sixteen realizations set average of roll STD in Case A. The dotted line indicates the ensemble average of running STDs

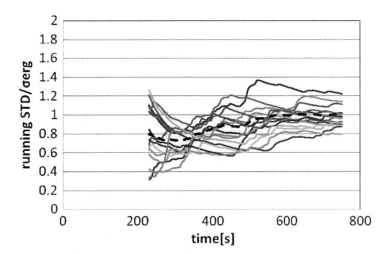

Fig. 3.11 Four realizations set average of roll STD in Case B. The dotted line indicates the ensemble average of running STDs

In Case B, it has more than 20% error at the same point of Case A. The error does not depend much on the number of set averages and time of realization. This is due to the fact that the condition of Case B is close to the threshold of parametric rolling for the car carrier. Comparing 12- and 16-realizations set averages in Case B, no remarkable difference exists. Therefore, it should be noted that such dispersion cannot be avoided for evaluation of parametric roll in irregular waves under the

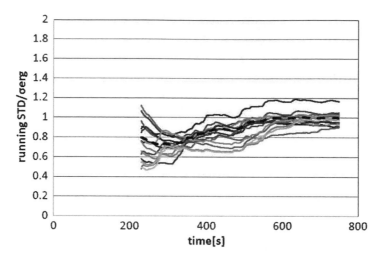

Fig. 3.12 Eight realizations set average of roll STD in Case B. The dotted line indicates the ensemble average of running STDs

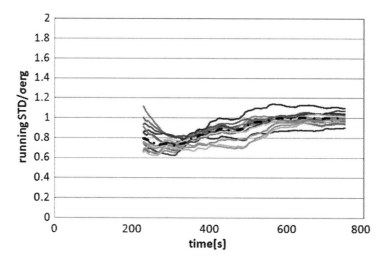

Fig. 3.13 Twelve realization set average of roll STD in Case B. The dotted line indicates the ensemble average of running STDs

condition close to the threshold. It is noteworthy that this guideline is obtained with this particular car carrier and only for these particular wave conditions. Thus further experiments with different ships and waves are required. Nevertheless, the guideline specification as well as the methodology used seems to be a base for further investigation using more comprehensive experimental data sets.

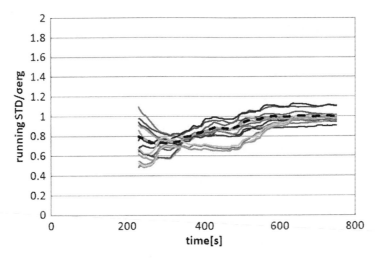

Fig. 3.14 Sixteen realizations set average of roll STD in Case B. The dotted line indicates the ensemble average of running STDs

3.6 Container Ship Case Study

To examine the applicability of the proposed guideline, physical and numerical experiments in irregular waves were conducted for a C11 class post-Panamax container ship modified by MARIN [20]. Its principal particulars and body plan are given in Table 3.3 and Fig. 3.15, respectively.

The physical model experiments for a 1/100 scaled model of the container ship in regular and irregular head waves were executed in a similar procedure for the car carrier model. The authors also executed numerical simulation for this container ship with a coupled heave–roll–pitch model given by [9]:

$$(m+A_{33}(\phi))\ddot{\zeta}+B_{33}(\phi)\dot{\zeta}+A_{34}(\phi)\ddot{\phi}+B_{34}(\phi)\dot{\phi}+A_{35}(\phi)\ddot{\theta}+B_{35}(\phi)\dot{\theta}$$
$$=F_3^{\mathrm{FK}+B}(\xi_G/\lambda,\zeta,\phi,\theta)+F_3^{\mathrm{DF}}(\phi) \tag{3.3}$$

$$(I_{xx}+A_{44}(\phi))\ddot{\phi}+N(\dot{\phi})+A_{43}(\phi)\ddot{\zeta}+B_{43}(\phi)\dot{\zeta}+A_{45}(\phi)\ddot{\theta}+B_{45}(\phi)\dot{\theta}$$
$$=F_4^{\mathrm{FK}+B}(\xi_G/\lambda,\zeta,\phi,\theta)+F_4^{\mathrm{DF}}(\phi) \tag{3.4}$$

$$(I_{yy}+A_{55}(\phi))\ddot{\theta}+B_{55}(\phi)\dot{\theta}+A_{53}(\phi)\ddot{\zeta}+B_{53}(\phi)\dot{\zeta}+A_{54}(\phi)\ddot{\phi}+B_{54}(\phi)\dot{\phi}$$
$$=F_5^{\mathrm{FK}+B}(\xi_G/\lambda,\zeta,\phi,\theta)+F_5^{\mathrm{DF}}(\phi) \tag{3.5}$$

In this model, the nonlinear Froude–Krylov forces are calculated by integrating wave pressure up to the wave surface. Dynamic components, that is, radiation and diffraction forces are calculated for an instantaneous submerged hull by

Table 3.3 Principal
particulars of the modified
C11 container ship

Item	Value
Length between perpendiculars: L	262.0 m
Breadth: B	40.0 m
Depth: D	24.45 m
Mean draught: T	11.5 m
Block coefficient: C_b	0.56
Metacentric height: $\overline{GM}$	1.965 m
Natural roll period: T_ϕ	25.1 s

Fig. 3.15 Body plan of the
modified C11 container ship

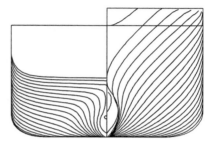

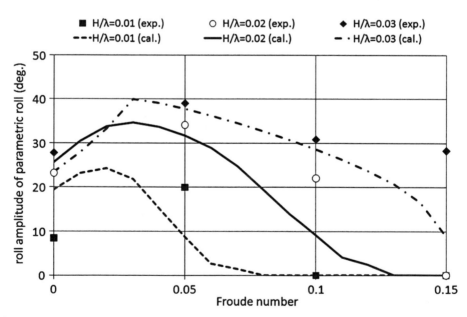

Fig. 3.16 Comparison of parametric roll amplitude for the modified C11 class container ship in regular waves between the physical and numerical experiments with the wave length to ship length ratio of 1.0 [9]

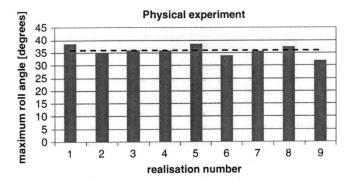

Fig. 3.17 Results of the maximum roll angle from the physical model experiments of the modified C11 class container ship in irregular head waves with the significant wave height of 10.43 m, the mean wave period of 9.99 s and the Froude number of 0.0. Here the dotted line indicates the mean of the maximum roll angles

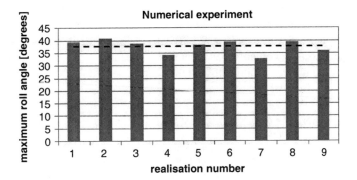

Fig. 3.18 Results of the maximum roll angle from the numerical experiments of the modified C11 class container ship in irregular head waves with the significant wave height of 10.43 m, the mean wave period of 9.99 s and the Froude number of 0.0. Here the dotted line indicates the mean of the maximum roll angles

considering a time-dependent roll angle. Two-dimensional hydrodynamic forces are calculated by solving the boundary integral equation for the velocity potential. Diffraction forces are calculated by the Salvesen–Tuck–Faltinsen (STF) method [21]. Hydrodynamic forces for vertical motion and diffraction modes are calculated with the mean wave frequency while for the sway and roll modes the natural roll frequency assuming parametric roll resonance are used. The roll damping moment is estimated from decay tests using the physical ship model.

In case of regular waves, good agreements between the physical and numerical experiments were already reported as shown in Fig. 3.16 by the authors [9]. Furthermore, the results shown in Figs. 3.17 and 3.18 show reasonably good agreements in the maximum roll angles between the physical and numerical experiments in irregular waves for several realizations. The duration of each realization is 500 s.

It is noteworthy that good agreements exist not only in the mean of the maximum roll angles but also for the dispersion of the roll angle. Thus it suggests that the tentatively proposed guideline could be applicable to both physical and numerical experiments.

3.7 Concluding Remarks

Model experiments of a car carrier were conducted with many realizations for long-time duration in order to establish guideline specifications for model experiment and numerical simulation of parametric rolling. It is confirmed that parametric rolling in irregular waves has practical nonergodicity. In order to evaluate parametric rolling, it is recommended to average many realizations. This chapter confirms that about 12-realizations and more than 360 encounter wave cycles should be used in the experiments. For cases close to the threshold for parametric rolling accurate estimation is difficult. Numerical results for parametric roll estimation should be sufficiently above the threshold when it is used in practical stability assessment. Reasonable good agreements between the physical and numerical experiments suggest that the above tentative guideline could be applicable to both physical and numerical experiments.

Acknowledgements This work was supported by a Grant-in Aid for Scientific Research of the Japan Society for Promotion of Science (No. 21360427). It was partly carried out as a research activity of Stability Project of Japan Ship Technology Research Association in the fiscal years of 2009–2010, funded by the Nippon Foundation. The author thanks Mr. Keiji Ohnishi for his assistance during the experiment and Dr. Gabriele Bulian for his fruitful discussion.

References

1. Munif, A., Umeda, N.: Modeling Extreme Roll Motions and Capsizing of a Moderate-Speed Ship in Astern Waevs, Journal of the Society of Naval Architects of Japan, Vol. 187, pp. 51–58 (2000)
2. Ribeiro e Silva, S., Santos, T., Guedes Soares, C.: Time Domain Simulations of a Coupled Parametrically Excited Roll Response in Regular and Irregular Head Seas, Proceedings of the 8th International Conference on the Stability of Ships and Ocean Vehicles, pp. 349–360 (2003)
3. Neves, M. A. S., Rodriguez, C.: A Non-Linear Mathematical Model of Higher Order for Strong Parametric Resonance of the Roll Motion of Ship in Waves, Marine Systems and Ocean Technology, Vol. 1, No. 2, pp. 69–81 (2005)
4. Brunswig, J., Pereira, R.: Validation of Parametric Roll Motion Predictions for a Modern Container Ship Design, Proceedings of the 9th International Conference on Stability of Ships and Ocean Vehicles, pp. 157–168 (2006)
5. Spanos, D., Papanikolaou, A.: Numerical Simulation of Parametric Roll in Head Seas, Proceedings of the 9th International Conference on Stability of Ships and Ocean Vehicles, pp. 169–180 (2006)

6. Taguchi, H., Ishida, S., Sawada, H., Minami, M.: Model Experiment on Parametric Rolling of a Post-Panamax Container Ship in Head Waves, Proceedings of the 9th International Conference on Stability of Ships and Ocean Vehicles, Vol. 1, pp. 147–156 (2006)
7. Ogawa, Y.: An Examination for the Numerical Simulation of Parametric Roll in Head and Bow Waves, Proceedings of the 9th International Ship Stability Workshop, Hamburg, pp. 4.2.1–4.2.8 (2007)
8. Rodriguez, C., Holden, C., Perez, T., Drummen, I., Neves, M. A. S., Fossen, T. I.: Validation of a Container Ship Model for Parametric Rolling, Proceedings of the 9th International Ship Stability Workshop, Hamburg, pp. 4.4.1–4.4.11 (2007)
9. Hashimoto, H., Umeda, N.: A Study on Quantitative Prediction of Parametric Roll in Regular Waves, Proceedings of the 11th International Ship Stabilityb Workshop, Wageningen, pp. 295–301 (2010)
10. Belenky, V., Weems, K. M., Paulling, J. R.: Probabilistic Analysis of Roll Parametric Resonance in Head Seas, Proc. STAB'2003, pp. 325–340 (2003)
11. Bulian, G., Francescutto, A., Umeda, N., Hashimoto, H.: Qualitative and Quantitative Characteristics of Parametric Roll in Random waves in the Light of Physical Model Experiments, Ocean Engineering Vol. 35, No. 17–18, pp. 1661–1675 (2008)
12. France, W. L., Levadou, M., Treakle, T. W., Paulling, J. R., Michel, K., Moore, C.: An Investigation of Head-Sea Parametric Rolling and Its Influence on Container Lashing Systems. Marine Technology Vol. 40, No. 1, pp. 1–19 (2003)
13. Sweden: Recordings of Head Sea Parametric Rolling on a PCTC, SLF 47/INF.5, IMO (2004)
14. Hua, J., Palmquist, M., Lindgren, G.: An Analysis of the Pasrametric Roll Events Measured Onboard the PCTC AIDA, Proc. STAB2006, Vol. 1, pp. 109–118 (2006)
15. Hashimoto, H., Umeda, N., Sakamoto, G.: Head-Sea Parametric Rolling of a Car Carrier, Proceedings of the 9th International Ship Stability Workshop, Hamburg, pp. 3.5.1–3.5.7 (2007)
16. ITTC: Recommended Procedures, Model Tests on Intact Stability, 7.5-02-07-04.1 (2008)
17. Blocki, W.: Ship Safety in Connection with Parametric Resonance of the Roll, International Shipbuilding Progress, Vol. 27, No. 306, pp. 36–53 (1980)
18. Themelis, N., Spyrou, K. J.: Probabilistic Assessment of Ship Stability, Transactions of the Society of Naval Architects and Marine Engineers, Vol. 97, pp. 139–168 (2007)
19. Jensen, J. J.: Efficient estimation of extreme non-linear roll motions using the first-order reliability method (FORM), Journal of Marine Science and Technology, Vol. 12, No. 4, pp. 191–202 (2007)
20. Levadou, M., Van't Veer, R.: Parametric Roll and Ship Design, Proceedings of the 9th International Conference of Ships and Ocean Vehicles, Vol. 1, pp. 191–206 (2006)
21. Salvesen, N., Tuck, E. O., Faltinsen, O. M.: Ship Motions and Sea Load, Transactions of the Society of Naval Architects and Marine Engineers, Vol. 78, pp. 250–287 (1970)

Part II
Parametric Roll

Chapter 4
Trimaran Vessels and Parametric Roll

Gabriele Bulian and Alberto Francescutto

4.1 Introduction

Multihull vessels are typically considered as a viable alternative to standard fast monohulls in particular when high speed and large deck areas are required. Typical extreme configurations can be considered as monohulls on one side and catamarans on the other side. However, between these two extremes, it is possible to find a category of ships, i.e., "ships with outriggers" [12], which are characterized by a central, usually slender, monohull and typically two (for trimarans) or four (for pentamarans) outriggers. In this paper we focus our attention to the case of trimaran, i.e., multihull vessels with two outriggers and a central slender hull. The pentamaran configuration can be considered along with the trimaran configuration, at least for what concerns linear hydrodynamics, because the two additional outriggers in pentamarans are typically outside water in calm water condition [10]. The very slender central hull is typically characterized by a limited, or even negative, initial stability and outriggers allow to compensate for this drawback by providing additional static stability. From the point of view of transport efficiency, trimaran/pentamaran vessels are potentially very competitive [14] especially when passengers and/or volume-based cargoes are of concern [12]. From the point of view of ship motions, favorable seakeeping characteristics of ships with outriggers have also been observed in bow/head waves [10, 12, 15], making these types of configurations potentially beneficial in terms of, e.g., passengers' comfort. Thanks to the fact that, with respect to a monohull, the positions, shapes, and dimensions of outriggers represent additional design parameters, ships with outriggers provide more degrees of freedom to the designer in the ship optimization procedure. This makes the design of ships with outriggers an interesting and challenging topic, especially in view of

G. Bulian (✉) • A. Francescutto
Department of Mechanical Engineering and Naval Architecture, University of Trieste,
Via A. Valerio 10, 34127 Trieste, Italy
e-mail: gbulian@units.it; francesc@units.it

T.I. Fossen and H. Nijmeijer (eds.), *Parametric Resonance in Dynamical Systems*,
DOI 10.1007/978-1-4614-1043-0_4, © Springer Science+Business Media, LLC 2012

the fact that the positions and shapes of the outriggers have direct consequences on resistance, static/dynamic stability, seakeeping, and manoeuvring. The total ship resistance can indeed be modified by changing the relative position between the main hull and the outriggers as a consequence of the wave interference [2]. Also the restoring moment of the ship in calm water is influenced by the positions and shapes of the outriggers and this also has direct consequences on the nonlinear rolling motion characteristics of ships with outriggers [6]. Of particular interest, from both the design and research point of view, it is the possibly strongly nonlinear roll behavior of ships with outriggers when the transversal separation of the outriggers is not very large and when the draught of the outriggers is limited. In such cases, indeed, the emergence of the outriggers at small heeling angles in calm water leads to a significant change in the slope of the restoring moment at relative small heeling [2, 6] and this leads also to a significant bending of the roll response curve in beam waves [6]. Designs with outriggers having limited draught can also lead to issues in longitudinal waves for what concerns parametric roll. Indeed, considering the case of a ship with outriggers in longitudinal regular waves, the possible emergence of outriggers as the wave crest moves along the ship can lead to significant variations of the metacentric height with respect to the calm water value. A sufficiently large variation of metacentric height with respect to the calm water value can eventually lead to the inception of parametric roll if the appropriate conditions are met. It would therefore be beneficial to design the outriggers with a sufficient draught [11]. However, in order to reduce the overall ship resistance, the draught of the outriggers, and hence their wetted surface, is often kept at minimum. This makes certain ships with outriggers potentially prone to the inception of parametric roll and this is exactly the topic covered by this paper for the particular case of a trimaran vessels tested at the Hydrodynamic Laboratories of the University of Trieste.

Investigations carried out on trimaran vessels have shown that the fluctuation of the metacentric height in longitudinal regular waves for this type of ships shows considerable differences with respect to the case of standard monohulls. In particular the position of the wave crest leading to the minimum of the metacentric height in waves depends on the position of the outriggers, while in case of monohulls the minimum of the metacentric height in waves is typically associated with a wave crest close to amidships. In addition, the fluctuation of the metacentric height in case of trimaran configurations contains a not negligible contribution from harmonics higher than the first harmonic, particularly from the second harmonic, even at small wave steepnesses [9], while in case of monohulls the fluctuation of the metacentric height is governed by its first harmonic even for relatively large wave steepnesses [4]. Due to the nonnegligible contribution from higher harmonics, particularly the second harmonic, in the fluctuation of the metacentric height, the Mathieu equation is no longer an appropriate model for checking the inception of parametric roll especially in the second (or higher) parametric resonance regions. As a natural extension it is therefore more appropriate to analyse the stability of the upright position by means of a more general Hill equation-based model.

The paper is structured as follows. Firstly, the nonlinear 1-DOF mathematical model used for describing parametrically excited rolling motion in longitudinal regular waves is presented, together with its corresponding linearized form, which is necessary for the linear stability analysis of the upright position. In this context the generic Floquet theory is described in view of its numerical implementation in this study. Indeed, for taking into account the complex nonpurely-sinusoidal behavior of the metacentric height in waves, the Floquet theory is applied numerically in order to obtain information on the stability of the upright position in generic ranges of model's parameters. The ship used for numerical simulations and experimental tests is then described. An example of computation of regions of instability of the upright position (stability map) is reported highlighting the importance of taking into account harmonics in the fluctuation of $\overline{\mathrm{GM}}$ up to at least the second one. The rolling amplitude in the nonlinear range is then addressed in order to show some peculiar features of the parametrically excited roll response for the considered trimaran vessel. In both cases relevant experimental data are also reported. Finally an example of determination of optimum positioning of the outriggers is shown where the target is the minimization of the variation of the metacentric height in waves.

4.2 Description of the 1-DOF Mathematical Model for Roll Motion in Longitudinal Regular Waves

The employed numerical model is a 1-DOF model for regular longitudinal waves [8], where roll restoring in waves is calculated using a quasi-static approach for heave and pitch. The model can be considered similar also when a dynamic description is given for heave and pitch if the coupling of roll into vertical motions is neglected. Smith effect is not considered and a purely hydrostatic pressure under the nonflat sea surface is assumed for the computation of the roll restoring moment in waves. This approximation is usually acceptable in case of long waves, where the ship's draught is small in comparison with the wave length. In case of shorter waves this approximation could be no longer valid, and the actual pressure under the wave should be considered. However, the same model can be applied also by calculating the restoring moment using the actual pressure field under the wave. In the following we assume that the ship, when upright in calm water, has positive metacentric height ($\overline{\mathrm{GM}}$), i.e., we implicitly assume that $\phi = 0$ is a stable equilibrium in calm water. Moreover we assume that the ship speed is constant, i.e., we neglect the surge motion. The nonlinear 1-DOF model for roll motion in longitudinal waves is written as:

$$\ddot{\phi} + d\left(\dot{\phi}, V\right) + \frac{\Delta}{J'_{xx}}\overline{\mathrm{GZ}}(\phi, V, x_{\mathrm{c}}(t), \lambda_{\mathrm{w}}, a_{\mathrm{w}}) = 0, \qquad (4.1)$$

where

- ϕ [rad] is the roll angle
- V [m/s] is the ship speed
- $d\left(\dot{\phi},V\right)$ [rad/s^2] is the damping function, explicitly considering a dependence on forward speed.
- Δ [N] is the ship displacement (assumed constant according to the quasi static assumption for heave)
- J'_{xx} [kg·m^2] is the roll moment of inertia comprising the effect of added inertia
- $\overline{GZ}(\phi,V,x_c(t),\lambda_w,a_w)$ [m] is the roll righting lever depending on

  - roll angle ϕ
  - ship speed V
  - instantaneous wave crest position $x_c(t)$ [m] along the ship
  - wave length λ_w [m]
  - wave amplitude a_w [m]

In the mathematical model (4.1) a speed dependence is considered on both the damping term and the restoring. Taking into account speed effects on damping is a quite standard and necessary procedure in order to reflect in particular the additional linear damping due to lift effects, but also the often observed reduction of nonlinear damping terms at forward speed [17]. However, forward speed lift effects also affect the ship restoring [27] and this effect is seldom considered. A complete description of the dependence of restoring on speed, heeling angle, and wave characteristics position is a quite complex problem which requires an extensive experimental or numerical investigation. Here we introduce lift effects in the model (4.1) in a simplified way on the basis of a superposition assumption. The forward speed righting lever is assumed to be decomposed as follows:

$$\overline{GZ}(\phi,V,x_c(t),\lambda_w,a_w) = \overline{GZ}_{zs}(\phi,x_c(t),\lambda_w,a_w) + \delta\overline{GZ}(\phi,V)$$

$$\overline{GZ}_{zs}(\phi,x_c(t),\lambda_w,a_w) = \overline{GZ}(\phi,V=0,x_c(t),\lambda_w,a_w), \tag{4.2}$$

where $\overline{GZ}_{zs}(\phi,x_c(t),\lambda_w,a_w)$ is the righting lever in waves at zero speed, while $\delta\overline{GZ}(\phi,V)$ is the forward speed contribution which, for simplification, is assumed to depend only on the ship speed and not on the wave characteristics. It is therefore worth underlining that the assumed simplified modeling neglects the direct effect of the wave on $\delta\overline{GZ}(\phi,V)$. In our experiments we have not measured directly the restoring moment at forward speed, but we have carried out roll decays at forward speed. Therefore the forward speed contribution to the restoring has been further simplified using a metacentric approximation as follows:

$$\delta\overline{GZ}(\phi,V) \approx \delta\overline{GM}(V)\sin(\phi), \tag{4.3}$$

where the variation of metacentric height $\delta\overline{GM}(V)$ is assumed to be representative of the speed dependent part of $\delta\overline{GZ}(\phi,V)$. The indirect determination of the speed dependent part $\delta\overline{GM}(V)$ of the metacentric height has been based on the analysis of the roll natural frequency $\omega_0(V)$ from roll decays at forward speed. Indeed, the increase or decrease of the metacentric height at forward speed can be associated with an increase or decrease, respectively, of the natural roll frequency with respect to the calm water value. Accordingly, it is possible to indirectly determine $\delta\overline{GM}(V)$ as follows:

$$\delta\overline{GM}(V) = \overline{GM}_{zs}\left(\frac{\omega_0^2(V)}{\omega_{0,zs}^2} - 1\right)$$

$$\omega_{0,zs} = \omega_0(V = 0), \tag{4.4}$$

where $\overline{GM}_{zs}$ [m] and $\omega_{0,zs}$ [rad/s] are the zero speed calm water metacentric height and roll natural frequency respectively. It must be underlined that the relation (4.4) implicitly assume that forward speed effect on the roll added inertia are negligible with respect to forward speed lift effects on restoring. A quite standard explicit modeling of the roll damping function $d(\dot{\phi}, V)$ can be considered to be given by a linear + quadratic + cubic model based on the roll velocity:

$$d(\dot{\phi}, V) = 2\mu(V)\dot{\phi} + \beta(V)\dot{\phi}|\dot{\phi}| + \delta(V)\dot{\phi}^3, \tag{4.5}$$

where the coefficients of the damping function, considered to be possibly dependent on the ship speed V, are as follows:

- $\mu(V)$ [1/s] is the linear roll damping coefficient;
- $\beta(V)$ [1/rad] is the quadratic roll damping coefficient;
- $\delta(V)$ $\left[s/rad^2\right]$ is the cubic roll damping coefficient;

Using a damping modeling as in (4.5), the model (4.1) can be rewritten as:

$$\ddot{\phi} + 2\mu(V)\dot{\phi} + \beta(V)\dot{\phi}|\dot{\phi}| + \delta(V)\dot{\phi}^3$$
$$+ \omega_{0,zs}^2 \frac{\overline{GZ}_{zs}(\phi, x_c(t), \lambda_w, a_w) + \delta\overline{GM}(V)\sin(\phi)}{\overline{GM}_{zs}} = 0. \tag{4.6}$$

The model (4.6) is a nonlinear one which is therefore suitable for direct time-domain simulations of roll motion in the nonlinear range. However, one of the important parts of parametric roll assessment, and often the initial part of the assessment procedure, is the determination of the regions of parameters where the upright position $\phi = 0$ becomes unstable under the parametric excitation induced by the considered longitudinal wave(s). To this end it is necessary to use the linearization of the model (4.6) close to $\phi = 0$, which is:

$$\ddot{\phi} + 2\mu(V)\dot{\phi} + \omega_{0,zs}^2 \frac{\overline{GM}_{zs}(x_c(t), \lambda_w, a_w) + \delta\overline{GM}(V)}{\overline{GM}_{zs}}\phi = 0, \tag{4.7}$$

where

$$\overline{GM}_{zs}\left(x_c(t),\lambda_w,a_w\right) = \left.\frac{\partial\overline{GZ}_{zs}\left(\phi,x_c(t),\lambda_w,a_w\right)}{\partial\phi}\right|_{(\phi=0,x_c(t),\lambda_w,a_w)}. \tag{4.8}$$

It is also worth noticing that, according to the models (4.6) and (4.7), and considering a constant forward speed, it is possible to define a reference average roll natural frequency at forward speed $\omega_0\left(V,\lambda_w,a_w\right)$ for a given wave length λ_w and wave amplitude a_w as follows:

$$\omega_0\left(V,\lambda_w,a_w\right) = \omega_{0,zs}\sqrt{\frac{\left\langle\overline{GM}_{zs}\left(\lambda_w,a_w\right)\right\rangle+\delta\overline{GM}\left(V\right)}{\overline{GM}_{zs}}}, \tag{4.9}$$

where

$$\left\langle\overline{GM}_{zs}\left(\lambda_w,a_w\right)\right\rangle = \frac{1}{\lambda_w}\int_0^{\lambda_w}\overline{GM}_{zs}\left(x_c,\lambda_w,a_w\right)dx_c. \tag{4.10}$$

Such reference average natural roll frequency is the roll frequency which governs the tuning between the ship and the waves for the inception of parametric roll. The special case of $a_w = 0$ corresponds to the calm water case. In this condition, in accordance with (4.4), the calm water natural roll frequency $\omega_0\left(V\right)$ at forward speed becomes:

$$\left\langle\overline{GM}_{zs}\left(\lambda_w,a_w=0\right)\right\rangle = \overline{GM}_{zs} \Rightarrow \omega_0\left(V\right) = \omega_{0,zs}\sqrt{1+\frac{\delta\overline{GM}\left(V\right)}{\overline{GM}_{zs}}}. \tag{4.11}$$

In general, the average zero speed metacentric height in waves $\left\langle\overline{GM}_{zs}\left(\lambda_w,a_w\right)\right\rangle$ is different from the calm water zero speed metacentric height $\overline{GM}_{zs}$. It therefore follows that the ratio between the reference natural roll frequency in waves $\omega_0\left(V,\lambda_w,a_w\right)$ and the encounter wave frequency (the so called tuning ratio) is not only a function of the speed, but it also depends on the wave length and wave amplitude. Since the inception of parametric roll is strongly influenced by the tuning ratio, it follows that, for a given wave length λ_w, the "most dangerous" ship speeds can significantly depend on the wave amplitude a_w, in particular when $\omega_0\left(V,\lambda_w,a_w\right)$ is strongly dependent on a_w.

For a given ship in given loading condition one of the interests is usually the determination of the set of parameters $\left(\lambda_w,a_w,V\right)$ for which $\phi = 0$ is unstable. The linear model (4.7) represents the basis for determining such set of parameters. It is important to note that the model (4.7) contains speed effects both in damping and in restoring, although with some approximations. Moreover the fluctuation of the metacentric height is, in general, nonlinearly dependent on the wave amplitude a_w for a given wave length λ_w. In addition, the time varying part $\overline{GM}_{zs}\left(x_c(t),\lambda_w,a_w\right)$ of the linearized restoring term is, in general, and in particular in case of trimaran vessels, a nonpurely-sinusoidal function of the time t (or, equivalently, of the wave crest position $x_c(t)$) [3, 4]. Instead, the Fourier decomposition of $\overline{GM}_{zs}\left(x_c(t),\lambda_w,a_w\right)$

contains, in general, and in particular for trimaran vessels, significant contributions from higher harmonics and particularly from the second harmonic component [9]. If $\overline{GM}_{zs}(x_c(t), \lambda_w, a_w)$ has, for a given wave length λ_w and a given wave amplitude a_w, a significant harmonic content, the widely used approach based on a Mathieu-like approximation of the model (4.7) (e.g., [1,7]) becomes unsuitable.

Considering a generic Fourier expansion of the restoring term, the model (4.7) can be assimilated to a generic Hill equation [16] with damping. Such model can show more complex and/or extended instability regions for the solution $\phi = 0$ with respect to its single harmonic, i.e. Mathieu-like, approximation. In order to handle this increased complexity and in order to address all the parametric resonance region the linear stability analysis of $\phi = 0$ is therefore carried out by exploiting the generic results from Floquet theory.

4.3 Floquet Theory and Its Direct Numerical Application

Floquet theory provides a criterion for determining whether the equilibrium solution of a linear system of first-order differential equations with periodic coefficients is unstable or not. Mathematical details of Floquet theory can be found in literature [16,18,19,22,23]. The intention of this section is to provide the essence of the ideas behind the numerical application of the Floquet theory as used in this paper.

As starting point we use the following generic linear first-order dynamic model in $\mathbb{R}^N$ with periodic matrix of coefficients $\mathbf{P}(t)$ with period T_P:

$$\begin{cases} \dot{\mathbf{u}} = \mathbf{P}(t)\mathbf{u} \\ \mathbf{P}(t + T_P) = \mathbf{P}(t), \end{cases} \tag{4.12}$$

where $\mathbf{u} \in \mathbb{R}^{N \times 1}$ and $\mathbf{P} \in \mathbb{R}^{N \times N}$. The vector $\mathbf{u}(t)$ is the state vector of the system under analysis. Since the system (4.12) is linear and homogeneous the origin of the state space is a solution, i.e., $\mathbf{u}(t) \equiv \mathbf{0}$ is always a solution of (4.12). The problem is to determine when such solution is stable and when it is unstable. It is worth highlighting already here that the linearized roll motion equation (4.7) can be put in the form (4.12) by using $\mathbf{u} = (\phi, \dot{\phi})^T$, where the superscript T represents the transpose operator.

In non-degenerate cases it is possible to construct for (4.12) a set of N independent fundamental solutions. Any other solution of (4.12) can be obtained by means of linear combination of the fundamental set of solutions. Although the way of determining the fundamental set of independent solutions is not unique, a choice for the generic jth fundamental solution $s^j(t)$ with $j = 1, \ldots, N$ could be:

$$\begin{cases} \dot{s}^j = \mathbf{P}(t)s^j \\ (s^j(t=0))_k = 1 \text{ for } k = j \\ (s^j(t=0))_k = 0 \text{ for } k \neq j. \end{cases} \tag{4.13}$$

When the fundamental set of solutions $s^j(t)$ according to (4.13) is known, it is possible to create an $N \times N$ matrix $\boldsymbol{\Sigma}(t)$ which contains, in the generic jth column, the fundamental solution $s^j(t)$. Accordingly, $\boldsymbol{\Sigma}(t)$ can be considered as solution of the following system:

$$\begin{cases} \dot{\boldsymbol{\Sigma}} = \mathbf{P}(t)\boldsymbol{\Sigma} \\ \mathbf{P}(t + T_P) = \mathbf{P}(t) \\ \boldsymbol{\Sigma}(t = 0) = \mathbf{I}_N \\ \boldsymbol{\Sigma}(t) \in \mathbb{R}^{N \times N} \; ; \; \mathbf{P} \in \mathbb{R}^{N \times N}, \end{cases} \tag{4.14}$$

where $\mathbf{I}_N$ is the $N \times N$ identity matrix. When the matrix $\boldsymbol{\Sigma}(t)$ is known it is possible to determine the generic solution of (4.12) for a given set of initial conditions $\mathbf{u}_0 = \mathbf{u}(t = 0)$ simply as follows:

$$\mathbf{u}(t) = \boldsymbol{\Sigma}(t)\mathbf{u}_0 \; \forall t \geq 0. \tag{4.15}$$

Thanks to the linearity of (4.12) and the T_P-periodicity of $\mathbf{P}(t)$ it follows that, for any generic $t \geq 0$, the following relation holds:

$$\mathbf{u}(t + T_P) = \boldsymbol{\Sigma}(t + T_P)\mathbf{u}_0 = \boldsymbol{\Sigma}(t)\mathbf{u}(T_P) = \boldsymbol{\Sigma}(t)\boldsymbol{\Sigma}(T_P)\mathbf{u}_0. \tag{4.16}$$

The matrix $\boldsymbol{\Sigma}(T_P)$ is called the monodromy matrix and it is the key element for the determination of the stability properties of the solution $\mathbf{u}(t) \equiv 0$. For the analysis of the stability of the system what is indeed sufficient to know is the behavior of the sequence $\mathbf{u}(t = nT_P)$ as $n \to +\infty$ with $n \in \mathbb{N}$. Indeed, according to (4.16) we have that:

$$\mathbf{u}(t = nT_P) = [\boldsymbol{\Sigma}(T_P)]^n \mathbf{u}_0. \tag{4.17}$$

The relation (4.17) allows to know the state vector $\mathbf{u}$ after a generic number n of characteristic periods T_P if the state vector is known at $t = 0$. The original continuous time problem (4.12) is transformed by (4.17) into a discrete time problem by means of a Poincaré mapping. From the geometrical point of view the discrete-time problem (4.17) can basically be considered as a linear mapping $\mathbb{R}^N \to \mathbb{R}^N$ with parameter n. The question now is how the mapping (4.17) can be used to decide on the stability of $\mathbf{u} = 0$. Basically the idea is that if some vector $\mathbf{u}_0$ exists such that at least one component of $\mathbf{u}(nT_P)$ grows indefinitely as $n \to \infty$, then the solution $\mathbf{u}(t) \equiv 0$ is unstable. To obtain information on the behavior of $\mathbf{u}(nT_P)$ for $n \to \infty$ it is sufficient to look at the eigenvalues λ_j of the monodromy matrix $\boldsymbol{\Sigma}(T_P)$. Such eigenvalues are called the Floquet multipliers and are related to the Floquet characteristic exponents σ_j by the relation:

$$\lambda_j = e^{\sigma_j T_P} \Leftrightarrow \sigma_j = \frac{\ln(\lambda_j)}{T_P}. \tag{4.18}$$

The Floquet multiplier λ_j, as well as the Floquet exponents σ_j, can be (and usually are) complex numbers. The solution $\mathbf{u}(t) \equiv \mathbf{0}$ is unstable when at least one Floquet multipliers has modulus larger than one, i.e.,

$$\exists j : |\lambda_j| > 1 \Rightarrow \mathbf{u}(t) \equiv \mathbf{0} \text{ is unstable.} \tag{4.19}$$

If the monodromy matrix $\boldsymbol{\Sigma}(T_\mathrm{P})$ is known it is then possible to decide upon the stability or instability of the system by simply checking the modulus of the eigenvalues of $\boldsymbol{\Sigma}(T_\mathrm{P})$.

Unfortunately closed form analytical expressions for the monodromy matrix can hardly be found in general, apart from simple cases. On the other hand the problem can always be approached from a direct numerical point of view [26] by determining from numerical integration the set of fundamental solutions according to (4.13) for one characteristic period T_P, i.e., for $t \in [0, T_\mathrm{P}]$. From the knowledge of $\mathbf{s}^j(t = T_\mathrm{P})$ for $j = 1, \ldots, N$ it is possible to construct the monodromy matrix $\boldsymbol{\Sigma}(T_\mathrm{P})$ from which an analysis of the eigenvalues can be performed to decide upon the stability of the solution $\mathbf{u}(t) \equiv \mathbf{0}$.

As anticipated, the linearized equation of motion (4.7) can be recast in the form (4.12) by using the standard transformation $\mathbf{u} = (\phi, \dot{\phi})^\mathrm{T}$. In case of parametric roll the characteristic period of the matrix of coefficients $\mathbf{P}(t)$ is equal to the ship-wave encounter period T_e (in our modeling the ship speed V is constant), i.e., $T_\mathrm{p} = T_\mathrm{e}$. The linear equation (4.7) is then numerically integrated for $t \in [0, T_\mathrm{e}]$ using two different initial conditions, namely, $\mathbf{u}_0^1 = (\phi = 1, \dot{\phi} = 0)^\mathrm{T}$ and $\mathbf{u}_0^2 = (\phi = 0, \dot{\phi} = 1)^\mathrm{T}$. From the knowledge of $\mathbf{u}^j(T_\mathrm{e}) = (\phi^j(T_\mathrm{e}), \dot{\phi}^j(T_\mathrm{e}))^\mathrm{T}$ $j = 1, 2$, i.e., from the knowledge of $\mathbf{s}^1$ and $\mathbf{s}^2$, see (4.13), the 2×2 monodromy matrix $\boldsymbol{\Sigma}(T_\mathrm{e})$ can be created and a calculation of eigenvalues allows to apply the criterion (4.19). The same method can be easily introduced in more complex analytical models as, e.g., 3-DOF models coupling dynamically heave, roll and pitch [20, 21].

The method based on the direct determination of the monodromy matrix is of quite straightforward application and, by using only N time domain numerical integrations in a short time window $[0, T_\mathrm{P}]$, it allows to obtain a formally correct stability check, although this check can be sometime influenced by the numerical accuracy of the integration scheme.

It is important to stress two characteristics associated with the direct application of the Floquet theory. First of all the stability check based on the eigenvalues of the monodromy matrix allows to address all the parametric resonance regions using the same approach/code. This means that we do not need to specify in advance whether we are interested in the first, or second, or some other parametric resonance region as it instead occurs when using approximate analytical expressions for the threshold boundaries in different parametric resonance region [16]. In addition, the direct determination and analysis of the monodromy matrix retains all the characteristics of the periodic parameters. In particular the method retains all the harmonics in the fluctuating restoring in (4.7), which is important in case of trimaran vessels and/or large wave steepness for both mono and multihulls especially when the second, or higher, parametric resonance regions are of concern.

4.4 Example Calculations and Experiments in Longitudinal Regular Waves

4.4.1 Ship Geometry and Forward Speed Effects on Linear Damping and Restoring

A trimaran vessel has been used in the numerical calculations and a 1:50 scale model has been used for experimental tests. An extended overview of the results from the experimental campaign and simulations in longitudinal regular waves for this ship has been given by [8], while results of experiments and simulations in beam regular waves have been provided by [6]. Herein we will only report some sample results from experiments and simulations in longitudinal regular waves by focussing our discussion on the highlighting of the peculiar characteristics of the trimaran behavior.

The trimaran geometry and the main ship data are shown in Fig. 4.1. The stagger is defined as the distance, in longitudinal direction, between the aft end of the main hull and the aft end of the outriggers. The clearance is defined as the distance, in transversal direction, between the ship centreplane and the centreplane of each

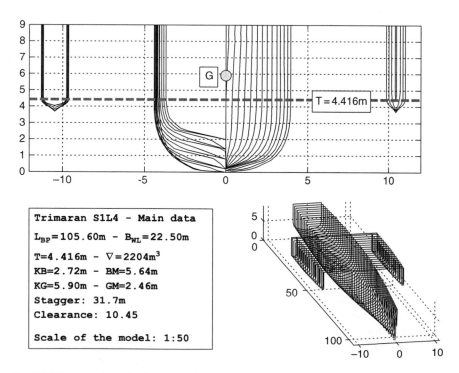

Fig. 4.1 Trimaran ship used for experiments and simulations

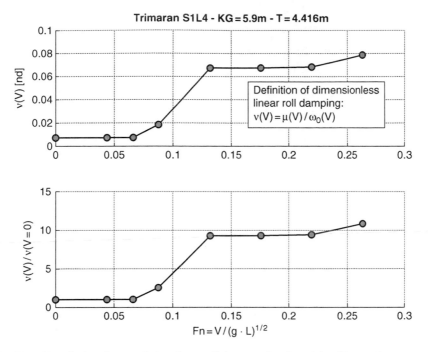

Fig. 4.2 Dimensionless linear roll damping coefficient as a function of the ship speed

outrigger. It can be seen that the outriggers have a very shallow draught (of the order of 0.7 m) which leads to their emergence at very small heeling (about 4°) in calm water. In addition, the outriggers can easily come (partially) out of water in case the ship is on a wave, especially if the wave trough is close to the position of the outriggers. Since the two outriggers significantly contribute to the metacentric radius, their emergence from water leads to a significant reduction of the initial stability and, eventually, to the whole restoring.

Roll decays at different forward speeds have been carried out on the model of the trimaran vessel in order to assess the influence of the speed on damping and on restoring. The linear roll damping coefficient typically increases with forward speed because of the lift effects on the hull while rolling. This has been observed also for the considered trimaran as shown in Fig. 4.2. The linear roll damping $\mu(V)$ is reported in Fig. 4.2 in dimensionless form, as $v(V)$, by dividing it using the forward speed calm water roll natural frequency $\omega_0(V)$. In addition, the ratio between the forward speed dimensionless linear roll damping $v(V)$ and its zero speed value $v(V=0)$ is also reported. It can be seen that the effect of the forward speed is significant, leading to almost a tenfold increase of $v(V)$ even at moderate Froude numbers.

As anticipated, also the roll natural frequency as measured from roll decays has been found to be affected by forward speed. This is something which occurs

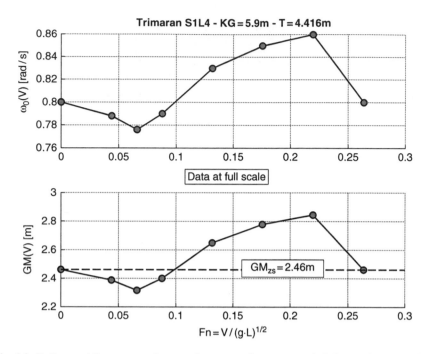

Fig. 4.3 Roll natural frequency and assumed corresponding metacentric height as functions of the forward speed

also in case of conventional monohulls [4]. In the present modeling we assume that the variation in the roll natural frequency is completely explained by restoring variations due to lift effects on the hull when heeled. This basically neglects different hydrodynamic effects such as, e.g., a possible speed dependence of the roll added inertia. The roll natural frequency from roll decays and the corresponding value of $\overline{GM}(V) = \overline{GM}_{zs} + \delta\overline{GM}(V)$ according to (4.4) and (4.11) are shown in Fig. 4.3. It can be seen that for small forward speed there is a small reduction in the roll natural frequency, which corresponds, under our assumptions, to a decrease of the forward speed metacentric height with respect to the zero speed value. However, as the forward speed increases, the roll natural frequency significantly increases above the zero speed value. According to our assumptions, the corresponding forward speed metacentric height also increases, reaching a maximum value which is, at full scale, about 0.38 m in excess of the calm water value. Finally, at the maximum tested forward speed an abrupt reduction is visible for the roll natural frequency. This final point is however doubtful, due to the difficulties involved in the execution and analysis of roll decays at high forward speed. Such difficulties at high forward speeds are associated with the strong linear damping and the limited usable length of the towing tank of the University of Trieste (about 35–40 m).

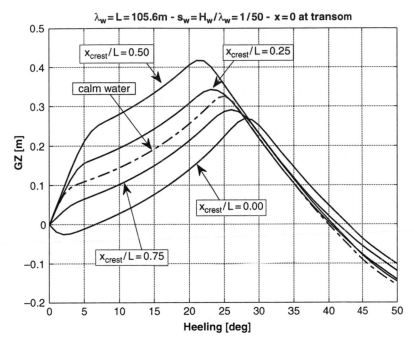

Fig. 4.4 Righting arm in calm water and in waves from hydrostatic calculations. Wave length: 105.6 m. Wave steepness: 1/50

4.4.2 Hydrostatic Calculations of Restoring and Metacentric Height in Waves (Zero Speed)

The calculation of the righting arm in wave has been carried out in the example case of a wave having length equal to the ship length and steepness equal to 1/50. This wave has been actually used also in the experiments. The resulting zero speed righting arm $\overline{GZ}_{zs}(\phi, x_c, \lambda_w = L, a_w = 0.5\lambda_w/50)$ from hydrostatic calculations is reported in Fig. 4.4. It can be seen that the fluctuation of restoring in waves is significant. For typical monohulls the minimum of the metacentric height occurs, for a wave having length close to the ship length, when the wave crest is close to amidships, while the maximum usually occurs when a wave trough is close to amidships. It can be seen that, for the considered trimaran configuration, the situation is actually opposite. Indeed the case with $x_{crest}/L = 0$ (trough amidships) is associated with an increased restoring with respect to calm water, while the case $x_{crest}/L = 0.5$ (crest amidships) is associated with a significant reduction of restoring which even leads to a negative metacentric height, i.e., a negative slope of $\overline{GZ}$ at $\phi = 0$. This result is a direct consequence of the emergence of the outriggers when the ship is on a wave through, with the bow and the stern on the wave crests (remember that in this example the wave length λ_w is equal to the ship length L) and the outriggers, which are positioned in the central part of the ship, out of water.

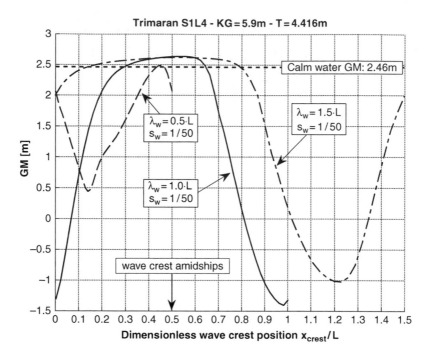

Fig. 4.5 Metacentric height in waves from hydrostatic calculations

It order to have a better understanding of the variation of restoring as the wave crest moves along the ship length, it is useful to look at the behavior of the metacentric height in waves for a few different wave lengths having the same wave steepness. Figure 4.5 shows $\overline{GM}_{zs}(x_c, \lambda_w, a_w)$ for three different wave lengths (0.5 L, 1.0 L, and 1.5 L) all having the same wave steepness ($s_w = 2a_w/\lambda_w = H_w/\lambda_w = 1/50$). It can be seen that the fluctuation of the metacentric height is characterized by a behavior which cannot be accurately described by a simple sinusoidal fluctuation. In addition, in all cases, the maximum of the metacentric height occurs when the wave crest is close to amidships, because in such condition the outriggers, which are close to the center of the main hull, remain completely in water and therefore contribute to the metacentric height. On the other hand, in case of wave trough close to amidships, the metacentric height is very significantly reduced particularly for the longer waves.

The significant departure of the fluctuation of $\overline{GM}$ in waves from a simple sinusoid is the main reason calling for the use of a theoretical approach for the assessment of the stability of the upright position which is able to take into account not only the first harmonic of the fluctuation of $\overline{GM}$ in waves, but actually its exact shape and hence all its harmonic components. Figure 4.6 shows the results of a Fourier analysis of the metacentric height in waves, without forward speed effects, i.e., $\overline{GM}_{zs}(x_c, \lambda_w, a_w)$, when $\lambda_w = L$ and $s_w = 1/50$. It can be noticed that the amplitude of the second harmonic is about 35% of the amplitude of

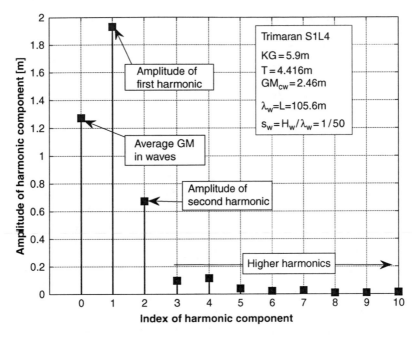

Fig. 4.6 Harmonic content of the fluctuation of the metacentric height. Wave length equal to the ship length and wave steepness equal to 1/50

the first harmonic and it cannot therefore be neglected. Higher harmonics have, instead, much smaller amplitudes and they could in principle be neglected in comparison with the first two harmonics. In addition it is important to highlight the significant loss for what concern the average metacentric height during a wave passage with respect to the calm water value. The calm water metacentric height of 2.46 m reduces to an average metacentric height $\langle \overline{GM}_{zs}(\lambda_w, a_w) \rangle$ of 1.27 m for the considered wave. From the point of view of restoring, the direct consequence can be considered as a reduction, on average, of static stability, at least for small heeling angles. The difference between the calm water righting arm and the average righting arm in waves has also been investigated in the past [4]. From a dynamical point of view the reduction of the average metacentric height in waves leads to a roll natural frequency in waves $\omega_0(V, \lambda_w, a_w)$ which becomes significantly different, and in this case reduced, with respect to the calm water roll natural frequency. The immediate consequence is a tendency towards a shift of the resonance conditions for the inception of parametric roll with respect to the nominal conditions estimated by using the calm water zero speed roll natural frequency. The additional consequence of the significant reduction of the average metacentric height in waves is an increase of the amplitude of the dimensionless parametric excitation (i.e., say, the amplitude of the fluctuation of $\overline{GM}$ divided by the average metacentric height) with respect to what would be determined by using the calm

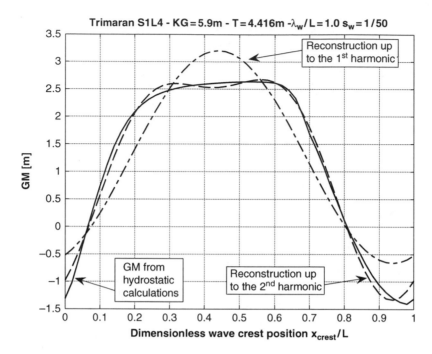

Fig. 4.7 Metacentric height in waves and reconstruction by a limited number of harmonics. Wave length equal to the ship length and wave steepness equal to 1/50

water metacentric height. The necessity of taking into account the variation of the average metacentric height in waves with respect to the calm water value has also been considered in the simplified methodology proposed by [7] for the assessment of vulnerability to parametric roll at early design stage.

In order to graphically show that the harmonics up to the second one are often sufficient to reproduce the behavior of the metacentric height in waves for the considered trimaran vessel, Fig. 4.7 compares, for $(\lambda_w = L, s_w = 1/50)$, the metacentric height in the considered wave from hydrostatic calculations and its approximation up to the first and up to the second harmonic. It can be seen that the approximation up to the second harmonic is a good approximation of the actual curve.

4.4.3 (In)Stability Maps for the Upright Position in Longitudinal Waves

The direct application of the Floquet theory allows to determine, for a given wave and a given ship speed, whether the upright position $\phi = 0$ is stable or unstable. The obtained results are so-called stability maps, which allow to identify

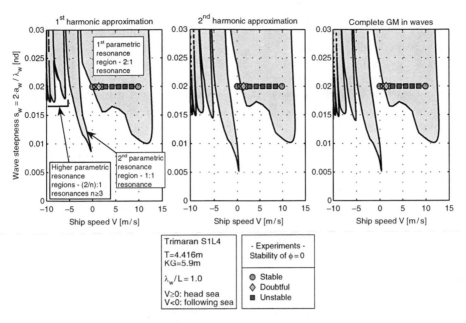

Fig. 4.8 Parametric resonance regions. Wave length equal to the ship length

those conditions where the upright position, which is assumed as stable in calm water, loses its stability. We have also shown that harmonics higher than the first one in the fluctuation of the metacentric height are important for correctly reproducing the behavior of the metacentric height in waves. In particular, the second harmonic component of the fluctuation of $\overline{GM}$ has been shown to provide the major correction to the first harmonic approximation of $\overline{GM}$ in waves in order to obtain a good approximation of the actual metacentric height in waves. This has a direct consequence also when dealing with the calculation of stability maps, as shown also by [24] when dealing with surge effects on the inception of parametric roll in following waves. Comparisons between calculated stability maps and results from experiments have been reported by [8]. Here we want to look more closely at how the approximation of the metacentric height in waves can lead to differences in the determined stability maps. For this reason we have determined the regions of instability for the upright position in longitudinal regular waves using the Floquet theory and three different approximation for the metacentric height in waves. On one side we have calculated the stability maps by using the metacentric height in waves as determined from hydrostatic calculations. This basically means we have retained all the harmonic components for the metacentric height in waves. On the other side, similarly to what has been shown in Fig. 4.7, we have used two approximations of the metacentric heights in waves by keeping a Fourier approximation up to the first and up to the second harmonic respectively. In case of a wave length equal to the ship length, the results from this analysis are shown in Fig. 4.8. The regions of instability for the upright position are shown as filled regions in the (V, s_w) plane. Results from experiments aimed at checking the stability of the solution $\phi = 0$ are also reported for comparison.

The results in Fig. 4.8 indicate that the approximation of the $\overline{GM}$ curve in waves up to the second harmonic practically leads to the same instability regions as obtained by using the complete $\overline{GM}$ curve in waves. On the other hand, the first harmonic approximation of $\overline{GM}$ in waves leads to predicted instability regions which are quite different from the other two cases. Differences are visible in particular in the second parametric resonance region, where neglecting the second harmonic component leads to a reduction in the extent of the predicted instability region for the upright position. This means that neglecting higher-order harmonics can lead to underestimations of the extent of the second, and higher, parametric resonance regions. This outcome can be qualitatively explained by the fact that, when the first harmonic of the fluctuation of $\overline{GM}$ is in the second parametric resonance region, its second harmonic falls in the first parametric resonance region, where the limiting threshold parametric excitation is smaller. Neglecting this aspect by neglecting the second harmonic in the fluctuation of $\overline{GM}$ leads to an underestimation of the extend of the second parametric resonance instability region. A similar outcome is also observed in case of higher parametric resonance regions, where the use of the first harmonic approximation of $\overline{GM}$ leads to a reduction in the predicted extent of the instability regions. Interestingly this outcome seems to be opposite to that described by [24]. The first parametric resonance region is not significantly affected by the number of harmonics considered in the Fourier expansion of the metacentric height in waves, apart from some effect in the region of speeds close to $V = 0$ m/s. This is also expectable because when the first harmonic of the fluctuation of the metacentric height is close to twice the roll natural frequency in waves, higher harmonics fluctuate at frequencies which are out of parametric resonance regions and therefore they have basically no effect in the determination of the threshold limits for the inception of parametric roll when the ship is close to the first parametric resonance region. Together with the numerical calculations of the instability regions for the upright position, experimental results are also reported, from which it can be seen that the predicted first parametric resonance instability region extends a little bit too much at the high speed limit in comparison to what has been experimentally determined. The predicted low speed limit for the first parametric resonance region better agrees with experiments. In general, however, there seems to be some overestimation of the instability zone for the upright position with respect to the experiments. This is of course expectable, due to the fact that the considered modeling contains significant simplifications with respect to reality, in particular for what concerns the simplification of hydrodynamic lift effects and the quasi-static approximation for the actually dynamic heave and pitch motions. More details on comparisons between experimentally and numerically determined stability maps are reported by [8].

4.4.4 Nonlinear Roll Motion Above Threshold

Nonlinearities of restoring are known to lead to bending of the roll response curve in regular beam waves [6, 13]. When nonlinearities of restoring are sufficiently strong and/or the excitation is sufficiently large, such bending leads to coexisting steady state solutions. Which solution the system will eventually be attracted to, depends on the initial conditions. A similar situation also occurs in longitudinal regular waves when parametric roll is excited [4, 5, 25]. In case of trimaran vessels the strongly nonlinear behavior of the righting moment in waves leads to an exacerbation of such nonlinear effects. It has indeed been observed, both from experiments and simulations [8], that quite wide ranges of parameters (ship speed, wave length, wave steepness, etc.) can be associated, in case of the considered trimaran configuration, to multiple coexisting solutions. In addition, large amplitude rolling can be observed outside the regions of instability for the upright position, i.e., a stable upright position can coexist with one (or more) large amplitude rolling solutions [4, 5, 25]. From a practical point of view, this usually means that large amplitude rolling motions, in this case in longitudinal waves, can be observed for trimaran vessels in quite unexpected regions of speed which can be quite far from the speeds associated with the 2:1 or 1:1 resonance based on nominal calm water rolling frequency. The bending, in accordance with the softening restoring, is usually towards the region of smaller ship–wave encounter frequencies, as it also happens in beam waves [6]. While the bending of the response curve is associated with the shape of the restoring (softening or hardening) the shifting of the, say,"center speed" of the response curve is due to the interaction of two effects. One point is that the average metacentric height in waves tends to be different, and for trimaran vessels usually smaller, than the calm water metacentric height. The other point is the variation of restoring due to hull lift effects. These two aspects contribute to the modification of the roll natural frequency in waves with respect to the calm water value, see (4.9), and are also responsible for the winding of the instability region for the upright position as reported in Fig. 4.8. Coexisting steady states have been quite extensively addressed in the past from different points of view [4,5,8,25]. Therefore, here we only report in Fig. 4.9 one example response curve with the intention of underlining that "nominal worst conditions", in terms of ship speed, can significantly differ from actual worst conditions. The understanding of nonlinear features can therefore play a dominant role in a correct assessment and interpretation of simulations and/or experimental results. Figure 4.9 compares the outcomes from experiments and from simulations based on the model (3.6) for the considered trimaran vessel in case of a short wave having length equal to half the ship length ($\lambda_w = 0.5\,L$) and wave steepness $s_w = 1/50$. The nonlinear roll damping has been modeled as a speed independent purely quadratic function of the roll velocity ($\delta(V) = 0$) with $\beta(V) = \beta = 0.3\,\text{rad}^{-1}$. In Fig. 4.9 the quantity ω_e is the frequency of encounter between the ship and the wave.

Looking at Fig. 4.9 it can be seen, first of all, that the numerical simulations based on the model (4.6) closely reproduce the experimental data, despite the evident

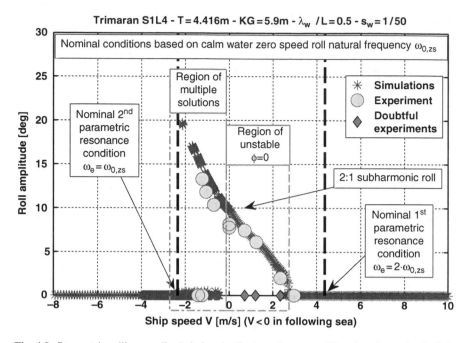

Fig. 4.9 Parametric rolling amplitude in longitudinal regular waves. Wave length equal to half the ship length and wave steepness equal to 1/50

simplicity of the 1-DOF modeling. More comparisons between experiments and simulations are given by [8]. It can also be noticed that the region of instability for the upright position is shifted towards lower speeds in head seas, i.e., lower encounter frequencies, when compared with the reported nominal first parametric resonance condition based on the calm water zero speed roll natural frequency $\omega_{0,zs}$. As anticipated, this shifting of the instability region for $\phi = 0$ is mostly due to the reduction of the average metacentric height in waves, as it is visible also from Fig. 4.5. In the region where $\phi = 0$ is unstable, a stable subharmonic rolling motion develops. However, the most dangerous condition for what concerns the rolling amplitude is observed, both numerically and experimentally, in the region of low speeds in following waves, well outside the speed range where $\phi = 0$ is unstable. The bending of the roll response curve towards the region of low encounter frequencies is a direct consequence of the softening restoring. A similar, but usually not so strong, behavior can also occur for conventional monohulls [4, 5]. Since we are out of the zone of instability for $\phi = 0$, the large amplitude rolling in such region coexists with a stable upright position. However due to the strongly nonlinear restoring which quite abruptly change derivative at small heeling angles, the domain of attraction of the solution $\phi = 0$ tends to be quite small, while the domain of attraction of the large amplitude rolling tends to be dominant [8]. This means that, for the majority of the initial conditions not leading to capsize, the roll motion is eventually attracted to the large amplitude solution.

It is also interesting to note that 2:1 subharmonic roll motions are numerically and experimentally observed up to speeds close to the region where a check based only on the nominal zero speed roll natural frequency would indicate, if any, an harmonic 1:1 response.

When addressing parametrically excited rolling motion for trimaran vessels it is therefore important to have clear in mind the possible occurrence of nonlinear phenomena such as those described in this example. Having them clear in mind could avoid checking for parametric roll in wrong regions of speeds and could also remind to check different initial conditions, or the influence of perturbations, for verifying the presence of multiple solutions.

4.5 Optimum Longitudinal Positioning of the Outrigger

In the design of multihull vessels the positioning of the outriggers gives additional degrees of freedom in the ship optimization process. A correct positioning of outriggers should take into account static stability in calm water and in waves, seakeeping, resistance, manoeuvring, the possibility of entering specific harbor facilities, etc.

In this section we show an example of how the positioning of the outriggers can significantly influence the variations of restoring in waves. We have indeed underlined in the previous sections that one of the main reasons for the significant reduction of initial stability in waves for the considered trimaran configuration is to be associated with the (partial) emergence of outriggers from water. Moreover, we have also underlined that the (partial) emergence of outriggers from water is the reason for having a maximum of the metacentric height when the wave crest is close to amidships for the considered trimaran configuration. On the contrary, for standard monohulls, a wave crest close to amidships usually leads to an almost minimum of the metacentric height in waves.

Changing the longitudinal positioning of the outriggers can have a significant influence on the variations of stability in waves, because the possibility of keeping the outriggers in water as the wave crest moves along the ship depends on the actual longitudinal position of the outriggers combined with the resulting ship trim in waves. We have therefore systematically changed the stagger of the outriggers, i.e., the distance between the transom of the main hull and the transom of the outriggers, in a range going from 1 m to 70 m. A stagger of 1 m corresponds to outriggers positioned at the stern region of the ship, while a stagger of 70 m corresponds to outriggers positioned close to the ship bow. For each different longitudinal position of the outriggers the variation of the metacentric height has been calculated for a reference wave having length equal to the ship length ($\lambda_w = L$) and steepness equal to 1/50. A Fourier analysis has been carried out on the resulting metacentric height in waves for each position of the outriggers. The results from the different design alternatives are shown in Fig. 4.10. From the results it can be seen that positioning the outriggers towards the aft or forward end of the ship is beneficial with respect to

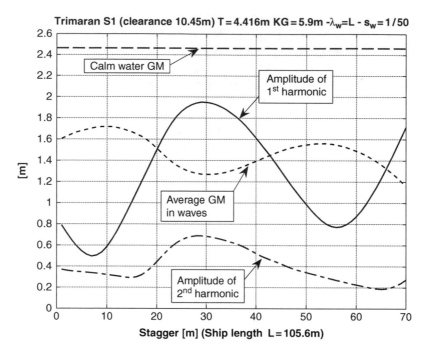

Fig. 4.10 Influence of stagger on mean value, first harmonic and second harmonic of the metacentric height in waves. Wave length equal to the ship length and wave steepness equal to 1/50

a central positioning for what concerns the average metacentric height and also for the amplitude of the first and second harmonics of the fluctuation.

In order to determine a unique "optimum stagger" it is necessary to reduce our problem to a single-objective problem, otherwise we would have to address a multi-objective optimization problem and hence a Pareto front. For this reason we have defined, as example objective function, the dimensionless first harmonic of the metacentric height in the considered reference wave. This quantity is defined as the ratio between the amplitude of the first harmonic of $\overline{GM}$ and the average metacentric height for the considered wave. The result is shown in Fig. 4.11.

It can be seen that the optimum configuration is determined as a stagger of about 7.25 m, i.e., a condition with outriggers close to the aft end of the ship. Of course the obtained optimum stagger depends on the particular definition of the objective function used in this exercise. Moreover, the obtained optimum depends on the reference wave which we have selected for the calculations. Different assumptions concerning the objective function and/or the reference wave can lead to, likely slightly, different results in term of optimum stagger. Some check carried out with waves having wave length equal to the ship length but different reference steepnesses have led to small variations in the identified optimum position of the outriggers. The original and the "optimum" configurations are compared in Fig. 4.12

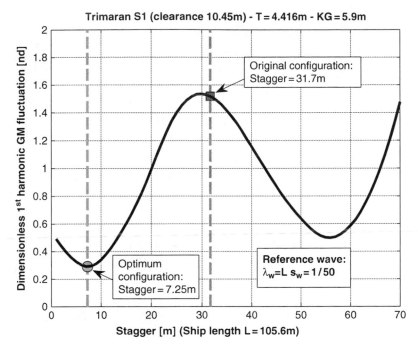

Fig. 4.11 Determination of "optimum stagger". Wave length equal to the ship length and wave steepness equal to 1/50

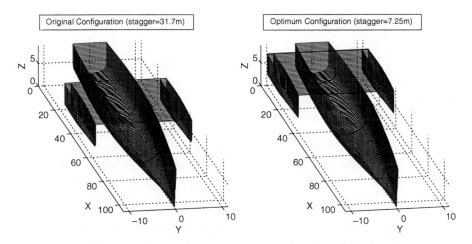

Fig. 4.12 Comparison between original and "optimum" configurations. Ship geometry

while the hydrostatically calculated metacentric heights are compared in Fig. 4.13. By looking at the metacentric height in waves reported in Fig. 4.13 it is evident how significant can be the effect of the longitudinal positions of the outriggers on the trimaran hydrostatic properties in waves.

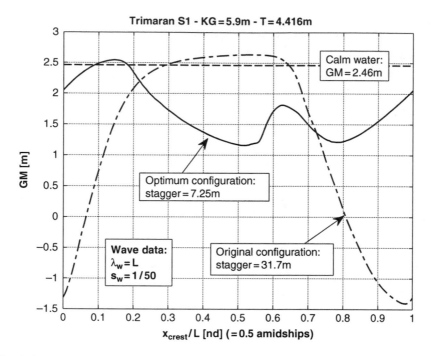

Fig. 4.13 Comparison between original and "optimum" configurations. Metacentric height from hydrostatic calculations. Wave length equal to the ship length and wave steepness equal to 1/50

4.6 Final Remarks

Ships with outriggers, and in particular trimaran vessels, with limited transversal separation between the outriggers and/or limited draught of the outriggers can be characterized by not extremely large metacentric heights and by significant restoring nonlinearities. Variations of restoring in waves for such configurations can also be significant. The combination of these factors leads to the possibility of inception of parametrically excited rolling motion with peculiar features which often represent exacerbation of what usually happens in case of standard monohulls. This paper addressed some of these peculiar features by assuming that their knowledge is important for those designers and researchers having to deal with, and hopefully eventually eliminate or at least limit, the risk of inception of parametric roll.

Parametric roll in longitudinal regular waves has been modeled by a simplified 1-DOF nonlinear mathematical model. The most peculiar and distinctive feature of the proposed model is the explicit, although simplified, consideration of forward speed effects not only for damping but also for restoring. It has been shown that the fluctuation of the metacentric height in waves in case of the considered trimaran vessel can be significantly different from a simple sinusoidal function, whereas the sinusoidal approximation is often sufficient for standard monohulls. For this reason the regions of instability for the upright position have been determined by means of a direct numerical application of the Floquet theory. Since the proposed methodology for the numerical determination of stability maps is independent of the form of the model containing periodic parameters, it can easily handle the unusual metacentric height variations of the considered trimaran ship which would be more difficult to address by more widely used semi-analytical approximate methods.

The analysis of the stability map for a sample case has shown that the first parametric resonance region is the dominant region in terms of extension, but also higher parametric resonance regions, and particularly the second one, could be a concern. Particular attention should probably be given to the often overlooked second parametric resonance region. It has been shown that the extent of this region is particularly sensitive to the number of harmonics taken into account in the description of the metacentric height in waves. For the considered sample case it has been found that the description of the metacentric height in waves should retain up to at least the second harmonic to obtain instability regions close to those obtained by using the exact shape of $\overline{GM}$ in waves.

For what concerns rolling motion in the nonlinear range, an example has been provided to show some characteristic features which should be borne in mind, especially when planning experiments and/or simulations. First of all the variation of the average metacentric height in waves with respect to the calm water value can lead to a significant shifting of the regions of speeds where the upright position becomes unstable in comparison to what would be predicted by using the nominal calm water roll natural frequency. This aspect is also visible when analyzing the stability maps. Some role in this shifting, though usually in the opposite direction, is also played by forward speed lift effects on the restoring. The, possibly large amplitude, roll motions in the region where the upright position is unstable can therefore occur at partially unexpected speeds. In addition, the significant softening nonlinearities of the restoring moment lead to a bending of the roll response curve towards the region of low encounter frequencies. The first consequence is that in wide ranges of speeds large amplitude rolling motions can coexist with a stable upright position. Therefore, simulations and/or experiments should be carried out in such a way to try to observe the coexisting steady states (e.g., by changing initial conditions). The second consequence is that in almost all cases the maximum of the rolling amplitude is observed at a speed where the upright position is stable, i.e., outside the regions of instability identified by the stability maps.

From the design point of view, changing the longitudinal position of the outriggers can provide significant benefits in terms of variation of restoring in waves. An example has been shown where the outriggers of the considered trimaran vessel has been systematically shifted from the extreme aft to the extreme forward of the ship. For each configuration the metacentric height in waves has been calculated for a reference wave having length equal to the ship length and steepness of 1/50. The ratio between the amplitude of the first harmonic of the fluctuation of $\overline{GM}$ and the average $\overline{GM}$ has been used as single objective function to be minimized for determining the optimum positioning of the outriggers. Outriggers positioned in the central part of the ship have been found to be associated with the largest dimensionless first harmonic of the fluctuation of $\overline{GM}$, whereas the positioning outriggers in the aft part of the ship has been found to lead to the absolute minimum of the objective function. An additional relative minimum has been found in case of outriggers positioned in the forward part of the ship. The selection of the longitudinal position of the outriggers, especially when they have a very limited draught as in the considered case, should therefore take into account not only resistance, manoeuvring, and general arrangement considerations, but also the problem of stability in waves.

Acknowledgements The authors wish to express their appreciation to those students which provided contributions for the present paper during their thesis works: Mr. Luca Boaro, Mr. Giovanni Dall'Aglio, Mr. Fabio Fucile, and Mr. Marco Sinibaldi.

References

1. American Bureau of Shipping (ABS): Guide for the assessment of parametric roll resonance in the design of container carriers, September 2004 (Updated June 2008)
2. Begovic, E., Bertorello, C., Caldarella, S., Cassella, P.: High Speed Multihull Craft for Medium distance Marine Transportation. Proc. High Speed Craft: Design & Operation, 17–18 November, RINA, London, UK (2004)
3. Bulian, G.: Nonlinear Parametric Rolling in Regular Waves – A General Procedure for the Analytical Approximation of the GZ Curve and Its Use in Time Domain Simulations. Ocean Engineering, Vol. 32, No. 3–4, March, pp. 309–330. doi:10.1016/j.oceaneng.2004.08.008 (2005)
4. Bulian, G.: Development of analytical nonlinear models for parametric roll and hydrostatic restoring variations in regular and irregular waves. Ph.D. Thesis, DINMA – University of Trieste, March (http://hdl.handle.net/10077/2518) (2006)
5. Bulian, G., Francescutto A.: Theoretical Prediction and Experimental Verification of Multiple Steady States for Parametric Roll. Proc. of the 10th International Ship Stability Workshop, 23–25 March, Daejeon, Republic of Korea, pp. 31–45 (2008)
6. Bulian, G., Francescutto A.: Experimental results and numerical simulations on strongly nonlinear rolling of multihulls in moderate beam seas. Proceedings of the Institution of Mechanical Engineers: Part M – Journal of Engineering for the Maritime Environment, Vol. 223, pp. 189–210, DOI: 10.1243/14750902JEME126 (2009)
7. Bulian, G., Francescutto A.: A simplified regulatory-oriented method for relative assessment of susceptibility to parametric roll inception at the early design stage. Proc. 4th International Maritime Conference on Design for Safety and 3rd Workshop on Risk-Based Approaches in the Marine Industries – Part I, 18–20 October, Trieste, Italy, pp. 93–106 (2010)
8. Bulian, G., Francescutto, A., Fucile, F.: Numerical and experimental investigation on the parametric rolling of a trimaran ship in longitudinal regular waves. Proc. 10th International Conference on Stability of Ships and Ocean Vehicles (STAB2009), St. Petersburg, 22–26 June, pp. 567–582 (2009)
9. Bulian, G., Francescutto, A., Fucile, F.: Study of trimaran stability in longitudinal waves. Ships and Offshore Structures, First published on: 31 August 2010 (iFirst), DOI: 10.1080/17445302.2010.507494 (2010)
10. Chacón, J. R., Ollero, P., Gee, N., Dudson, E.: The Pentamaran Ro-Ro: the Solution for Modal Shift from Land to Sea. Proc. Design and Operation of Trimaran Ships, 29–30 April, RINA, London, UK, ISBN 0-903055-99-6 (2004)

11. Courts, M.: Keynote Presentation: Trimaran Warship - The issues. Proc. Design and Operation of Trimaran Ships, 29–30 April, RINA, London, UK, ISBN 0-903055-99-6 (2004)
12. Dubrovsky, V. A., Matveev, K. I.: Concept design of outrigger ships. Proc. Design and Operation of Trimaran Ships, 29–30 April, RINA, London, UK, ISBN 0-903055-99-6 (2004)
13. Francescutto, A., Contento, G.: Bifurcations in ship rolling: experimental results and parameter identification technique. Ocean Engineering, Vol. 26, pp. 1095–1123 (1999)
14. Gee, N.: Keynote presentation: Stabilised Monohulls – Trimarans and Pentamarans – History, Evolution and the Future. Proc. Design and Operation of Trimaran Ships, 29–30 April, RINA, London, UK, ISBN 0-903055-99-6 (2004)
15. Gee, N., Dudson, E., González, J. M.: The IZAR Pentamaran – Tank Testing, Speed Loss and Parametric Rolling. Downloaded on 16 February 2009 from http://www.ngal.co.uk/Documents%20&%20Resources/?/1/2876/2876 (2002)
16. Hayashi, C.: Nonlinear Oscillations in Physical Systems. New York, McGraw Hill (1964)
17. Ikeda, Y., Himeno, Y., Tanaka, N.: A Prediction Method for Ship Roll Damping. Report No. 00405 of Department of Naval Architecture – University of Osaka Prefecture, December (1978)
18. José, J. V., Saletan, E. J.: Classical Dynamics - A Contemporary Approach. Cambridge University Press, Cambridge, UK, ISBN 0-521-63636-1 (1998)
19. Nayfeh, A. H., Mook, D. T.: Nonlinear Oscillations. John Wiley & Sons, Inc. (1979)
20. Neves, M. A. S., Rodríguez, C. A.: Nonlinear Aspects of Coupled Parametric Rolling in Head Seas. Proc. 10th International Symposium on Practical Design of Ships and Other Floating Structures, Houston (2007)
21. Oh, I. G., Nayfeh, A. H., Mook, D. T.: A theoretical and experimental investigation of indirectly excited roll motion in ships. Philosophical Transactions of the Royal Society, Mathematical, Physical and Engineering Sciences 358, pp. 1853–1881 (2000)
22. Simakhina, S. V.: Stability Analysis of Hill's equation. Department of Mathematics, Statistics and Computer Science – University of Illinois – Chicago, July 20, available at http://www.math.fsu.edu/~ssimakhi/Svetlana%20Simakhina_files/Sveta_thesis_pdf.pdf (2003)
23. Simakhina, S. V., Tier, C.: Computing the stability regions of Hill's equation. Applied Mathematics and Computation, Vol. 162, pp. 639–660 (2005)
24. Spyrou, K. J.: Designing against parametric instability in following seas. Ocean Engineering, Vol. 27, pp. 625–653 (2000)
25. Spyrou, K. J.: Design Criteria for Parametric Rolling. Ocean Engineering International, Vol. 9, No. 1, pp. 11–27 (2005)
26. Turhan, Ö., Bulut G.: Dynamic stability of rotating blades (beams) eccentrically clamped to a shaft with fluctuating speed. Journal of Sound and Vibration, Vol. 280, pp. 945–964 (2005)
27. Van Walree, F., de Jong, P.: Time domain simulations of the behaviour of fast ships in oblique seas. Proc. 6th Osaka Colloquium on Seakeeping and Stability of Ships (OC2008), 26–29 March, Osaka, Japan. pp. 383–391 (2008)

Chapter 5
Probability of Parametric Roll in Random Seaways

Jørgen Juncher Jensen

5.1 Introduction

The roll motion of ships can lead to various types of failures ranging from seasickness over cargo shift and loss of containers to capsize of the vessel. Hence, it is important to minimize the roll motion during a voyage. Currently, on-board decision support systems, e.g., [28, 32, 35], are being installed in vessels with the aim to provide the officer on watch with guidance on the best possible route, taking into account the weather forecast, the time constraints for the voyage and limiting criteria for motions etc.

A main problem is real-time estimation of the sea state. Here two approaches are being tested in full scale. The first is based on the use of a wave radar, e.g., the WAVEX system, e.g., [3], and the second uses ship responses (e.g. motions, accelerations, and strains) measured real-time by sensors installed on board together with linear transfer functions to estimate the sea state, e.g., [29], where also a comparison between the two approaches can be found.

After estimation of the sea state, a real-time estimation of the maximum ship responses within the next few hours as function of ship speed and course is needed to guide the officer on the action to take if excessive responses are foreseen with the present course and speed. To linear responses the standard frequency domain approach using transfer functions can easily be applied. For nonlinear responses most often time domain simulations are performed to obtain short-term statistics for the nonlinear roll response of ships, e.g., [6, 23, 34]. However, less time-consuming stochastic procedures have also been suggested. Most of them are based on simplifying, but often reasonable assumptions like equivalent linear damping, e.g., [4], second- or third-order perturbation procedures, e.g., [27], Melnikov functions, e.g.,

J.J. Jensen (✉)
DTU Mechanical Engineering, Department of Mechanical Engineering, Technical University
of Denmark, Bygning 403, Niels Koppels Alle, DK 2800 Kgs. Lyngby, Denmark
e-mail: jjj@mek.dtu.dk

T.I. Fossen and H. Nijmeijer (eds.), *Parametric Resonance in Dynamical Systems*,
DOI 10.1007/978-1-4614-1043-0_5, © Springer Science+Business Media, LLC 2012

[13, 36] and moment closure techniques, [26]. More recently, different procedures based on identification of critical wave episodes related to the roll motion have been suggested, see [19, 20, 37]. The present note is largely based on [20] but deals only with prediction of parametric rolling in head sea leaving out forced rolling.

5.2 Roll Motion of a Ship

A very comprehensive discussion of intact stability, including parametric rolling can be found in the ITTC report on ship stability in waves, [15]. The report discusses various modes of failure and the prediction procedures available. The report is partly based on the result of a questionnaire distributed to a large number of organizations and thus reflects very well the current status. To cover all modes of failure (static loss of stability, parametric excitation, dynamic rolling, resonance excitation, and broaching) a general nonlinear 6-DOF time domain procedure including viscous effects and maneuvering models must be applied. Some codes, e.g., LAMP, [11] and [33], seem to be able to do so with reasonable accuracy, but are very time-consuming to run, restricting the application to regular waves or very short stochastic realizations.

Another nonlinear 6-DOF procedure is GL-SIMPEL, see e.g., [31], based on a nonlinear strip theory formulation. The frequency dependence of the added mass and damping is taken into account using a higher differential equation formulation. FREDYN, see e.g., [11], is another nonlinear code based on a strip theory formulation. Generally, these codes are much faster than nonlinear 3-D procedures like LAMP and, as the 3-D effects on the roll motion is usually not that important, to be preferred for design work and onboard decision support system.

Other procedures have more limiting capabilities as some of the capsize modes are excluded. An example is the ROLLS procedure, [24], where the following nonlinear differential equation is used to estimate the roll angle ϕ, (omitting the terms due to wind and fluids in tanks):

$$\ddot{\phi} = \frac{M_\phi + M_{sy} - M_d - \Delta(g - \ddot{w})\overline{GZ}(\phi) - I_{xz}\left[\left(\ddot{\theta} + \theta\dot{\phi}^2\right)\sin\phi - \left(\ddot{\psi} + \psi\dot{\phi}^2\right)\cos\phi\right]}{I_{xx} - I_{xz}\left(\psi\sin\phi + \theta\cos\phi\right)}.$$

$$(5.1)$$

Here M_ϕ, M_{sy}, and M_d are the roll moments due to waves, sway and yaw, and hydrodynamic damping, respectively. Furthermore, I_{xx} and I_{xz} are the mass moment of inertia about the longitudinal axis and the cross term mass moment of inertia. The displacement of the ship is denoted by Δ and g is the acceleration of gravity. The instantaneous value of the righting arm $\overline{GZ}$ is in irregular waves calculated approximately using the so-called Grim's effective wave. The heave w, pitch θ, and yaw ψ motions are determined by standard strip theory formulations, whereas the surge motion is calculated from the incident wave pressure distribution. The advantage of this formulation compared to full nonlinear calculations is the much faster computational speed, still retaining a coupling between all six-degrees-of-freedom, [23]. The model can, however, not deal with broaching due to the assumption of a linear yaw motion. Both the ROLLS and the GL-SIMBEL procedures are described and validated in [14, IMO-SLF submission by Germany 2007].

Here a simplified version of (5.1) is considered. The heave motion w is taken to be a linear function of the wave elevation using the closed-form expression given by [16]. The cross term mass moment of inertia is assumed to be small, and pitch is thus only included through the static balancing of the vessel in waves in the calculation of the $\overline{GZ}$ curve. Furthermore, the sway and yaw motions can be ignored as only head sea is considered. The effect of speed variations due to the time varying added resistance in waves was investigated in [41], showing some reduction in the overall parametric roll behavior. Hence, the surge motion can be important, but will be omitted in the present treatment. The damping term M_d is modeled by a standard combination of a linear, a quadratic and a cubic variation in the roll velocity. Finally, the wave excitation roll moment M_ϕ is equal to zero in long-crested head sea. With these simplifications (5.1) reads

$$\ddot{\phi} = -2\beta_1 \omega_\phi \dot{\phi} - \beta_2 \dot{\phi}|\dot{\phi}| - \frac{\beta_3 \dot{\phi}^3}{\omega_\phi} - \frac{(g-\ddot{w})\overline{GZ}(\phi)}{r_x^2}, \qquad (5.2)$$

where r_x is the roll radius of gyration. The roll frequency ω_ϕ is given by the metacentric height $\overline{GM}_{sw}$ in still water:

$$\omega_\phi = \frac{\sqrt{g\overline{GM}_{sw}}}{r_x}. \qquad (5.3)$$

The instantaneous $\overline{GZ}$ curve in irregular waves is estimated from numerical results for a regular wave with a wave length equal to the length L of the vessel and a wave height equal to $0.05\,L$. These numerical results are fitted with analytical approximations of the form

$$\overline{GZ}(\phi, x_c) = \left(C_0 \sin\phi + C_1\phi + C_3\phi^3 + C_5\phi^5\right)\cos^4\left(\frac{\pi x_c}{L_e}\right)$$

$$+ \left(D_0 \sin\phi + D_1\phi + D_3\phi^3 + D_5\phi^5\right)\sin\left(\frac{\pi x_c}{L_e}\right), \qquad (5.4)$$

where the wave crest position x_c is measured relative to the aft end of the vessel. Similarly, the $\overline{GZ}$ curve in still water is fitted by:

$$\overline{GZ}_{sw}(\phi) = \left(\overline{GM}_{sw} - A_1\right)\sin\phi + A_1\phi + A_3\phi^3 + A_5\phi^5. \qquad (5.5)$$

The coefficients $(A_1, A_3, A_5, C_0, C_1, C_3, C_5, D_0, D_1, D_3, D_5, L_e)$ in (5.4)–(5.5) are found by the least square method. Other polynomial or Fourier series representations have been suggested, e.g., [5, 36], and generally a very good fit can be achieved for the range of roll angles of interest.

An alternative, albeit rather time-consuming, procedure to (5.4)–(5.5) is to calculate the instantaneous value of the righting arm by static balancing the ship in the instantaneous wave elevation, [41]. It was found that it did not change the results significantly compared to more simplified approaches including (5.4)–(5.5). Thus, (5.4)–(5.5) are assumed accurate enough for the present discussion focusing on effective stochastic procedures for estimation of the probability of occurrence of parametric roll.

In a stochastic seaway the following approximation to the instantaneous value of the righting arm $\overline{GZ}(t)$ is then:

$$\overline{GZ}(\phi,t) = \overline{GZ}_{sw}(\phi) + \frac{h(t)}{0.05L}\left(\overline{GZ}(\phi,x_c(t)) - \overline{GZ}_{sw}(\phi)\right). \qquad (5.6)$$

The instantaneous wave height $h(t)$ along the length of the vessel and the position of the crest x_c in head sea are determined by an equivalent wave procedure somewhat similar to the one used by [24]:

$$a(t) = \frac{2}{L_e}\int_0^{L_e} H\left(X\left(x,t\right),t\right)\cos\left(\frac{2\pi x}{L_e}\right)dx$$

$$b(t) = \frac{2}{L_e}\int_0^{L_e} H\left(X\left(x,t\right),t\right)\sin\left(\frac{2\pi x}{L_e}\right)dx$$

$$X\left(x,t\right) = -(x+Vt)$$

$$h(t) = 2\sqrt{a^2(t)+b^2(t)}$$

$$x_c(t) = \begin{cases} \frac{L_e}{2\pi}\arccos\left(\frac{2a(t)}{h(t)}\right) & \text{if } b(t) > 0 \\ L_e - \frac{L_e}{2\pi}\arccos\left(\frac{2a(t)}{h(t)}\right) & \text{if } b(t) < 0 \end{cases} \qquad (5.7)$$

Stationary sea conditions are assumed and specified by a JONSWAP wave spectrum with significant wave height H_s and zero-crossing period T_z. The frequency range is taken to be $\pi \leq \omega T_z \leq 3\pi$ covering the main part of the JONSWAP spectrum.

The next step in the solution procedure is to account for the stochastic behavior of the sea. The straight forward procedure is to generate time series of random waves and use them as input to the ship motion code and then extract extreme values by simple counting and subsequent fitting to a proper extreme value distribution, e.g., the Gumbel distribution. This, however, requires long simulation time and also CPU time to get sufficient reliable results. A solution is to use cluster of computers. Other methods seek to identify the most probable wave episodes leading to a specified large roll angle. Spyrou and Themelis [37] describe such an approach in which a specific ship motion parameter, e.g., a large roll angle, is calculated for a range of wave heights, wave periods, and number of adjacent high waves. Thereafter, the probabilities of encountering these wave groups are determined and used to estimate the corresponding probability of exceeding the prescribed ship motion response. The feasibility of the method has been documented in [37] and [38] in the EU FP7 sponsored integrated project SAFEDOR.

A related procedure for calculation exceedance probabilities and associated critical wave episodes has been developed in [19] for parametric roll in head sea and extended in [20] to cover other types of roll motions. This procedure uses the First-Order Reliability Method (FORM) to determine the mean out-crossing rate of the ship response considered. The procedure also identifies a design point with a corresponding most probably wave episode leading to the prescribed response value. Thereby, the tedious task to identify critical wave episodes is done automatically by the procedure and the user (i.e., the designer) only has to select or program a proper time domain procedure able to model the ship response in question. All the statistical estimates are then done within a standard FORM. In the present treatment, the time domain simulation routine, (5.4), has been linked to the FORM software PROBAN [8]. It is clear that (5.2) has a rather limited accuracy, but anyway contains the main features needed to model parametric rolling. It is, however, straight forward to replace (5.2) with (5.1) or another more general time domain ship motion code.

Apart from the FORM results, which is only asymptotically correct for very low probability of occurrence, a very useful property can be derived from the FORM analysis, namely that the reliability index is strictly inversely proportional to the significant wave height. This observation can be used to accelerate MCS using artificially increased significant wave heights. These results can then afterwards be scaled down to the actual (real) wave height, resulting in several order of magnitudes reduction in simulation time.

In the following, the FORM procedure is first described in general terms and then results for a container ship are presented and compared with MCS using artificially increased wave heights.

5.3 First-Order Reliability Method Applied to Wave Loads

5.3.1 Design Point and Reliability Index

In FORM, the excitation or input process is a stationary stochastic process. Considering in general wave loads on marine structures, the input process is the wave elevation and the associated wave kinematics. For moderate sea states the wave elevation can be considered as Gaussian distributed, whereas for severer wave conditions corrections for nonlinearities must be incorporated. Such corrections are discussed and accounted for by using a second-order wave theory in a FORM analysis of a jack-up platform [17]. In the present paper dealing with the roll motion of a ship, linear, long-crested waves are assumed and hence the normal distributed wave elevation $H(X, t)$ as a function of space X and time t can be written as:

$$H(X,t) = \sum_{i=1}^{n} \left(u_i c_i(X,t) + \bar{u}_i \bar{c}_i(X,t) \right), \qquad (5.8)$$

where the variables $u_i, \bar{u}_i$ are statistical independent, standard normal distributed variables to be determined by the FORM procedure and with the deterministic coefficients given by:

$$c_i(x,t) = \sigma_i \cos(\omega_i t - k_i X),$$
$$\bar{c}_i(x,t) = -\sigma_i \sin(\omega_i t - k_i X),$$
$$\sigma_i^2 = S(\omega_i)\mathrm{d}\omega_i, \tag{5.9}$$

where ω_i and $k_i = \omega_i^2/g$ are the n discrete frequencies and wave numbers applied. Furthermore, $S(\omega)$ is the wave spectrum and the increment between the discrete frequencies. It is easily seen that the expected value $E[H^2] = \int S(\omega)\mathrm{d}\omega$, thus the wave energy in the stationary sea is preserved. Short-crested waves could be incorporated, if needed, but require more unknown variables $u_i, \bar{u}_i$ and thus a larger computational effort.

From the wave elevation, (5.8)–(5.9), and the associated wave kinematics, any nonlinear wave-induced response $\phi(t)$ of a marine structure can in principle be determined by a time domain analysis using a proper hydrodynamic model:

$$\phi = \phi(t\,|u_1, \bar{u}_1, u_2, \bar{u}_2, \ldots, u_n, \bar{u}_n). \tag{5.10}$$

Each of these realizations represents the response for a possible wave scenario. In the present case the realization is the roll angle and the realization which exceeds a given threshold ϕ_0 at time $t = t_0$ with the highest probability is sought. This problem can be formulated as a limit state problem, well-known within time-invariant reliability theory [7]:

$$G(u_1, \bar{u}_1, u_2, \bar{u}_2, \ldots, u_n, \bar{u}_n) \equiv \phi_0 - \phi(t_0\,|u_1, \bar{u}_1, u_2, \bar{u}_2, \ldots, u_n, \bar{u}_n) = 0. \tag{5.11}$$

An approximate solution can be obtained by using FORM. The limit state surface G is given in terms of the statistical independent, standard normal distributed variables $\{u_i, \bar{u}_i\}$, and hence determination of the design point $\{u_i^*, \bar{u}_i^*\}$, defined as the point on the failure surface $G = 0$ with the shortest distance to the origin, is rather straightforward. A linearization about this point replaces (5.11) with a hyper plane in $2n$-space. The distance

$$\beta_{\mathrm{FORM}} = \min \sqrt{\sum_{i=1}^{n} (u_i^2 + \bar{u}_i^2)} \tag{5.12}$$

from the hyper plane to the origin is denoted the FORM reliability index. The calculation of the design point $\{u_i^*, \bar{u}_i^*\}$ and the associated value of β_{FORM} can be performed by standard reliability codes, e.g., [8]. Alternatively, standard optimization codes using (5.12) as the objective function and (5.11) as the constraint can be applied.

The time integration in (5.11) must cover a sufficient time period $\{0,t_0\}$ to avoid any influence on $\phi(t_0)$ of the initial conditions at $t = 0$, i.e., to be longer than the memory in the system. Proper values of t_0 would usually be 1–3 min, depending on the damping in the system. However, for bifurcation problems with low damping as e.g., parametric roll a larger transient period can be expected. To avoid repetition in the wave system and for accurate representation of typical wave spectra $n = 25 - 50$ would thus be needed.

The deterministic wave profile:

$$H^*(X,t) = \sum_{i=1}^{n} \left(u_i^* c_i(X,t) + \bar{u}_i^* \bar{c}_i(X,t) \right) \qquad (5.13)$$

can be considered as a design wave or a critical wave episode. It is the wave scenario with the highest probability of occurrence that leads to the exceedance of the specified response level ϕ_0. For linear systems the result reduces to the standard Slepian model, see e.g., [1, 9, 25, 40]. The critical wave episode in itself is a useful result as it can be used as input in more elaborate time domain simulations to correct for assumptions made in the hydrodynamic code, (5.10), applied in the FORM calculations. Such a model correction factor approach can provide an effective tool of accounting for even very complicated nonlinear effects [10].

It should be noted that other definitions of design waves based on a suitable nonuniform distribution of phase angles have been applied, especially for experimental application in model basins. The selection of the phase angle distribution is, however, not obvious, see e.g., [2].

5.3.2 Mean Out-Crossing Rates and Exceedance Probabilities

The time-invariant peak distribution follows from the mean out-crossing rates. Within a FORM approximation the mean out-crossing rate can be written as follows [17]:

$$\nu(\phi_0) = \frac{1}{2\pi\beta_{\text{FORM}}} e^{-\frac{1}{2}\beta_{\text{FORM}}^2} \sqrt{\sum_{i=1}^{n} \left(u_i^{*2} + \bar{u}_i^{*2} \right) \omega_i^2} \qquad (5.14)$$

based on a general formula given by [22]. Thus, the mean out-crossing rate is expressed analytically in terms of the design point and the reliability index. For linear processes it reduces to the standard Rayleigh distribution. Often the gradient vector $\{\alpha_i^*, \bar{\alpha}_i^*\} = \{u_i^*, \bar{u}_i^*\}/\beta_{\text{FORM}}$ to the design point does not vary much with exceedance level ϕ_0. This is so for e.g., parametric rolling, [19]. Hence, (5.14) reduces to

$$\nu(\phi_0) = \nu_0 e^{-\frac{1}{2}\beta_{\text{FORM}}^2}, \qquad (5.15)$$

where v_0 can be viewed as an effective mean zero out-crossing rate. It value can be taken as the inverse of the roll period in calm water without notable error. Finally, on the assumption of statistically independent peaks and, hence, a Poisson distributed process, the number of exceedance of the level ϕ_0 in a given time T can be calculated from the mean out-crossing rate $v(\phi_0)$:

$$P\left[\max_T \phi > \phi_0\right] = 1 - e^{-v(\phi_0)T}. \tag{5.16}$$

The present procedure can be considered as an alternative to the random constrained simulation, see e.g., [9]. The present method has, however, the advantage that the number of time domain simulations is much smaller due to the very efficient optimization procedures within FORM, and that it does not require the curve-fitting of lines of constant probabilities needed in the other procedure. Furthermore, the present procedure does not rely on a mean wave conditional from a linear response and can hence be applied also to bifurcation types of problems like parametric roll.

For bifurcation type of problems the optimization procedure used in the FORM analysis must be chosen appropriately, i.e., of the nongradient type and here a circle step approach is used [8]. Furthermore, to facilitate the convergence of the optimization procedure, the limit state surface, (5.11), is replaced by a logarithm transformation:

$$\tilde{G}(u_1,\bar{u}_1,u_2,\bar{u}_2,\ldots,u_n,\bar{u}_n) \equiv \log t(\phi_0) - \log t(\phi(t_0 | u_1,\bar{u}_1,u_2,\bar{u}_2,\ldots,u_n,\bar{u}_n)) = 0,$$
$$\tag{5.17}$$

where

$$\log t(y) \equiv \begin{cases} -1 - \log(-y); & y < -1 \\ y; & -1 \leq y \leq 1 \\ 1 + \log(y); & 1 < y \end{cases} \tag{5.18}$$

Finally, an arbitrary starting point different from zero is used and a range monotonically increasing threshold values ϕ_0 are applied in order to get convergence.

The FORM is significantly faster than direct MCS, but most often very accurate. In a study by [19] dealing exclusively with parametric rolling of ships in head sea the FORM approach was found to be two orders of magnitude faster than direct simulation for realistic exceedance levels and with results deviating less than 0.1 in the reliability index. However, a scaling property derived from the FORM procedure can be used to accelerate very effectively the MCS. This will be investigated later in this chapter.

5.4 Numerical Example

The same Panmax container ship as used in [20] is considered here. The pertinent data including the coefficients in (5.4) and (5.5) can be found in [20], but it is noted that the damping coefficients, β_1–β_3, are taken, quite arbitrarily, from a study

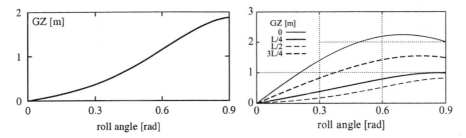

Fig. 5.1 *Left*: $\overline{GZ}$ curve in still water. *Right*: $\overline{GZ}$ curves in regular waves with wave length equal to the ship length L and a wave height equal to 0.05 L. Wave crest positions at $x_c = 0$, 0.25 L, 0.5 L, 0.75 L, and 1.0 L [19]

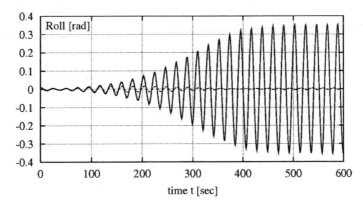

Fig. 5.2 Parametric roll in a regular wave (*solid line*) and the roll response for a slightly smaller regular wave (*dashed line*) [19]

considering a different vessel [5], and corresponds to about 0.05 in equivalent linear roll damping. The ship is sailing in head sea with a forward speed of 6 m/s.

In the following the roll angles are given in radians. The approximate $\overline{GZ}$ curves, (5.4)–(5.5) are accurate for roll angles up to 0.9 rad [18]. The $\overline{GZ}$ curves are shown in Fig. 5.1 and it is clear that a significant reduction in righting lever occurs when the wave crest moves from the aft perpendicular (AP) to 0.25 L forward of AP. This is quite typical for ships with fine hull forms like container ships.

By use of the closed-form expressions given in [16] for the heave w, all pertinent data for calculation of the roll angle as function of time is defined. In order to show that (5.2) can model parametric roll, calculations have been performed with a regular wave with an encounter frequency close to twice the roll frequency [19]. Two wave heights were used: one (3.65 m) where parametric roll is not triggered and one slightly higher (3.7 m) where parametric roll develops. The roll motions for the two wave heights are shown in Fig. 5.2. The onset of parametric roll and its saturation level are clearly noticed.

The regular wave height needed to trigger parametric roll is thus about 3.7 m for the present vessel. If the wave height is increased above this value, parametric roll develops faster and to a higher saturation level. These results are consistent with both model test results and numerical calculations using more elaborate hydrodynamic codes [11].

In the following, some results are shown for parametric roll motions in head sea in a stochastic seaway. More results can be found in [19], whereas results for other headings are given in [20]. A parameter study is included in [20] quantifying the sensitivity of the reliability index β_{FORM} to the zero-crossing wave period T_z and the forward speed V. Here it is only noted that the probability of parametric roll decreases if the speed is either lowered or increased for the present example.

The sea state has a significant wave height $H_s = 12$ m and a zero-crossing wave period $T_z = 11.7$ s. The zero-crossing period is chosen such that parametric roll can be expected due to occurrence of encounter frequencies in the range of twice the roll frequency. Of course neither the encounter frequency nor the roll frequency is constant in irregular waves.

The time simulations are carried out from $t = 0$ to $t = t_0 = 300$ s with a time step of 0.5 s. The effect of the initial condition ($\phi(t = 0) = 0.01$ rad) is negligible after about 50 s, but in order to build up parametric roll a longer duration is needed. With $n = 50$ equidistant frequencies, the wave repetition period relative to the ship is about 400 s with a forward speed of 6 m/s.

5.4.1 Results by the First-Order Reliability Method

A detailed analysis using the present approach is given in [19]. As an example the most probable roll response and the associated critical wave episode, (5.13), corresponding to exceedance of a prescribed roll angle, i.e., ϕ_0, of 0.5 rad, are shown in Fig. 5.3.

The interesting observation is as stated in [19]:

> "The critical wave episode is basically a sum of two contributions: firstly, a regular wave with encounter frequency close to twice the roll frequency and a wave height just triggering parametric roll and, secondly, a transient wave with magnitude depending on the prescribed roll response ϕ_0".

The last part resembles the critical wave episodes as obtained from quasi-static response analyzes, e.g., [1] and has basically the shape of the autocorrelation function. The first term, which is independent of the prescribed response level, is unique for parametric roll, but is needed to initiate parametric roll. After the peak in roll angle has been reached (i.e., for $t > 300$ s) the first part is seen to disappear. This is consistent with an unconditional mean wave equal to zero after $t = t_0 = 300$ s.

As the wave spectrum does not change shape with H_s the critical wave episode, (5.13), becomes independent of H_s. A change of H_s by a factor μ will then just change the design point $\{u_i^*, \bar{u}_i^*\}$ and hence β_{FORM} by a factor $1/\mu$. This behavior has

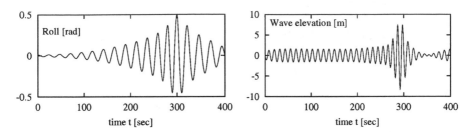

Fig. 5.3 *Left*: Most probable roll response yielding $\phi_0 = 0.5$ rad at $t_0 = 300$ s. *Right*: Corresponding critical wave episode [20]

previously been noted by [39], based on a different argument, and is also mentioned and discussed in the [14, IMO-SLF submission by Germany 2007]. Clearly this property greatly facilitates the long-term convolution of the heeling angle, but is also useful in MCS as will be discussed below.

5.4.2   Results by Monte Carlo Simulations

The FORM solution is asymptotically correct, i.e., only strictly valid for very low probabilities of occurrence. Furthermore, the FORM solution does not always converge, most probably due to parametric rolling being a bifurcation type of problem. A comparison between FORM results and MCS has been made in [30]. Generally, fairly good agreement was found. However, if the simulation length, t_0 in (5.11), exceeded 600 s then the FORM solution did not converge using the present FORM code. This might constitute a problem in some cases as parametric rolling typically has a long transient run-in period due to the relative low roll damping.

The reliability index β_{FORM} calculated by the FORM analysis is strictly inversely proportional to the square root of the intensity of the excitation spectrum *irrespectively of the nonlinearity* in the system as shown in [20] and [12]: Hence, for wave excitation in stationary stochastic sea based on the JONSWAP or the Pierson–Moskowitz spectrum, β_{FORM} is always inversely proportional to the significant wave height as in a purely linear analysis:

$$\beta_{FORM}(\phi_0 \,|H_s, T_z, V, \dots) = \frac{C(\phi_0 \,|T_z, V, \dots)}{H_s}. \tag{5.19}$$

If this property can be assumed valid also in the MCS the out-crossing rates can be increased with a corresponding reduction of the necessary length of the time domain simulations by using a larger significant wave height than relevant from a design point-of-view in the simulations. The reliability index and the corresponding mean out-crossing rate, (5.15), thus obtained can then afterwards be scaled down to the actual significant wave height. The accuracy of this approach has been investigated

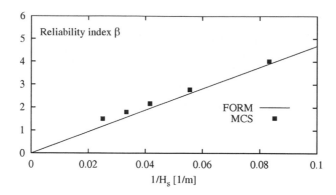

Fig. 5.4 Reliability index from First-Order Reliability Method (FORM) and Monte Carlo simulations (MCS) as function of $1/H_s$

by [21] considering the overturning of a jack-up rig and the extreme wave bending moment in a ship as examples. It was found that a relation of the type $\beta = a + b/H_s$, where a and b are constants depending on the threshold level ϕ_0, very accurately modeled the MCS results.

The example is the same as above except that the threshold roll angle is increased to $\phi_0 = 0.6$ rad in order to lower the probability of occurrence and thereby increase the computational time for direct simulations. With a significant wave height $H_s = 12$ m the reliability index was found in [19] to be $\beta_{FORM} = 3.9047$ whereas a MCS simulation for the same problem using 1,000,000 simulations resulted in a mean reliability index of 4.0208 with a 90% confidence interval of $[3.9576, 4.1048]$ and a CoV $= 0.186$. The agreement is good with FORM being slightly on the conservative side, but there were two orders of magnitude in difference in the computational time between the FORM analysis and the MCS, so an improvement in the MCS is strongly needed. Therefore the MCS have now been performed again but with significant wave heights $H_s = 18$ m, 24 m, 30 m and 40 m using a CoV $= 0.05$ as stopping criteria. The results for the reliability index thus calculated are shown in Fig. 5.4 as function of $1/H_s$. Clearly, a relation of the type $\beta = a + b/H_s$ will provide a very accurate fit of the MCS results using the reliability index β at $H_s = 18$ m, 24 m and/or 30 m as input. The result for $H_s = 40$ m falls a little outside this fit in agreement with other studies [21, 39] showing that the linearity in the reliability index β with $1/H_s$ usually only holds if β is greater than about 2.

It is noted that the simulation time for the results with the increased wave heights only are small fractions of what was needed in the simulation with $H_s = 12$ m. More specifically, the number of simulations required to get results with the same CoV is inversely proportional to the corresponding probability of exceedance. Hence, if the values at $H_s = 30$ m ($\beta = 1.8001$) and $H_s = 24$ m ($\beta = 2.1597$) are used in the extrapolation scheme $\beta = a + b/H_s$, then the reduction in simulation time is about 300. However, if simulation results for $H_s = 10$ m was requested then the extrapolation scheme will provide a four order of magnitude reduction in simulation time. Finally, it is noted that the FORM results seems to be asymptotically correct as expected, but also that the extrapolation scheme for the MCS does not need any input from a FORM analysis. Thus the two methods supplement each other rather nicely, providing accurate results for low and high values of the reliability index.

5.5 Conclusions

The Monte Carlo simulation (MCS) results presented for the reliability index β for parametric rolling of a ship suggest that the relation $\beta = a + b/H_s$ can provide an accurate scaling of the reliability index with significant wave height. This could be a useful procedure in case direct MCS for the design sea states are not feasible due to too long computational time.

It is also noted that the result from First-Order Reliability Method (FORM) is quite accurate, especially for high values of the reliability index.

From the reliability index the mean out-crossing rate and the probability of exceedance directly follow from (5.15) and (5.16).

References

1. Adegeest, L. J. M, Braathen, A. Løseth, R. M.: Use of non-linear sea-loads nimulations in design of ships. Proc. PRADS'1998, Delft, pp. 53–58 (1998)
2. Alford, K. A., Troesch, A. W., McCue, L.: LS design wave elevations leading to extreme roll motions. Proc. STAB'2005, Istanbul, Turkey (2005)
3. Borge, J., Gonzáles, R., Hessner, K., Reichert, K., Soares, C. G.: Estimation of sea state directional spectra by using marine radar imaging of sea surface. Proc. ETCE/OMAE2000 Joint Conference, New Orleans, Louisiana, USA (2000)
4. Bulian, G., Francescutto, A. A.: Simplified modular approach for the prediction of the roll motion due to the combined action of wind and waves. Proc Instn Enggrs, Part M: J. Engineering for the Maritime Environment 218:189–212 (2004)
5. Bulian, G.: Nonlinear parametric rolling in regular waves – a general procedure for the analytical approximation of the GZ curve and its use in time domain simulations. Ocean Engineering 32:309–330 (2005)
6. Daalen, E. F. G., Boonstra, H., Blok, J. J.: Capsize probability analysis of a small container vessel. Proc- 8th Int. Workshop on Stability and Operational Safety of Ships, Istanbul, October 6–7 (2005)
7. Der Kiureghian, A.: The geometry of random vibrations and solutions by FORM and SORM. Probabilistic Engineering Mechanics 15:81–90 (2000)
8. Det Norske Veritas: General purpose probabilistic analysis program, Version 4.4 (2003)
9. Dietz, J. S., Friis-Hansen, P., Jensen, J. J.: Most likely response waves for estimation of extreme value ship response statistics. Proc PRADS'2004, Travemünde, September, Germany (2004)
10. Ditlevsen, O., Arnbjerg-Nielsen, T.: Model-correction-factor method in structural reliability. Journal of Engineering 120:1–10 (1994)
11. France, W. N., Levadou, M., Treakle, T. W., Paulling, J. R., Michel, R. K., Moore, C.: An investigation of head-sea parametric rolling and its influence on container lashing systems. Marine Technology 40:1–19 (2003)

12. Fujimura, K., Der Kiureghian, A.: Tail-equivalent linearization method for nonlinear random vibration. Probabilistic Engineering Mechanics 22: 63–76 (2007)

13. Hsieh, S.-R., Troesch. A. W., Shaw, S. W.: A nonlinear probabilistic method for predicting vessel capsize in random beam seas. Proc Royal Society London, Part A, 446:195–211 (1994)

14. IMO submission (SLF 50) by Germany: Proposal for an additional intact stability regulations. Subcommittee on Stability, Loadlines and on Fishing Vessels (2007)

15. de Kat, J. O.: ITTC specialist committee on stability in waves (2005)

16. Jensen, J. J., Mansour, A. E., Olsen, A. S.: Estimation of ship motions using closed-form expressions. Ocean Engineering 31:61–85 (2004)

17. Jensen, J. J., Capul, J.: Extreme response predictions for jack-up units in second-order stochastic waves by FORM. Probabilistic Engineering Mechanics 21:330–337 (2006)

18. Jensen, J. J., Olsen, A. S.: On the assessment of parametric roll in random sea. Proc. World Maritime Technology Conference, London, 6–10 March (2006)

19. Jensen, J. J., Pedersen, P. T.: Critical wave episodes for assessment of parametric roll Proc. IMDC'06, Ann Arbor, May, pp. 399–411 (2006)

20. Jensen, J. J.: Efficient estimation of extreme non-linear roll motions using the first-order reliability method (FORM). J. Marine Science and Technology 12:191–202 (2007)

21. Jensen, J. J.: Extreme value predictions using Monte Carlo simulations with artificially increased load spectrum. Probabilistic Engineering Mechanics, 26:399–404 (2010)

22. Koo, H., Der Kiureghian, A., Fujimura, K.: Design point excitation for nonlinear random vibrations. Probabilistic Engineering Mechanics. 20:136–147 (2005)

23. Krüger, S., Hinrichs, R., Cramer, H.: Performance based approaches for the evaluation of intact stability problems. Proc. PRADS'2004, Travemünde, September, Germany (2004)

24. Kroeger, H.-P.: Roll Simulation von Schiffen im seegang. Schiffstechnik 33:187–216 (in German) (1986)

25. Lindgren, G.: Some properties of a normal process near a local maximum. Ann. Math. Stat. 41:1870–1883 (1970)

26. Ness, O. B., McHenry, G., Mathisen, J., Winterstein, S. R.: Nonlinear analysis of ship rolling in random beam waves. Proc. STAR Symposium on 21st Century Ship and Offshore Vessel Design, Production and Operation, April 12–15, pp. 49–66 (1989)

27. Neves, M. A. S., Rodriquez, C. A.: A coupled third-order model of roll parametric resonance. Proc Maritime Transportation and Exploitation of Ocean and Coastal Resources, pp. 243–253, Taylor and Francis, London, UK (2005)

28. Nielsen, J. K., Hald, N. H., Michelsen, J., Nielsen, U. D., Baatrup, J., Jensen, J. J., Petersen, E. S.: Seasense – real-time onboard decision Support. Proc. World Maritime Technology Conference, London, 6-10 March (2006)

29. Nielsen, U. D.: Estimations of on-site directional wave spectra from measured ship responses. Marine Structures 19:33–69 (2006)

30. Nielsen, U. D., Jensen, J. J.: Numerical simulations of the rolling of a ship in a stochastic sea – Evaluations by Use of MCS and FORM. Proc of 28th International Conference on Offshore Mechanics and Arctic Engineering (OMAE'09), Honolulu, USA, June, Paper No. 79765 (2009)

31. Pereira, R.: Simulation nichtlinearer Seegangslasten. Schiffstechnik 35:173-193 (in German) (1988)

32. Rathje, H.: Impact of Extreme Waves on Ship Design and Ship Operation. Proc Design & Operation for Abnormal Conditions, Royal Institution of Naval Architects, London, 26–27 January (2005)

33. Shin, Y. S., Belenky, V. L., Paulling, J. R., Weems, K. M., Lin, W. M.: Criteria for parametric roll of large container ships in head seas. Transactions of SNAME 112:14–47 (2004)

34. Spanos, D., Papanikolaou, A. A numerical simulation of a fishing vessel in parametric roll in head sea. Proc. 8th Int. Workshop on Stability and Operational Safety of Ships, Istanbul, October 6–7 (2005)

35. Spanos, D., Papanikolaou, A., Papatzanakis, G.: Risk-based onboard guidance to the master for avoiding dangerous seaways. 6th OSAKA Colloquium on Seakeeping and Stability of Ships, Osaka University, Japan, March 26–28 (2008)
36. Spyrou, K. J.: Designing against parametric instability in following seas. Ocean Engineering 26:625–653 (2000)
37. Spyrou, K., Themelis, N.: Development of probabilistic procedures and validation – Alternative 2: Capsize mode analysis. Deliverables D235, SAFEDOR-D-235-2006-11-9-NTUA–rev-1 (2006)
38. Themelis, N., Spyrou, K., Niotis, S.: Implementation and application of probabilistic Procedures. Deliverables D236, SAFEDOR-D-236-2007-06-15-NTUA–rev-1 (2007)
39. Tonguc, E., Söding, H.: Computing capsizing frequencies of ships in a seaway. Proc. 3rd Int. Conf. on Stability of Ships and Ocean Vehicles STAB'86, Gdansk (1986)
40. Tromans, P. S., Anaturk, A. R., Hagemeijer, P.: A New Model for the Kinematics of Large Ocean Waves - Application as a Design Wave. Proc 1st Offshore and Polar Engineering Conference (ISOPE), Vol. 3, pp. 64–71 (1991)
41. Vidic-Perunovic, J., Jensen, J. J.: Parametric roll due to instantaneous volumetric changes and speed variations. Ocean Engineering 36:891–899 (2009)

Chapter 6
Domains of Parametric Roll Amplification for Different Hull Forms

Claudio A. Rodríguez and Marcelo A.S. Neves

6.1 Introduction

Wave loads on specific hull forms under certain loading conditions can trigger excessive roll motions on intact vessels. Due to these excessive motions loss of carglo might occur or even capsizing. This is typically the situation when parametric roll is involved [29]. This is specially true for small fishing vessels and large container ships, [11, 12, 19], and hence of significant importance in the safe design of such vessels. Various methods which are employed to simulate parametric rolling are described in the literature, [4, 5]. Most of these methods are based on uncoupled mathematical models and the Mathieu equation and its variations are usually taken as the governing equation of the resonant amplification process, [2, 7, 9, 31, 37]. Bulian [8] proposed a 1.5-DOF model; following another perspective, Spyrou [38] investigated the coupling of surge motion with roll. For practical reasons more commonly investigation focus on the first region of instability, which corresponds to encounter frequency close to twice the roll natural frequency. A singular exception is reported by Obreja et al. [27].

In following seas, parametric rolling may be reasonably conceived as an uncoupled process but in extreme head seas the coupling of the three restoring modes (heave–pitch–roll) is essential, as coupling becomes significantly nonlinear [20]. Paulling and Rosenberg [28] and Blocki [7] are examples of limited formulations involving more than one DOF. A third-order model coupled in heave, roll, and pitch was first introduced in Neves and Rodríguez [21].

In order to allow incorporation of safety measures into ship design against parametric rolling in head seas, the influence of hull forms and operating conditions (including wave length and height) on the inception of this phenomenon and on the resultant motion amplitudes should be well understood. Clarification of these issues

C.A. Rodríguez (✉) • M.A.S Neves
LabOceano, COPPE, Universidade Federal do Rio de Janeiro, Rio de Janeiro, Brazil
e-mail: claudiorc@peno.coppe.ufrj.br; masn@peno.coppe.ufrj.br

T.I. Fossen and H. Nijmeijer (eds.), *Parametric Resonance in Dynamical Systems*,
DOI 10.1007/978-1-4614-1043-0_6, © Springer Science+Business Media, LLC 2012

is within the scope of the present work. This is intended to be achieved making use of the concept of parametric amplification domain (PAD for short), a generalization of the more usual concept of diagrams of limits of stability.

Spyrou and Tigkas [39], McCue et al. [18] and Bulian et al. [10] have employed different techniques of analyzes related to limits of stability. Neves and Rodríguez [25] introduced analytical expressions for the limits of stability of the roll linear variational equation of the third-order model based on the technique developed by Hsu [15]. An interesting aspect of these limits of stability is that they may display upper frontiers, as a consequence of the bending to the right of the limits of stability. This feature implies that at a given frequency, above a certain level of wave amplitude, further increasing the wave amplitude may result in the complete disappearance of parametric amplification.

Neves and Rodríguez [22, 23] proposed a 3-DOF mathematical model based on multivariable Taylor series expansions of hydrostatic and incident wave pressure fields with terms in heave, roll, and pitch expressed in terms of derivatives. They showed that the 3-DOF algorithm efficiently reproduced intense parametric rolling. That mathematical model has recently been extended to 6-DOF, Rodríguez [32]. This more robust new model will be employed in the present investigation. Examples of the capability of the 6-DOF model to adequately reproduce roll amplification will be presented for two hulls for which experimental results are available in the literature. These are a transom stern fishing vessel and a container vessel. It is worth mentioning that the same hulls have been numerically investigated by Ahmed et al. [3] employing a distinct algorithm. They compared the roll time series for both hulls for different tested conditions, achieving in general adequate agreement. However, the main target of the present investigation is not only to demonstrate the good agreement of the numerical and experimental results. The more ample objective here is to use the new model as a tool to, through systematic mapping of PADs and a structured comparative analysis for different hulls, build up a better understanding of the complexities of nonlinear parametric rolling, identify tendencies in the responses and to achieve a consistent perception of the limitations of the usual Mathieu modeling of parametric rolling as applied to the design of safe ships.

Employing numerical techniques, it is possible to determine the mentioned PADs for a given ship at different conditions (Neves and Rodríguez [24]). These correspond to realistic mappings of roll amplifications, since the essential coupled nonlinearities are kept in the approach. They are also more informative than the classical Ince–Strut diagram (Belenky and Sevastianov [6]) or the more elaborate Hsu [15] limits of stability, since all the regions inside the limits of stability are mapped. Numerical PADs are here employed as a tool to investigate how distinct hull forms may display substantially different standards of responses. These aspects deserve careful examination in the context of the present IMO initiatives towards the development of simple and reliable models for new generation intact ship stability criteria.

A brief outline of the paper is as follows. In the next section a new 6-DOF mathematical model is introduced. Subsequently, time series obtained from the

6-DOF mathematical model are compared with experimental results. The nonlinear responses of two ship hulls with marked hull form differences are considered, a fishing vessel and a container ship. Additionally, a third hull, one of a large vertical cylinder is taken as a counter example of parametric rolling weakly affected by nonlinearities. Subsequently, the concept of parametric rolling amplification domain is used to demonstrate the influence of relevant dynamical characteristics such as: (a) coupling between modes; (b) influence of third-order terms on the topology of the PADs; (c) influence of roll restoring curve; (d) influence of wave amplitude; and (e) influence of initial conditions. Finally, main conclusions are drawn.

6.2 6-DOF Mathematical Model

The 6-DOF mathematical model is presented below in a concise form. More details are given in Rodríguez [32]. Let the vector $s(t) = [x(t), y(t), z(t), \phi(t), \theta(t), \psi(t)]^T$ represent rigid-body motions in 6-DOF with respect to an inertial frame of reference. These are the surge, sway, and heave translational motions and roll, pitch, and yaw angular motions. The nonlinear ship equations of motions may be represented as:

$$\left(\tilde{\mathbf{M}} + \tilde{\mathbf{A}}\right)\ddot{\mathbf{s}} + \tilde{\mathbf{B}}(\dot{\phi})\dot{\mathbf{s}} + \mathbf{C}_r(\mathbf{s}, \zeta) = \mathbf{C}_{\text{ext}}\left(\zeta, \dot{\zeta}, \ddot{\zeta}\right). \tag{6.1}$$

In (6.1), $\tilde{\mathbf{M}}$ is a 6×6 matrix which describes hull inertial characteristics:

$$\tilde{M} = \begin{bmatrix} m & 0 & 0 & 0 & mz_G & 0 \\ 0 & m & 0 & -mz_G & 0 & 0 \\ 0 & 0 & m & 0 & 0 & 0 \\ 0 & -mz_G & 0 & J_{xx} & 0 & -J_{zx} \\ mz_G & 0 & 0 & 0 & J_{yy} & 0 \\ 0 & 0 & 0 & -J_{xz} & 0 & J_{zz} \end{bmatrix}. \tag{6.2}$$

Its elements are: m, the ship mass, mz_G a first moment, J_{xx}, J_{yy} and J_{zz} the mass moments of inertia in the roll, pitch, and yaw modes, respectively, and J_{xz}, the roll–yaw product of inertia, all moments taken with reference to the chosen origin. Matrices $\tilde{\mathbf{A}}$ and $\tilde{\mathbf{B}}$ are defined as:

$$\tilde{A} := \begin{bmatrix} X_{\dot{x}} & 0 & X_{\dot{z}} & 0 & X_{\ddot{\theta}} & 0 \\ 0 & Y_{\dot{y}} & 0 & Y_{\dot{\phi}} & 0 & Y_{\dot{\psi}} \\ Z_{\dot{x}} & 0 & Z_{\dot{z}} & 0 & Z_{\ddot{\theta}} & 0 \\ 0 & K_{\dot{y}} & 0 & K_{\dot{\phi}} & 0 & K_{\dot{\psi}} \\ M_{\dot{x}} & 0 & M_{\dot{z}} & 0 & M_{\ddot{\theta}} & 0 \\ 0 & N_{\dot{y}} & 0 & N_{\dot{\phi}} & 0 & N_{\dot{\psi}} \end{bmatrix}, \tag{6.3}$$

$$\tilde{\mathbf{B}} := \begin{bmatrix} X_{\dot{x}} & 0 & X_{\dot{z}} & 0 & X_{\dot{\theta}} & 0 \\ 0 & Y_{\dot{y}} & 0 & Y_{\dot{\phi}} & 0 & Y_{\dot{\psi}} \\ Z_{\dot{x}} & 0 & Z_{\dot{z}} & 0 & Z_{\dot{\theta}} & 0 \\ 0 & K_{\dot{y}} & 0 & K_{\dot{\phi}}(\dot{\phi}) & 0 & K_{\dot{\psi}} \\ M_{\dot{x}} & 0 & M_{\dot{z}} & 0 & M_{\dot{\theta}} & 0 \\ 0 & N_{\dot{y}} & 0 & N_{\dot{\phi}} & 0 & N_{\dot{\psi}} \end{bmatrix}. \tag{6.4}$$

$\tilde{\mathbf{A}}$ is also a 6×6 matrix, whose elements represent hydrodynamic generalized added masses. Following the reasoning of Abkowitz [1], these hydrodynamic reactions may be taken as linear. $\tilde{\mathbf{B}}$ describes the coefficients of the hydrodynamic reactions dependent on ship velocities (damping). These velocity-dependent reactions are also taken as linear, except for the roll damping moment, which incorporates a quadratic term in the roll equation:

$$K_{\dot{\phi}}(\dot{\phi})\dot{\phi} := K_{\dot{\phi}}\dot{\phi} + K_{\dot{\phi}|\dot{\phi}|}\dot{\phi}|\dot{\phi}|. \tag{6.5}$$

$\mathbf{C}_r(\mathbf{s}, \zeta)$ is a 6×1 vector which describes nonlinear restoring forces and moments dependent on the relative motions between ship hull and wave elevation $\zeta(t)$. On the right-hand side of (6.1), the generalized 6×1 vector $\mathbf{C}_{ext}(\zeta, \dot{\zeta}, \ddot{\zeta})$ represents linear wave external excitation, defined as the sum of the linear Froude–Krilov plus diffraction wave forcing terms, dependent on wave heading χ, encounter frequency ω_e, wave amplitude A_w and time t. Its components represent the wave exciting forces (X_W, Y_W and Z_W for surge, sway, and heave, respectively) and moments (K_W, M_W and N_W for roll, pitch, and yaw):

$$\mathbf{C}_{ext}(\zeta, \dot{\zeta}, \ddot{\zeta}) = [X_W(t), Y_W(t), Z_W(t), K_W(t), M_W(t), N_W(t)]^T \tag{6.6}$$

Elements in (6.3), (6.4), and (6.6) are derived assuming potential theory. WAMIT code [42] is employed to compute zero-speed contributions of the added masses, damping coefficients and exciting forces. Corrections due to ship speed of advance are introduced approximately considering the 2-D speed-correction terms of Salvesen et al. [30] to be applicable to the 3-D panel methodology, a procedure that was proposed by Inglis [16].

In the following the mathematical form of the nonlinear restoring actions $\mathbf{C}_r(\mathbf{s}, \zeta)$ will be presented, more details may be found in Rodríguez [32]. Considering that these actions are modeled with terms defined to third order in Taylor, they may be grouped as constituted of first-, second-, and third-order contributions:

$$\mathbf{C}_r(\mathbf{s}, \zeta) = \mathbf{C}_{r(\mathbf{s})}^{(1)} + \left(\mathbf{C}_{r(\mathbf{s})}^{(2)} + \mathbf{C}_{r(\mathbf{s}, \zeta)}^{(2)} \right) + \left(\mathbf{C}_{r(\mathbf{s})}^{(3)} + \mathbf{C}_{r(\mathbf{s}, \zeta)}^{(3)} \right). \tag{6.7}$$

First-order terms are purely hydrostatic (depend only on ship displacements, $\mathbf{s}$). Second- and third-order terms have contributions from both hydrostatic (subscript $(\mathbf{s})$) and incident wave pressure field (subscript $(\mathbf{s}, \zeta)$):

$$
\mathbf{C}_{r(s)}^{(1)} = \begin{bmatrix} Z_z & 0 & Z_\theta \\ 0 & K_\phi & 0 \\ M_z & 0 & M_\theta \end{bmatrix} \begin{bmatrix} z \\ \phi \\ \theta \end{bmatrix}
\tag{6.8}
$$

$$
\mathbf{C}_{r(s)}^{(2)} = \frac{1}{2} \begin{bmatrix} Z_{zz}z + 2Z_{z\theta}\theta & Z_{\phi\phi}\phi & Z_{\theta\theta}\theta \\ 0 & 2\left(K_{z\phi}z + K_{\phi\theta}\theta\right) & 0 \\ M_{zz}z + 2M_{z\theta}\theta & M_{\phi\phi}\phi & M_{\theta\theta}\theta \end{bmatrix} \begin{bmatrix} z \\ \phi \\ \theta \end{bmatrix}
\tag{6.9}
$$

$$
\mathbf{C}_{r(s)}^{(3)} = \frac{1}{6} \begin{bmatrix} (Z_{zzz}z + 3Z_{zz\theta}\theta)z^2 + 3(Z_{\phi\phi z}z + Z_{\phi\phi\theta}\theta)\phi^2 + (Z_{\theta\theta\theta}\theta + 3Z_{\theta\theta z}z)\theta^2 \\ 3K_{zz\phi}z^2\phi + (K_{\phi\phi\phi}\phi^2 + 6K_{z\phi\theta}z\theta)\phi + 3K_{\theta\theta\phi}\theta^2\phi \\ (M_{zzz}z + 3M_{zz\theta}\theta)z^2 + 3(M_{\phi\phi z}z + M_{\phi\phi\theta}\theta)\phi^2 + (M_{\theta\theta\theta}\theta + 3M_{\theta\theta z}z)\theta^2 \end{bmatrix}
$$

$$
\tag{6.10}
$$

$$
\mathbf{C}_{r(s,\zeta)}^{(2)} = \begin{bmatrix} 0 & 0 & 0 & 0 & 0 & X_{\zeta\psi}(t) \\ 0 & 0 & 0 & 0 & 0 & Y_{\zeta\psi}(t) \\ 0 & 0 & Z_{\zeta z}(t) & 0 & Z_{\zeta\theta}(t) & Z_{\zeta\psi}(t) \\ 0 & 0 & 0 & K_{\zeta\phi}(t) & 0 & K_{\zeta\psi}(t) \\ 0 & 0 & M_{\zeta z}(t) & 0 & M_{\zeta\theta}(t) & M_{\zeta\psi}(t) \\ 0 & 0 & 0 & 0 & 0 & N_{\zeta\psi}(t) \end{bmatrix} \begin{bmatrix} x \\ y \\ z \\ \phi \\ \theta \\ \psi \end{bmatrix}
\tag{6.11}
$$

$$
\mathbf{C}_{r(s,\zeta)}^{(3)} = \begin{bmatrix} X_{\zeta\psi z}(t)\psi z + X_{\zeta\psi\phi}(t)\psi\phi + X_{\zeta\psi\theta}(t)\psi\theta + X_{\zeta\psi\psi}(t)\psi^2 \\ Y_{\zeta\psi z}(t)\psi z + Y_{\zeta\psi\phi}(t)\psi\phi + Y_{\zeta\psi\theta}(t)\psi\theta + Y_{\zeta\psi\psi}(t)\psi^2 \\ Z_{\zeta\zeta z}(t)z + Z_{\zeta zz}(t)z^2 + Z_{\zeta z\theta}(t)z\theta + Z_{\zeta\phi\phi}(t)\phi^2 + Z_{\zeta\zeta\theta}(t)\theta \\ \quad + Z_{\zeta\theta\theta}(t)\theta^2 + Z_{\zeta\psi z}(t)\psi z + Z_{\zeta\psi\phi}(t)\psi\phi + Z_{\zeta\psi\theta}(t)\psi\theta + Z_{\zeta\psi\psi}(t)\psi^2 \\ K_{\zeta\zeta\phi}(t)\phi + K_{\zeta z\phi}(t)z\phi + K_{\zeta\phi\theta}(t)\phi\theta + K_{\zeta\psi z}(t)\psi z + K_{\zeta\psi\phi}(t)\psi\phi \\ \quad + K_{\zeta\psi\theta}(t)\psi\theta + K_{\zeta\psi\psi}(t)\psi^2 \\ M_{\zeta\zeta z}(t)z + M_{\zeta zz}(t)z^2 + M_{\zeta z\theta}(t)z\theta + M_{\zeta\phi\phi}(t)\phi^2 + M_{\zeta\zeta\theta}(t)\theta \\ \quad + M_{\zeta\theta\theta}(t)\theta^2 + M_{\zeta\psi z}(t)\psi z + M_{\zeta\psi\phi}(t)\psi\phi + M_{\zeta\psi\theta}(t)\psi\theta \\ \quad + M_{\zeta\psi\psi}(t)\psi^2 + N_{\zeta\psi z}(t)\psi z + N_{\zeta\psi\phi}(t)\psi\phi + N_{\zeta\psi\theta}(t)\psi\theta + N_{\zeta\psi\psi}(t)\psi^2 \end{bmatrix}
$$

$$
\tag{6.12}
$$

The restoring coefficients nomenclature follows the logic for the derivatives coefficients in Taylor series expansions in which X, Y, Z, K, M, and N identify the DOF of the restoring action and their subscripts represent the order of the dependency of restoring forces and moments on ship motions and wave elevation. It should be noticed that purely hydrostatic terms have no components in the horizontal modes. Second and third-order terms due to wave passage also incorporate dependency on the instantaneous yaw angle (terms with ψ subscripts). In the case of longitudinal waves this dependency may not be strong, but may be significant in oblique seas. Each coefficient is obtained by means of polynomial fitting to the hydrostatic and wave pressure variations obtained for displaced positions of the hull in heave, roll, and pitch, as described in Rodríguez et al. [33] and Holden et al. [14].

6.3 Tested Hulls

Three hulls have been selected for the purpose of discussing the influence of coupling on the development of parametric rolling:

(a) TS – a transom stern fishing vessel ($L_{pp} = 22.09$ m, $\overline{GM} = 0.37$ m). Experimental model tests in head seas at different speeds and tuning $\omega_e/\omega_{n4} = 2$ have shown that this is a hull capable of developing intense parametric excitation in very few cycles. Details of the hull characteristics and test program can be found in Neves et al. [19].

(b) SAFEDOR – a container ship ($L_{pp} = 150$ m, $\overline{GM} = 1.38$ m). Details of the hull characteristics and experimental test program for this hull, also referred in the literature as ITTC-A1 hull, can be found in SAFEDOR [34], Spanos and Papanikolaou [35].

c) SPAR – a large cylindrical offshore platform (diameter $D = 37.5$ m; draft $T = 202.5$ m, $\overline{GM} = 4.0$ m) numerically and experimentally tested by Haslum and Faltinsen [13]. Unfortunately their experimental results do not allow any comprehensive validation of the model in this case. Neves et al. [26] numerically investigated the possible causes of parametric rolling amplification of this platform.

Figure 6.1 shows 3-D images of the three hulls (not to scale).

Fig. 6.1 3-D views:
(**a**) fishing vessel TS;
(**b**) containership SAFEDOR;
(**c**) SPAR platform

6.4   Verification of the 6-DOF Mathematical Model

It is not a simple task to numerically reproduce all cases of parametric rolling in head seas when a set of different wave conditions is considered, as demonstrated by the recent SAFEDOR benchmark exercise, Spanos and Papanikolaou [35].

Samples of comparisons of roll motion, simulated numerical results against experimental results are presented in Figs. 6.2 and 6.3 for the TS and SAFEDOR hulls, respectively. It may be observed that the mathematical model in general correctly captures the tendency for roll amplification for both hulls, although some quantitative differences in roll amplitude may be observed in some cases. A larger number of comparisons are given in Rodríguez [32], which additionally shows that the nonoccurrence of parametric rolling is also adequately reproduced. It may be concluded that, in general, the third-order mathematical model gives reliable results when compared with the experimental results for both hulls for different speeds and wave amplitudes, despite their marked differences in inertia and geometry.

Another successful verification exercise for this mathematical model in head seas may be found in Rodríguez et al. [33] and Holden et al. [14], corresponding to a large container vessel; in both references the original 3-DOF algorithm of Neves and Rodríguez [22, 23] was employed.

6.5   Parametric Amplification Domains (PADs)

6.5.1   General

Analytical limits of stability have been discussed by Neves and Rodríguez [25] based on the roll linear variational equation. The analytical approach, due to its relatively easy implementation, may be easily applied in the ship preliminary design stage. Two limitations are apparent: (a) the limits of stability are those of a linearized mathematical model; (b) they do not provide information on the magnitude and distribution of parametric rolling amplifications within the unstable region, only the borders of the PADs are determined. To overcome this inconvenience, Neves and Rodríguez [24] proposed a completely numerical approach considering the set of nonlinearly coupled equations of the heave, roll, and pitch motions. By means of direct integration of the set of equations for varying values of encounter frequency and wave amplitude, the steady state roll amplitude may be mapped, therefore adding, by means of appropriately defined color scaling, relevant quantitative information to the area inside the limits of stability.

Considering the previously demonstrated strength of the new 6-DOF model to simulate intense parametric rolling, in the following sub-sections the PADs for the three hulls will be obtained. In the next five sub-sections a comparative analysis of PAD topologies is addressed aiming at revealing global characteristics of the rolling

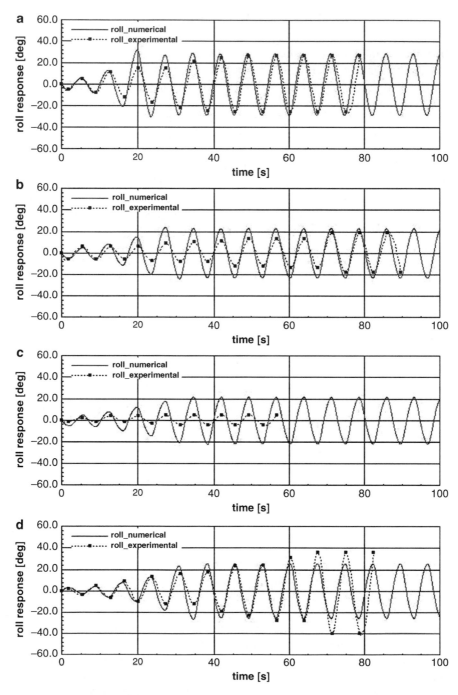

Fig. 6.2 Ship TS: (**a**) Fn = 0.11, $H_W/\lambda = 1/24$; (**b**) Fn = 0.20, $H_W/\lambda = 1/33$; (**c**) Fn = 0.30, $H_W/\lambda = 1/40$; (**d**) Fn = 0.30, $H_W/\lambda = 1/31$

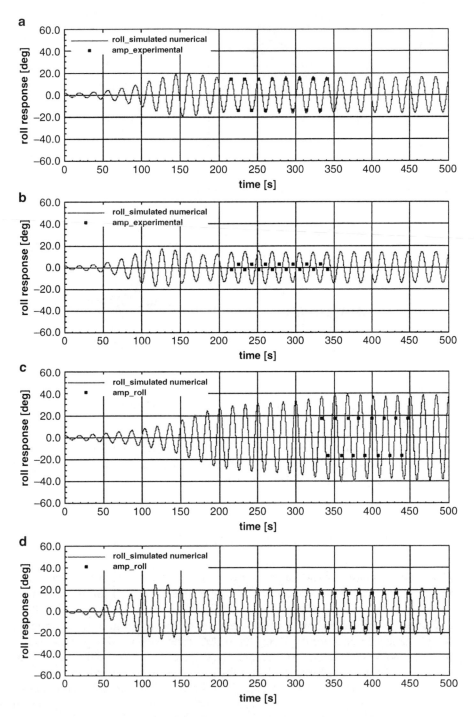

Fig. 6.3 Ship SAFEDOR: (**a**) Fn = 0.08, $H_W/\lambda = 1/49$; (**b**) Fn = 0.08, $H_W/\lambda = 1/31$; (**c**) Fn = 0.12, $H_W/\lambda = 1/49$; (**d**) Fn = 0.12, $H_W/\lambda = 1/31$

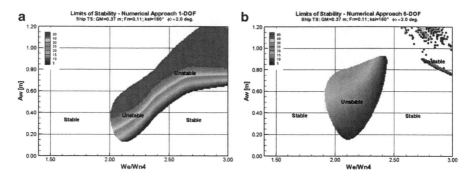

Fig. 6.4 PADs for ship TS: $\overline{GM} = 0.37$ m, $\phi_0 = 2°$, Fn $= 0.11$: (**a**) 1-DOF; (**b**) 6-DOF

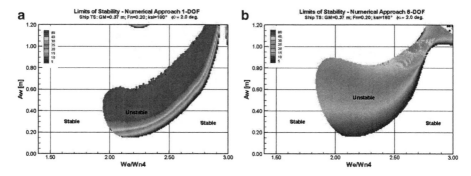

Fig. 6.5 PADs for ship TS: $\overline{GM} = 0.37$ m, $\phi_0 = 2°$, Fn $= 0.20$: (**a**) 1-DOF; (**b**) 6-DOF

responses of substantially different hull forms and exploring influences of different nonlinearities. The expected outcome of the following discussions is an increased understanding of the complexities of parametric rolling.

6.5.2 Influence of Coupling

What essential characteristics of global responses are lost when a simple uncoupled model is assumed? Are such losses comparable as distinct hull forms are considered?

Figures 6.4 and 6.5 show a comparison of PADs for TS hull considering on the left-hand side diagram the PADs for 1-DOF (uncoupled) roll equation and on the right the 6-DOF model. Results are given for two different ship speeds. For both speeds, it is noticed that the PADs for uncoupled and coupled responses are substantially different. In the case of the uncoupled model not only the topology is different: the intensity of roll responses shown in the dark areas (indicating ship capsizing) are much higher than the experimental results previously discussed in

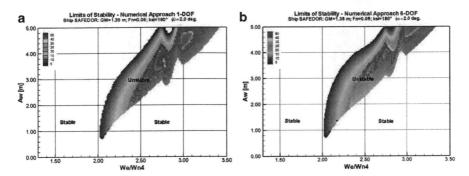

Fig. 6.6 PADs for ship SAFEDOR: $\overline{GM} = 1.38$ m, $\phi_0 = 2°$, Fn = 0.08: (**a**) 1-DOF; (**b**) 6-DOF

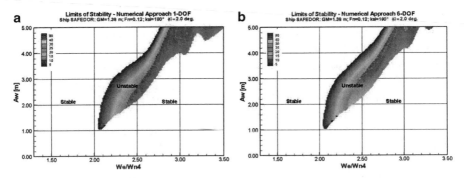

Fig. 6.7 PADs for ship SAFEDOR: $\overline{GM} = 1.38$ m, $\phi_0 = 2°$, Fn = 0.12: (**a**) 1-DOF; (**b**) 6-DOF

Sect. 6.4. Thus, it is clear that consideration of coupling is essential for this type of hull, since PADs for uncoupled model indicate ship capsizing for the majority of the domains. When 6-DOF model is applied, the large capsize area disappears and results accordingly conform to the experimental results. It is important to notice that in all cases the PADs display the occurrence of upper frontiers due to the existence of a backbone curve inclined to the right. Figures 6.6 and 6.7 show comparisons of PADs for SAFEDOR hull considering on the left-hand side of each figure the 1-DOF (uncoupled) roll equation and on the right the 6-DOF model.

Firstly, it is noticed that the PADs for this hull are much *slimer* than in the case of the TS hull. Secondly, PADs are not too different for the two speeds considered. It may also be observed that for both speeds the PADs are almost invariant, irrespective of simulating with 1-DOF or 6-DOF. Thus, in general, this hull is somewhat insensitive to coupling. Yet, if this hull do not depend much on nonlinear coupling with the vertical modes, it still displays upper frontiers, which is a roll nonlinear characteristic, essentially dependent on the heave and pitch responses. Noticing the existence of a strong backbone bending the PADs to the right, it is concluded that it is necessary for this hull to take into account the nonlinear additional stiffness R_0 term defined in (28) of the Neves and Rodríguez [25] (or the k term defined in (6) of

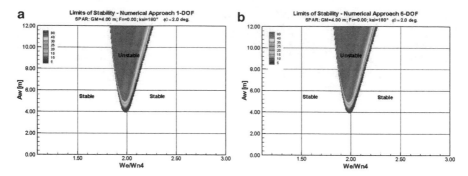

Fig. 6.8 PADs for SPAR platform: $\overline{\mathrm{GM}} = 4.00\,\mathrm{m}$, $\phi_0 = 2°$, Fn $= 0.00$: (**a**) 1-DOF; (**b**) 6-DOF

the paper Spanos and Papanikolaou [36]) to the uncoupled roll equation. Further to be observed is that, as a consequence of the strong backbone bending to the right, in all cases the PADs for this hull are located totally to the right of the exact Mathieu tuning, which corresponds to encounter frequency equal to twice the roll natural frequency.

After examination of the PADs for the two ships, both of them quite prone to parametric rolling, it is now time to compare their results with those corresponding to the SPAR platform, here defined as a simple vertical cylinder. But, before that, it is important to notice that Neves et al. [26] have demonstrated that hull forms with vertical walls are not parametrically excited by variations of the hydrostatic pressure field. Instead, it was shown that it is the incident wave potential that may induce time-dependent characteristics to the system, proportional to its oscillatory motions, thus creating an internal excitation at wave frequency. Therefore, a weak parametric excitation is expected to result for this type of hull. As will be seen below, this feature makes this hull an interesting counter example to the previous cases. Here, it is possible to notice that the behavior of this type of hull may be perfectly modeled as a linear Mathieu equation.

Figure 6.8 shows the 1-DOF and 6-DOF PADs for the SPAR hull. In practice, nothing changes. As anticipated, the 1-DOF and 6-DOF models give the same results. Therefore, a 1-DOF model may be adequate for modeling parametric rolling in this case. Also important to realize is that the PADs have no upper frontiers. Quite distinct from the previous cases presented for the TS and SAFEDOR hulls.

In order to summarize the above discussion, it may be seen that: (a) TS hull, nonlinear coupling terms are crucially relevant, uncoupled roll equation is not capable of taking into account the complexity of responses in this case; (b) SAFEDOR hull, heave and pitch equations may be taken as uncoupled from the roll mode, but the roll equation must include nonlinear terms responsible for the bending to the right (backbone curve) discussed by Neves and Rodríguez [25] and Spanos and Papanikolaou [36]; (c) SPAR: this may be seen as a physical system nicely modeled by the linear Mathieu equation.

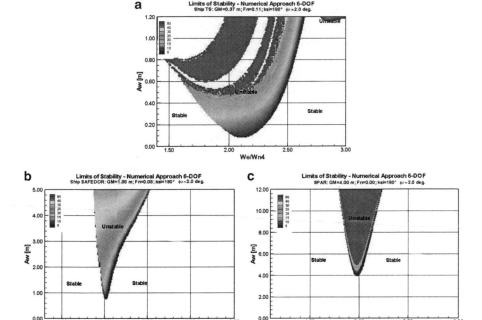

Fig. 6.9 PADs for second-order modeling: (**a**) TS, Fn = 0.11; (**b**) SAFEDOR, Fn = 0.08; (**c**) SPAR

6.5.3 Influence of Third-Order Terms

This analysis is relevant for the establishment of the concept of nonlinear additional stiffness. It is convenient to address the influence of nonlinear couplings separately from the analysis of roll/roll nonlinearities, which will be discussed in the next sub-section. So, here the roll/roll nonlinearities will be kept in the simulations. By means of an analytical methodology, Neves and Rodríguez [25] demonstrated that the heave–pitch–roll coupling induce a bending to the right of the limits of stability. Now, we wish to numerically verify this relevant characteristic, again following the comparative approach. For this purpose Fig. 6.9 (upper) for TS hull should be compared to Fig. 6.4 (right); Fig. 6.9 (lower left) for the SAFEDOR hull should be compared to Fig. 6.6 (right); Fig. 6.9 (lower right) should be compared to Fig. 6.8 (right).

When third-order coupling terms are not considered, the resulting roll responses for the TS hull are extremely (and unrealistically) high and the PAD has no upper frontier, as shown in Fig. 6.9 (upper). This is indicative that the roll responses for this hull are very much dependent on the third-order nonlinearities. When these terms are deleted the equations of motions remain nonlinearly coupled but are not capable of reproducing the experiments. Therefore, there is a direct connection between the existence of a backbone bent to the right and the efficient numerical control of the excessive second-order results. In fact, this aspect had been highlighted in the analytical analysis of Neves and Rodríguez [25]; this is now numerically confirmed in the case of the nonlinear model.

Now it is interesting to check whether the other hulls tend to be also so much dependent on this characteristic. Figure 6.9 (lower left) shows that the container vessel is affected in its tendency to have the PAD bent to the right thus losing the upper frontiers, as is the case with the classical Ince–Strutt diagram. Finally, it is observed that in the case of the SPAR the PAD with the second-order terms (Fig. 6.9 (lower right)) and with third-order terms (Fig. 6.8 – right) are practically the same, there is no noticeable influence of third-order terms, the amplifications having a great resemblance to the Ince–Strutt diagram.

6.5.4 Influence of Roll/Roll Nonlinearities

Naval architects traditionally apply the rationale that the static restoring curve is of fundamental importance in the characterization of the capability of a ship to avoid the risks associated with unstable conditions. Accepting this view, it may be relevant to point out the limited role that the restoring curve plays in the development of unstable motions associated with parametric rolling in head seas.

The purpose here is to investigate what is the role of the nonlinearities of the restoring curve when compared with the nonlinearities associated with the coupling of roll with heave and pitch. Referring again to the analytical limits of stability developed by Neves and Rodríguez [25], it was established that the limits of stability, derived on the basis of the roll linear variational equation suffered no influence from the shape of the restoring curve. The present nonlinear numerical approach allows one to verify if the limits are changed as the restoring curve is considered linear or not. The SAFEDOR hull will be used here as example. Figure 6.10 (left) shows the PAD for simulations in which the 6-DOF had all nonlinearities up to third-order, but the restoring curve was taken as linear. Figure 6.10 (right) corresponds to the complete model, with restoring curve adjusted by a ninth-order polynomial. It should be noticed that Fig. 6.10 (right) is the same as Fig. 6.6 (right). When comparing the two PADs, it is noticed that the area of parametric amplification is exactly the same in both cases, what is different are the roll amplitudes inside the domains of amplification. These tend to be distributed inside the PAD in an inverted topology. That is, in Fig. 6.10 (right) the lower roll amplitudes are distributed along the upper frontiers of the PAD, whereas in Fig. 6.10 (left) these are located along the lower frontiers of the PAD.

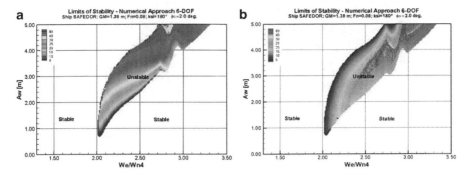

Fig. 6.10 Influence of roll/roll nonlinearities on third-order modeling for ship SAFEDOR at $Fn = 0.08$: (**a**) linear; (**b**) nonlinear

6.5.5 Influence of Wave Amplitude

It has been shown that wave amplitude plays a significant role on the development of nonlinear characteristics of the coupling between roll motion and the other modes. The aim here is to get a clear perception of the quite distinct bifurcations that each hull would undergo as wave amplitude is increased. Figure 6.11 shows the steady roll amplitudes inside the PAD as the tuning between encounter frequency and roll natural frequency varies, for values close to the exact Mathieu tuning of $\omega_e = 2\omega_{n4}$, that is, encounter frequency equal to twice the roll natural frequency. It is observed in Fig. 6.11 (upper left) that for the fishing vessel, increasing wave amplitude the roll amplitude starts to amplify at a given threshold wave amplitude and grows up continuously and smoothly. It then reaches a maximum, and subsequently diminishes in the form of an abrupt jump down to a condition of no amplification response.

On the other hand, as shown in Fig. 6.11 (lower left) for the SAFEDOR container ship, for low wave amplitudes the roll responses to parametric excitation start with a jump bifurcation, reaching very rapidly high amplitude values, of the order of 24° for some frequencies. After this initial bifurcation, roll responses for increasing wave amplitudes tend to decrease in a smooth trend, finally disappearing for higher wave amplitudes. This trend had been observed in the experiments conducted by Taguchi et al. [40]. It is concluded that the fishing vessel and container vessel display quite distinct paths to bifurcations.

For both ship hulls a tendency for the roll responses to disappear for increased wave amplitudes is observed. A markedly distinct variation of roll responses with increasing wave amplitude is observed in the case of the SPAR hull, as shown in Fig. 6.11 (lower right). In this case there is no upper limit of stability; hence roll amplitudes are ever increasing with wave amplitude, following a well known trend of linear systems.

There is here a point that deserves further discussion. If in all the previous discussions it became apparent that the fishing vessel tends to display more complex

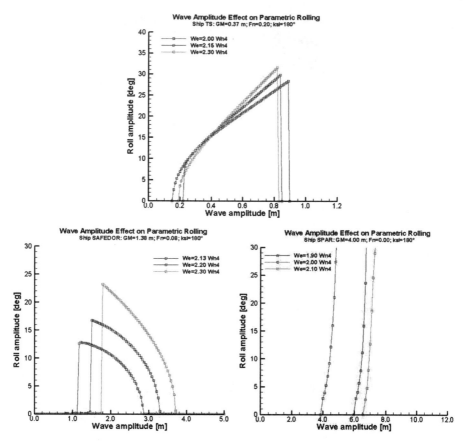

Fig. 6.11 Influence of wave amplitude

dynamical responses than the container vessel, it is necessary to ponder that the latter undergoes a sudden bifurcation to high roll amplitudes for small values of wave amplitude. On the contrary, the fishing vessel has the same type of bifurcation but located at the end of the range of larger wave amplitudes. This is a result that must be taken into account when alert systems are to be designed for the on-line prediction and detection of parametric rolling of container ships.

6.5.6 Influence of Initial Conditions

It is well known that numerically simulated roll parametric amplification may develop earlier or later in time in case a different initial roll amplitude is assumed. Matusiak [17] showed a case of numerical simulation in which a larger initial roll amplitude resulted in parametric amplification developing earlier but with the same final roll amplitude; this tendency has also been addressed by

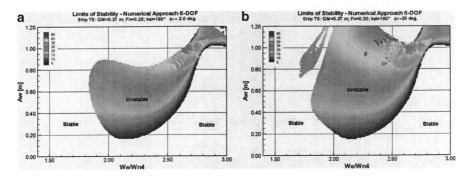

Fig. 6.12 Influence of initial conditions ship TS, Fn = 0.20: (a) $\phi_0 = 2°$; (b) $\phi_0 = 20°$

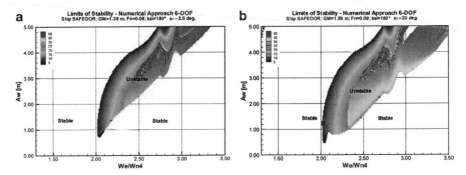

Fig. 6.13 Influence of initial conditions ship SAFEDOR, Fn = 0.08: (a) $\phi_0 = 2°$; (b) $\phi_0 = 20°$

Paulling [29]. Umeda et al. [41] applied the same reasoning when comparing numerical simulations to experimental time series. However, there are conditions in which the influence of different initial conditions may result in responses migrating to substantially different attractors.

Figure 6.12 compares the PADs for the TS hull for two initial roll amplitudes. Clearly, in this case the first noticeable influence of a distinct initial condition is to radically modify the topology of the amplification region. In the case illustrated in Fig. 6.12 the amplification region become much larger (growing mainly upwards and to the left). Equivalent topological changes had already been observed in Neves and Rodríguez [24] for the same TS hull employing the original 3-D mathematical model. It is now interesting to verify in Fig. 6.13 that the SAFEDOR hull will not undergo the same level of topological metamorphosis. In fact, in this case the PAD is only marginally increased, particularly in the region of higher encounter frequencies. On the other hand, it is possible to identify zones internal to both PADs having distinct levels of roll amplification. This is exemplified in Fig. 6.14 which shows time series and phase diagrams indicating that for the same wave conditions but distinct initial heel angles, for the SAFEDOR hull substantially different attractors may be reached.

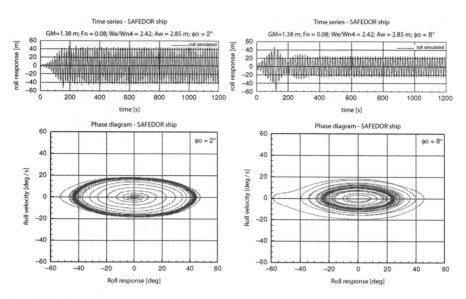

Fig. 6.14 Time series and phase diagrams for distinct initial conditions

It is concluded that the SAFEDOR hull is substantially less affected by initial conditions than the TS hull, thus confirming that the TS hull tends to display more complex dynamical responses. It goes without demonstration that in the case of the SPAR hull no dependency on initial conditions takes place.

6.6 Conclusions

A new 6-DOF mathematical model has been introduced and verified. The model has been employed for generating numerical domains of roll amplification for three different types of floating vessels. The following conclusions have been reached:

(a) Influence of coupling: PADs have been obtained for two mathematical models, one with uncoupled roll, the second one with complete 6-DOF coupling. It was shown that in the case of the fishing vessel the nonlinear couplings are essential, resulting in quite distinct PADs for the distinct numerical models. On the other hand, it was shown that the container ship dynamics may be approximately modeled by means of an uncoupled model, but retaining a tendency to display PADs with upper frontiers. Therefore, it may be concluded that the uncoupled equation may be assumed to be an acceptable model for the container ship if a detuning factor is considered; but this is not the case of the highly unstable fishing vessel. On the other hand it has been shown that the SPAR platform behaves as a typically uncoupled linear Mathieu-type solution.

(b) Influence of third-order terms: it has been shown that a coupled second-order model cannot reproduce adequately the roll amplifications for the two ships.

(c) Influence of roll/roll nonlinearities: it has been confirmed that the nonlinear roll restoring curve affects only the intensity of roll responses inside the PAD, but the borders of the PAD are not affected by these nonlinearities.

(d) Influence of wave amplitude: steady roll amplitudes simulated within the amplification area have shown quite distinct characteristics when results for the fishing vessel are compared to those corresponding to the container ship. The appearance of bifurcations is quite distinct in each case, as wave amplitude is increased in each case. This is an important dynamical characteristic displayed by the present investigation: distinct hulls may display quite distinct responses to roll parametric excitation. These aspects should be carefully examined when on-line detection of parametric rolling is a design objective.

(e) Influence of initial conditions: by applying different initial conditions it has been shown that different attractors may be reached, a dynamical feature of relevance for the modeling of parametric rolling in irregular seas. This is an important subject for future investigations.

Acknowledgements The present investigation is supported by CNPq within the STAB Project (Nonlinear Stability of Ships). The Authors also acknowledge financial support from LabOceano, CAPES, and FAPERJ.

References

1. Abkowitz, M. A.: *Stability and motion control of ocean vehicles.* The MIT Press, MA (1969)
2. ABS: Guide for the assessment of parametric roll resonance in the design of container carriers. American Bureau of Shipping, Houston, TX (2004)
3. Ahmed, T. M., Hudson, D. A., Temarel, P.: An investigation into parametric roll resonance in regular waves using a partly non-linear numerical model. Ocean Engineering vol. 37, issues 14–15, pp. 1307–1320, doi 10.1016/j.oceaneng.2010.06.009 (2010)
4. Arnold, L., Chueshov, I., Ochs, G.: Stability and capsizing of ships in random sea – a survey. Report No. 464 (June), Institut für Dynamische Systeme, Universität Bremen, Germany (2003)
5. Bassler, C. C., Belenky, V., Bulian, G., Francescutto, A., Spyrou, K., Umeda, N.: Review of available methods for application to second level vulnerability criteria. In Proc. of the 10th Int. Conference on Stability of Ships and Ocean Vehicles (STAB'2009), St. Petersburg, Russia (2009)
6. Belenky, V. L., Sevastianov, N. B.: Stability and safety of ships – risk of capsizing (second edition). The Society of Naval Architects and Marine Engineers (SNAME). Jersey City, NJ (2007)
7. Blocki, W.: Ship safety in connection with parametric resonance of the roll. Int. Shipbuilding Progress 27(306):36–53 (1980)
8. Bulian, G.: Development of analytical nonlinear models for parametric roll and hydrostatic restoring variations in regular and irregular waves. PhD Thesis. University of Trieste, Italy (2006)
9. Bulian, G., Francescutto, A., Lugni, C.: On the nonlinear modelling of parametric rolling in regular and irregular waves. In Proc. of the 8th International Conference on the Stability of Ships and Ocean Vehicles (STAB'2003). Madrid: 305–323 (2003)

10. Bulian, G., Francescutto, A., Fucili, F.: Numerical and experimental investigation on the parametric rolling of a trimaran ship in longitudinal regular waves. In Proc. of the 10th International Conference on the Stability of Ships and Ocean Vehicles (STAB'2009), St. Petersburg, Russia, pp. 567–582 (2009)
11. Carmel, S. M.: Study of parametric rolling event on a panamax container vessel, Journal of the Transportation Research Board, No. 1963, Transportation Research Board of the National Academy, Washington DC, pp. 56–63 (2006)
12. France, W. G., Levadou, M., Treakle, T. W., Paulling, J. R., Michel, R. K., Moore, C.: An investigation of head seas parametric rolling and its influence on container lashing systems. Marine Technology 40(1):1–19 (2003)
13. Haslum, H. A., Faltinsen, O. M.: Alternative shape of spar platform for use in hostile areas. In: Proc. of the Offshore Technology Conference, Paper No. OTC10953, Houston, TX (1999)
14. Holden, C., Galeazzi, R., Rodríguez, C. A., Perez, T., Fossen, T. I., Blanke, M., Neves, M. A. S.: Nonlinear container ship model for parametric resonance. Modeling, Identification and Control. 28: 87–103 (2008)
15. Hsu, C. S.: On the parametric excitation of a dynamic system having multiple degrees of freedom. Trans. of the ASME Journal of Applied Mechanics 30(3):367–372 (1963)
16. Inglis, R. B.: A three dimensional analysis of the motion of a rigid ship in waves. PhD Thesis, Dept. of Mechanical Engineering, University College London, UK (1980)
17. Matusiak, J.: On the effects of wave amplitude, damping and initial conditions on the parametric roll resonance. In Proc. of the 8th International Conference on the Stability of Ships and Ocean Vehicles (STAB'2003). Madrid: 341–347 (2003)
18. McCue, L., Campbell, B. L., Belknap, W. F.: On the parametric resonance of tumblehome hullforms in a longitudinal seaway. Naval Engineers Journal 119(3):35–44 (2007)
19. Neves, M. A. S., Pérez, N., Lorca, O.: Experimental analysis on parametric resonance for two fishing vessels in head seas. In Proc. of the 6th International Ship Stability Workshop, Webb Institute, New York (2002)
20. Neves, M. A. S., Pérez, N. A., Lorca, O., Rodríguez, C. A.: Hull design considerations for improved stability of fishing vessels in waves. In Proc. of the 8th International Conference on Stability of Ships and Ocean Vehicles (STAB'2003). Madrid: 147–165 (2003)
21. Neves, M. A. S., Rodríguez, C. A.: Limits of stability of ships subjected to strong parametric excitation in longitudinal waves. In Proc. of the 2nd International Maritime Conference on Design for Safety, Sakai, Japan v1: 81–89 (2004)
22. Neves, M. A. S., Rodríguez, C. A.: A non-linear mathematical model of higher order for strong parametric resonance of the roll motion of ships in waves. Marine Systems & Ocean Technology, Journal of Sociedade Brasileira de Engenharia Naval 1(2):69–81 (2005)
23. Neves, M. A. S., Rodríguez, C. A.: On unstable ship motions resulting from strong nonlinear coupling. Ocean Engineering 33(14–15):1853–1883 (2006)
24. Neves, M. A. S., Rodríguez, C. A.: An investigation on roll parametric resonance in regular waves. International Shipbuilding Progress 54:207–225 (2007)
25. Neves, M. A. S., Rodríguez, C. A.: Influence of nonlinearities on the limits of stability of ships rolling in head seas. Ocean Engineering 34:1618–1630 (2007)
26. Neves, M. A. S., Sphaier, S., Mattoso, B., Rodríguez, C. A., Santos, A., Villeti, V., Torres, F.: Parametric resonance of mono-column structures In: Proceedings of the 6th Osaka Colloquium on Seakeeping and Stability of Ships (OC'2008), pp. 405–411, Osaka, Japan (2008)
27. Obreja, D. C., Nabergoj, R., Crudu, L. I., Pacuraru-Popoiu, S.: Transverse stability of a cargo ship at parametric rolling on longitudinal waves. In Proc. of the 27th International Conference on Offshore Mechanics and Arctic Engineering (OMAE2008), paper no. 57768, Estoril, Portugal (2008)
28. Paulling, J. R., Rosenberg, R. M.: On unstable ship motions resulting from nonlinear coupling. Journal of Ship Research 3(1):36–46 (1959)
29. Paulling, J. R.: Parametric resonance – now and then. In Proc. of the 9th International Conference on Stability of Ships and Ocean Vehicles (STAB'2006). Rio de Janeiro, Brazil (2006)

30. Salvesen, N., Tuck, O. E., Faltinsen, O. M.: Ship motions and sea loads, Transactions of SNAME, vol. 78 pp. 250–287 (1970)
31. Shin, Y. S., Belenky, V. L., Paulling, J. R., Weems, K. M., Lin, W. M.: Criteria for parametric roll of large container ships in longitudinal seas. Transactions SNAME 112 (2004)
32. Rodríguez, C. A.: On the nonlinear dynamics of parametric rolling. Doctoral thesis, COPPE/UFRJ, Brazil (in Portuguese) (2010)
33. Rodríguez, C. A., Holden, C., Perez, T., Drummen, I., Neves, M. A. S., Fossen, T. I.: Validation of a container ship model for parametric rolling. In Proc. of the 10th International Ship Stability Workshop, Hamburg, Germany (2007)
34. SAFEDOR: Ship model tests on parametric rolling in waves. Research Project SP.7.3.9, of the FP6 Sustain Surface Transportation Programme.
35. Spanos, D., Papanikolaou. A.: SAFEDOR international benchmark study on numerical simulation methods for the prediction of parametric rolling of ships in waves. NTUA-SDL report, Revision 1.0, National Technical University of Athens, Greece (2009)
36. Spanos, D., Papanikolaou, A.: On the decay and disappearance of parametric roll of ships in steep head waves. In Proc. of the 10th International Conference on Stability of Ships and Ocean Vehicles (STAB'2009). St. Petersburg, Russia, pp. 259–270 (2009)
37. Spyrou, K. J.: Designing against parametric instability in following seas. Ocean Engineering 27(6):625–653 (2000)
38. Spyrou, K. J.: On the parametric rolling of ships in a following sea under simultaneous nonlinear periodic surging. Philosophical Transactions of the Royal Society, London, A 358:1813–1834 (2000)
39. Spyrou, K. J., Tigkas, I. G.: Principle and application of continuation methods for ship design and operability analysis. PRADS 2007, vol. 1, pp. 388–395, Houston, TX (2007)
40. Taguchi, H., Ishida, S., Sawada, H., Minami, M.: Model experiment on parametric rolling of a post-panamax container ship in head waves. In: Proc. of the 9th International Conference on Stability of Ships and Ocean Vehicles (STAB'2006), Rio de Janeiro, Brazil, pp. 147–156 (2006)
41. Umeda, N., Hashimoto, H., Minegaki, S., Matsuda, A.: Preventing parametric roll with use of devices and their practical impact. Proc. of the 10th International Symposium on Practical Design of Ships and Other Floating Structures, PRADS'2007, pp. 693–698 (2007)
42. WAMIT Inc.: WAMIT's user manual. WAMIT Inc., MA (2006)

Chapter 7
Probabilistic Properties of Parametric Roll

Vadim Belenky and Kenneth Weems

7.1 Introduction

Parametric roll in irregular waves is characterized by a number of specific properties making it different from other nonlinear ship motions in irregular seas. The parametric excitation has a relatively narrow frequency band and a threshold that depends on roll damping. The narrow frequency band makes autocorrelation to decay slowly. The existence of damping thresholds adds statistical uncertainty; a wave group slightly below the threshold does not add any energy into the dynamical system; that is why parametric roll "comes and goes".

7.2 Numerical Simulation

The driving force of parametric roll is variation of stability in waves. Therefore a numerical tool for parametric roll must be based on body-nonlinear formulation for hydrostatic and Froude–Krylov forces and moments.

Roll damping is another phenomenon that needs to be included as accurate as possible. The only way to calculate roll damping completely is CFD/RANS. However high computational cost prevents it from being a practical tool for Monte Carlo simulation where long record may be required. A hybrid model is more

V. Belenky (✉)
David Taylor Model Basin, 9500 MacArthur Blvd., West Bethesda, MD 20817, USA
e-mail: vadim.belenky@navy.mil

K. Weems
Science Application International Corporation, Ste 250, 4321 Collington Rd, Bowie,
MD 20716, USA
e-mail: kenneth.m.weems@saic.com

T.I. Fossen and H. Nijmeijer (eds.), *Parametric Resonance in Dynamical Systems*,
DOI 10.1007/978-1-4614-1043-0_7, © Springer Science+Business Media, LLC 2012

practical, where potential forces of roll damping (wave damping) are computed while skin friction and vortex forces are taking from the roll decay test. The model and calculation procedure are described by Belenky et al. [4].

Influence of other degrees of freedom, namely pitch and heave may be important especially in head or bow quartering seas where these motions are not insignificant. As it was shown in [11], including heave and pitch may decrease the magnitude of the change of stability in waves.

7.3 Configuration of a Sample Ship and Setup for Numerical Simulations

Following the line of the previous studies from France et al. [7], Shin et al. [11], Belenky et al. [3], C11-class post panamax container carrier was used as a sample ship for the study. The sketch of the body plan and a panel model taken from Shin et al. [11] are shown in Fig. 7.1.

Following Belenky et al. [3] JONSWAP spectrum was used. Fourier series are used to represent the stochastic process of wave elevation.

$$\zeta_w(t) = \sum_{i=1}^{N} a_i \cos(\omega_i t + \varphi_i), \tag{7.1}$$

where w_i is the frequency set, amplitudes a_i are defined from the spectrum, and phase shift j_i are random numbers with uniform distribution. Each realization of waves is generated with a new set of random phase shift.

Discretization with 200 equally spaced frequencies provides 1500 s of simulation time free from the self-repeating effect. Absence of the self-repeating effect is verified by evaluation of the autocorrelation function using cosine Fourier transform:

$$R(\tau_j) = \int_0^\infty s(\omega)\cos(\omega\tau_j)d\omega = \sum_{i=1}^{N_t} S_{Wi}\cos(\omega_i\tau_j), \quad \tau_j = \Delta t \cdot j. \tag{7.2}$$

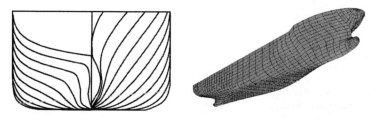

Fig. 7.1 Body plan and panel model of C11-class containership

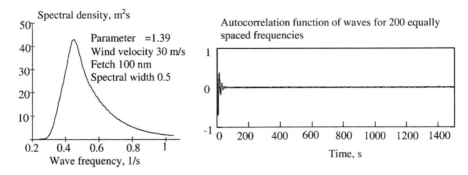

Fig. 7.2 Spectral density and autocorrelation function for wave elevations

Table 7.1 Parameters for numerical simulation

Length BP, m	262	Speed, kn	10
Breadth, m	40	Number of records	50
Draft, m	12.36	Length of record, s	1500
Depth, m	27.4	Time step, s	0.8
$\overline{KG}$, m	17.55	Number of points	1875

Here s is value of spectral density, S_{Wi} is the power spectrum, t_j is current moment of time and Dt is a time step. If the value of the autocorrelation function remains near zero after the initial decay, then the self-repeating effect is not present and formula (7.1) with the current frequency discretization is the valid model of the stochastic process on the considered simulation time. On the details of self-repeating effect (see Belenky [1, 2]. The spectral density and the autocorrelation functions are shown in Fig. 7.2. Key numerical parameters of the sample ship and the numerical simulation setup are summarized in Table 7.1.

Configuration and conditions of simulation described above exactly correspond to those used by Belenky et al. [3], where parametric roll in long-crested head seas was detected. Numerical simulation included three degrees of freedom: heave, roll, and pitch. Forward speed was assumed constant, influence of wave of ship motions on speed and heading was not simulated. Hybrid solver LAMP-2 was used: it included 3-D body-nonlinear formulation of hydrostatic and Froude–Krylov forces. Hydrodynamic forces including diffraction, radiation, wave damping, and added mass was simulated over a mean waterline, i.e., body-linear formulation was used.

This formulation allows utilization of Impulse Response Function (IRF) option of LAMP-2. This lead to dramatic decrease of computational costs as the kernel of the convolution integrals are pre-calculated and reused for each record.

Figure 7.3 shows samples of records of roll motions. Since the simulations were performed in exact head long-crested seas, there is no direct roll excitation. Therefore all roll motions are caused by parametric resonance; to provoke its development all records were started with $10°$ initial angle.

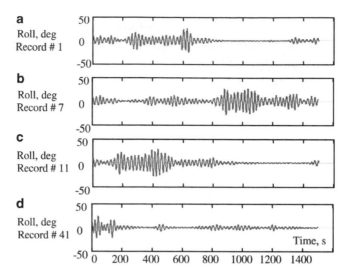

Fig. 7.3 Sample records of parametric rolls

Similar to the picture described in Belenky et al. [3], parametric roll comes and goes in random instances of time, as the waves capable generating of parametric excitation are not encountered all the time. This results in practical nonergodicity of parametric roll, as generation of a sufficiently long wave record may be prohibitively expensive due to self-repeating effect.

7.4 Spectrum of Parametric Roll

Figure 7.4 shows a power spectrum of the selected record produced by the numerical simulations. These spectra were evaluated with FFT with 512 frequencies, so only 1024 points were used from each record.

As it can be seen from Fig. 7.4, the record power spectra of parametric roll are highly variable from record to record. Sum of the spectral value presents a variance estimate over the length of record that was used for spectral evaluation. This estimate varies quite significantly, which can be seen as a manifestation of practical nonergodicity.

Figure 7.5 presents a power spectrum averaged over the ensemble. As it can be expected it is smoother than that any of the spectra calculated over a record. The variance estimated over the averaged spectrum is closer to the statistical estimate over the ensemble.

The value of spectral width e shown in Fig. 7.4 wave calculated using linear interpolation of the spectral density $s(\omega)$ evaluated from the power spectrum $S(\omega)$, followed by calculation of spectral moments of the second and fourth order:

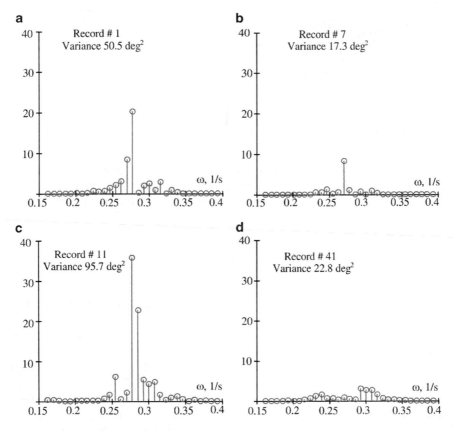

Fig. 7.4 Power spectra of different records

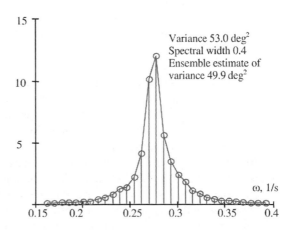

Fig. 7.5 Ensemble-averaged power spectrum

$$s(\omega) = \frac{S(\omega)}{\Delta\omega}, \quad m_r = \int_0^\infty s(\omega)\omega^r d\omega, \quad \varepsilon = \sqrt{1 - \frac{m_2^2}{m_0 m_4}}. \quad (7.3)$$

Here m_r is a spectral moment of rth order.

7.5   Autocorrelation Function

The autocorrelation function is another important characteristic of a stochastic process showing the dependence of the current value from the previous history. It is related with the spectral density through cosine Fourier transform (7.2), but also can be estimated directly from each record:

$$R_j^* = \frac{1}{N-j} \sum_{i=1}^{N-j} (\phi_i - m_\phi^*)(\phi_{i+j} - m_\phi^*) r_j^* = \frac{R_j^*}{V^*}. \quad (7.4)$$

Here f are values of roll angle, m_ϕ^* is estimate of the mean value, r_j^* is estimate of relative autocorrelation function.

Figure 7.6 shows autocorrelation functions estimate for a selected sample records. The most evident difference between these autocorrelation functions and the one shown in Fig. 7.2 is that the statistical estimates do not decay. They not only show some values at the end of the record, but even grow. This is a result of rising statistical uncertainty. Once j in the formula (7.4) becomes large, there are fewer and fewer data points left, the last values of the estimate (7.4) cannot be trusted. The evaluation of the autocorrelation function with cosine Fourier transform does not carry this type of uncertainty.

The artificial character of the increase of the estimate of autocorrelation function can also be stated from general considerations; there is no physical reason why the dependence should be stronger with time. Belenky and Weems [4] described a procedure of cutting the autocorrelation function and demonstrated that it is equivalent to smoothing of the spectrum.

The main idea of that procedure was to find a point where the estimate of autocorrelation starts to grow again and cut it there. Looking at Fig. 7.6 it is possible to visually define these points (shown as black arrows). These points appear to be quite far from each other; this may be another manifestation of practical nonergodicity of parametric roll.

Prior to cutting the autocorrelation function, it makes sense to find ensemble average estimate of autocorrelation, shown in Fig. 7.7. As the ensemble-average spectrum (Fig. 7.5) is smooth, the ensemble-averaged estimate of the autocorrelation function has significantly reduced numerical error caused by statistical uncertainty. Visually the cutting point (shown by the black arrow) is around 70 s.

Autocorrelation

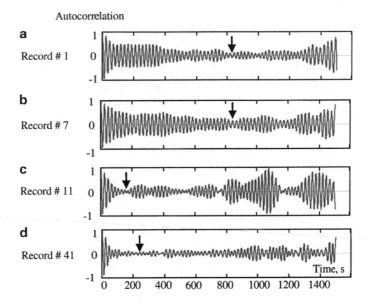

Fig. 7.6 Autocorrelation functions estimated over selected records

Autocorrelation

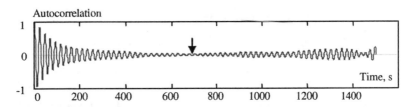

Fig. 7.7 Ensemble-averaged estimate of autocorrelation function

Indeed, the mean value of this error must be zero; once the number of records tends to infinity, the estimate of autocorrelation function becomes the true theoretical autocorrelation function. Since there is no reason for an increase of the autocorrelation function, its values beyond the virtual cut-off point must be small.

Belenky and Weems [5] proposed a formal procedure to find the cut-off point using the piecewise linear envelope of the autocorrelation function. This procedure can be improved if the formal definition of the envelope is used:

$$E_R(\tau) = \sqrt{(R_a^*(\tau))^2 + (Q_a^*(\tau))^2} \tag{7.5}$$

Here Q_a^* is the result of Hilbert transform of ensemble-averaged autocorrelation function. It is defined as follows:

$$Q_a^*(\tau) = \int_0^\infty (C_I(\omega)\cos(\omega\tau) - C_R(\omega)\sin(\omega\tau))\,d\omega, \tag{7.6}$$

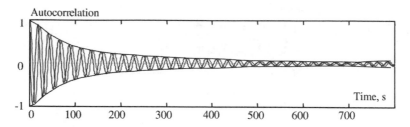

Fig. 7.8 Envelope for the ensemble-averaged autocorrelation function

Fig. 7.9 Comparison of power spectra calculated with FFT and from the estimate of autocorrelation function

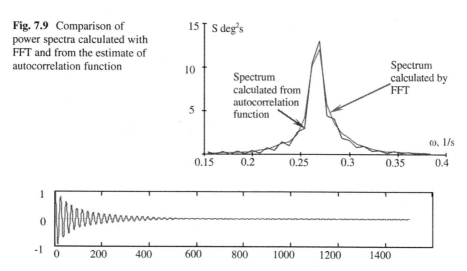

Fig. 7.10 Autocorrelation function averaged between statistical and spectral estimates

where C_R is real and C_I is imagined part of Fourier coefficients of the R_a^*. So the Hilbert transform produces a complimentary function to R_a^* by shifting it 90°. Numerically it was performed with the combination of direct and inverse FFT. The results are shown in Fig. 7.8. The minimum of the envelope is the cut-off point.

The autocorrelation function can be used to calculate the spectrum:

$$s(\omega_j) = \frac{2}{\pi} \int_0^\infty R(\tau)\cos(\omega_j\tau)\mathrm{d}\tau = \frac{2}{\pi} \sum_{i=1}^N R^*(\tau_i)\cos(\omega_j\tau_i)\Delta\tau. \qquad (7.7)$$

The spectrum calculated with formula (7.7) may contain a systematic numerical error: instead of zero it tends to some small value both in low and high frequency. This error can be easily corrected by subtraction of the spectral value corresponding to highest frequency.

A comparison between the spectra calculated with both methods is shown in Fig. 7.9. They are very close, which demonstrates validity of the considered procedure. The autocorrelation function for the further analysis can be taken as an average between statistical and spectral estimates, see Fig. 7.10.

7.6   Uncertainty of Statistical Estimates of a Single Record

Estimates of the average and the variance of the record are expressed with well-known formulae. The variance can be estimated using the true value of the average m_f (if known) or its estimate m_ϕ^* based on the volume of the sample N:

$$m_\phi^* = \frac{1}{N}\sum_{i=1}^{N}\phi_i, \quad V_\phi^* = \frac{1}{N}\sum_{i=1}^{N}(\phi_i - m_\phi)^2 = \frac{1}{N}\sum_{i=1}^{N}(\phi_i - m_\phi^*)^2. \qquad (7.8)$$

The mean value and the variance estimated over a sample of a finite volume are random numbers. Really, what is the estimate of the mean value? It is a sum of random numbers divided by a constant. The result is, indeed, a random number.

As any other random numbers, the estimates have their own distribution and all other properties. So, it is appropriate to introduce into consideration mean and variance of the estimates.

As per the laws of big numbers, increasing volume of sample leads to more accurate estimates; therefore the variance of a statistical estimate is expected to decrease with the increase of the sample volume.

The mean value of the average equals to itself as the average is an unbiased estimate. While the mean value of the variance does not equal to itself as the variance estimate is slightly biased. This bias is removed with another well-known formula.

$$m\left(m_\phi^*\right) = m_\phi^*, \quad m\left(V_\phi^*\right) = \frac{N}{N-1}V_\phi^*. \qquad (7.9)$$

Evaluation of the variance of the mean and the variance of the variance is more complex as the variance of estimates bears information on statistical uncertainty. For independent data points, the variances of the mean and the variance are given as:

$$V\left(m_\phi^*\right) = \frac{V_\phi}{N} \approx \frac{V_\phi^*}{N}, \quad V\left(V_\phi^*\right) = \frac{M_4}{N} - \frac{N-3}{N-1}\frac{V_\phi^2}{N}. \qquad (7.10)$$

Here M_4 is the fourth central moment of the roll angle. It is defined as:

$$M_4 := \int_{-\infty}^{\infty}(\phi - m_\phi)^4 f(\phi)\mathrm{d}\phi = \frac{1}{N}\sum_{i=1}^{N}(\phi_i - m_\phi^*)^4 = \mathrm{Kt}\cdot\left(V_\phi^*\right)^2, \qquad (7.11)$$

where Kt is the kurtosis. This estimation, however, may be difficult, as it requires impractical amount of data; due to the fourth power, the estimate is very sensitive to outliers. As a result, better accuracy can be achieved if it is expressed through an assumed kurtosis Kt of the distribution deemed suitable for the considered case. The usual practice is to assume Kt = 3 for most periodic processes, i.e., the kurtosis is taken from normal distribution. This leads to the following approximate formula for the variance of the variance of the random variable:

$$V\left(V_\phi^*\right) = \frac{2}{N-1}\left(V_\phi^*\right)^2 \qquad (7.12)$$

Formulae (7.10) and (7.12) are written for a set of independent data points. It is not the case when considered a response of nonlinear dynamical system, which is characterized by a certain degree of dependency of the data points. A correction needs to be introduced that will "fine" the data for dependency. The estimate of autocorrelation was needed exactly for this purpose. While the mean value itself is not affected by the dependence, the variance of the mean for a set of dependent data points x_i, is expressed as [10]:

$$V\left(m_\phi^*\right) = \frac{1}{N} \sum_{i=-N}^{N} R\left(\tau_{|i|}\right)\left(1 - \frac{|i|}{N}\right). \tag{7.13}$$

Here the dependency is defined by the autocorrelation function $R(t)$, symbol $|i|$ is used as "absolute value of i".

The formula (7.13) can be seen as generalization of (7.10) for the case of dependent data points. If the data points are independent (the case of white noise), the autocorrelation function equals to zero everywhere except of the origin:

$$\text{White noise: } R(\tau) = \begin{cases} V_\phi & \tau = 0 \\ 0 & \tau > 0 \end{cases} \tag{7.14}$$

Substitution of autocorrelation (7.14) into (7.13) leads to the first formula (7.10). While the estimate of the variance itself is not affected by the dependence, the variance of the variance is expressed as [10]:

$$V\left(V_\phi^*\right) = \frac{2}{N} \sum_{i=-N}^{N} \left(R\left(\tau_{|i|}\right)\right)^2 \left(1 - \frac{|i|}{N}\right). \tag{7.15}$$

This formula uses the assumption that the process has normal distribution, so $Kt = 3$. The formula (7.15) can be also seen as a generalization for independent data points. Substitution of (7.14) into (7.15) leads to

$$V\left(V_\phi^*\right) = \frac{2}{N}\left(V_\phi\right)^2. \tag{7.16}$$

For the large values of N, the formula (7.16) is practically identical to the second formula (7.10).

Distribution of these estimates, strictly speaking, depends on the distribution of the process (or a random variable) itself. For example, if the random variable is normal, estimate of its mean value has Student t distribution while, the estimate of the variance has chi-square distribution. The distribution of parametric roll is quite far from normal, see Belenky et al. [3], Hashimoto et al. [8]. As can be seen from Fig. 7.11, the distribution is leptokurtic (more "pointy" than normal). This also means that the tail of the distribution is "thicker" than normal. This is a sign of danger, as the ship tends to spend more time with large roll angles. Fitting a distribution for parametric roll is a separate problem that falls beyond the scope of this text.

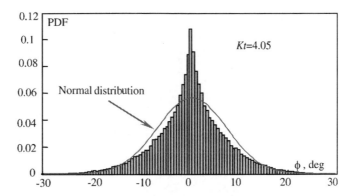

Fig. 7.11 Distribution of parametric roll

Distribution of both mean value and variance estimate must tend to normal distribution with the increase of the volume of the sample. Central limit theorem provides that a sum of independent comparable random variables tends to normal distribution with the increase of number of components irrespective to the distribution of each individual component.

Roll motions are presented with a set of dependent data point. In the considered case of parametric roll, this dependence lasts for almost half of the record length. Therefore application of the normal distribution for mean value and variance estimates of a record is approximate.

Once the decision of approximation is made, calculation of confidence interval for the mean value estimate is straight forward:

$$m_\phi^* \pm \Delta m_\phi^*, \quad \Delta m_\phi^* = K_\beta \sqrt{V\left(m_\phi^*\right)} \tag{7.17}$$

$$K_\beta = Q_N\left(1 - 0.5(1 - \beta)\right), \quad Q_N = \Phi^{-1}. \tag{7.18}$$

Here Q_N is the inverse function to a normal cumulative distribution function with zero mean and unity variance, β is accepted confidence probability.

Calculation of confidence interval for the variance estimate for a record is more complex. By definition, the variance is a positive value, however, normal distribution allows negative values as well. It means that application of formula (7.17) for variance may, in principle, extend lower boundary of the confidence interval below zero which does not carry any meaning.

An attempt to use chi-square distribution encounters difficulties. The mean value and variance are known for the variance estimate. Chi-square distribution has only one parameter – number of degrees of freedom. Therefore it is not possible to use it for a random variable with known mean value and variance.

Another possibility is to try noncentral chi-square distribution. In addition to the number of degrees of freedom it has a noncentric parameter, which must be positive as well as the number of degrees of freedom (that is actually not necessarily integer).

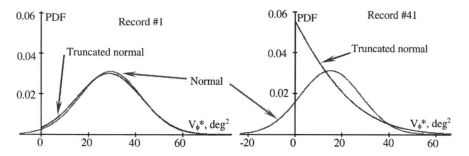

Fig. 7.12 Zero-truncated normal distributions for variance estimates

These limitations may prevent the use of noncentral chi-square distribution for the considered case. The zero-truncated normal distribution, however, does not have these limitations. It is defined as:

$$f(v; k, p_1, p_2) = \begin{cases} 0 & v < 0 \\ \frac{k}{\sqrt{2\pi p_2}} \exp\left(-\frac{(v-p_1)^2}{2p_2}\right) & v \geq 0. \end{cases} \tag{7.19}$$

Here v is the random variable, k is a normalization factor needed to make sure that the area under the PDF equals 1. The parameters p_1 and p_2 take the place of mean value and variance: the truncated normal distribution no longer directly use the mean value and the variance as parameters.

As a result, the factor k as well as the parameters p_1 and p_2 need to be found from the following system of algebraic equations:

$$\begin{cases} \int_0^\infty f(v; k, p_1, p_2) dv = 1 \\ \int_0^\infty f(v; k, p_1, p_2) v \, dv = m\left(V_\phi^*\right) \\ \int_0^\infty f(v; k, p_1, p_2) \left(v - m\left(V_\phi^*\right)\right)^2 dv = V\left(V_\phi^*\right). \end{cases} \tag{7.20}$$

Example distributions for records # 1 and 41 are shown in Fig. 7.12.

Once the distribution has been accepted, further calculation of the boundaries of the confidence interval V_{Low}^* and V_{Up}^* does not encounter any principal difficulties.

$$V_{Low}^* = Q_T(0.5(1-\beta)); \quad V_{Up}^* = Q_T(0.5(1+\beta)). \tag{7.21}$$

Here β is the confidence probability, while Q_T is an inverse function of CDF of truncated normal distribution F:

$$F(v; k, p_1, p_2) = \begin{cases} 0 & v < 0 \\ \int_0^v \frac{k}{\sqrt{2\pi p_2}} \exp\left(-\frac{(z-p_1)^2}{2p_2}\right) dz & v \geq 0. \end{cases} \tag{7.22}$$

7.7 Uncertainty of Statistical Estimates of an Ensemble

Consider an ensemble set containing N_R independent records/subsets, each consisting of N_i dependent data points. The estimate of mean value is expressed as:

$$m_{A\phi}^* = \frac{1}{\sum_{i=1}^{N_R} N_i} \left(\sum_{i=1}^{N_R} \sum_{j=1}^{N_i} \phi_{ij} \right) = \sum_{i=1}^{N_R} \left(\frac{N_i}{\sum_{i=1}^{N_R} N_i} \frac{1}{N_i} \sum_{j=1}^{N_i} \phi_{ij} \right) = \sum_{i=1}^{N_R} W_i m_{\phi i}^*. \quad (7.23)$$

Here $m_{\phi i}^*$ is the mean value estimate of a record i defined by the formula (7.8). In the considered case all of the records are of the same length, so $W_i = 1/N_R$.

Evaluation of the confidence interval requires variance of the mean. It can be calculated by applying variance operator to both sides of (7.23) taking into account that random values $m_{\phi i}^*$ are deemed independent (as they were estimated over independent records) and weights W_i are deterministic numbers. Using the formula for a variance of the linear combination of independent random variables, the variance of the mean of the ensemble can be expressed as:

$$V\left(m_{A\phi}^*\right) = V\left(\sum_{i=1}^{N_R} W_i m_{\phi i}^* \right) = \sum_{i=1}^{N_R} W_i^2 V\left(m_{\phi i}^*\right). \quad (7.24)$$

Results of calculations for the numerical example are given in Fig. 7.13. Normal distribution was assumed for the ensemble estimate, while confidence probability β was taken as 95%. The width of the confidence interval for the ensemble is narrower than for any of the records. This is expected, as the ensemble contains more statistical information.

Obviously, the true value for the mean of roll motion in the case of symmetric problem is zero. However the zero is not even included in the ensemble confidence interval. This is a result that all the records were started with the same initial condition of $10°$ and the initial transient was kept in the records. As can be seen

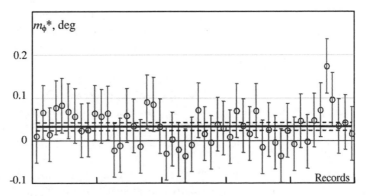

Fig. 7.13 Estimates of mean values of records and ensemble

from Fig. 7.10, it took almost one-third of a simulation time for the autocorrelation function to decay. Therefore it could been expected that the initial conditions may have some influence on the ensemble mean. However, this bias is pretty small ($0.032°$) and the mean value is not as important as variance for the purposes of practical analysis of parametric roll.

The variance estimate of the ensemble can be expressed in a similar way, but it is more convenient to consider the estimate of the second raw moment. Its definition, estimate and relation with variance is expressed as:

$$\alpha_2 = \int_{-\infty}^{\infty} \phi^2 f(\phi) d\phi; \quad \alpha_{2i}^* = \frac{1}{N_i} \sum_{j=0}^{N_i} \phi_{ij}^2, \quad V_\phi = \alpha_2 - m_\phi^2. \tag{7.25}$$

Here $f(\phi)$ is the PDF of roll angles. Consider the estimate of second raw moment for the ensemble:

$$\alpha_{2A}^* = \frac{1}{\sum_{i=1}^{N_R} N_i} \left(\sum_{i=1}^{N_R} \sum_{j=1}^{N_i} \phi_{ij}^2 \right) = \sum_{i=1}^{N_R} \left(\frac{N_i}{\sum_{i=1}^{N_R} N_i} \frac{1}{N_i} \sum_{j=1}^{N_i} \phi_{ij}^2 \right) = \sum_{i=1}^{N_R} W_i \alpha_{2i}^*. \tag{7.26}$$

The estimate of the variance of the ensemble is expressed taking into account that the estimate of mean value is small:

$$V_A^* = \sum_{i=1}^{N_R} W_i \left(V_{\phi i}^* + m_{\phi i}^2 \right) - \left(m_{A\phi}^* \right)^2 \approx \sum_{i=1}^{N_R} W_i V_{\phi i}^*. \tag{7.27}$$

Evaluation of confidence interval for the variance estimate requires evaluation of the variance of the variance. It can be done by applying variance operator to both parts of (7.27):

$$V \left(V_{A\phi}^* \right) = V \left(\sum_{i=1}^{N_R} W_i V_{\phi i}^* \right) = \sum_{i=1}^{N_R} W_i^2 V \left(V_{\phi i}^* \right) \tag{7.28}$$

Figure 7.14 shows the results of calculations of the confidence interval for the variance estimate of the ensemble together with estimates of each individual record. Normal distribution was assumed for the ensemble estimate. This assumption seems to be more reasonable for the ensemble as it contains significantly more independent data in comparison with a single record.

Figure 7.14 shows a dramatic decrease of the width of the confidence interval in comparison with a singe record. The figure also shows the scale of variability of the variance estimate of the records. For example, the first two estimates shows difference more than twofold: $29 \, deg^2$ versus $72 \, deg^2$. The difference between the largest and the smallest estimates in the ensemble is more than a factor of 5 ($15 \, deg^2$ versus $77 \, deg^2$).

The reason for such a dramatic difference is the nature of parametric excitation. It is only capable of producing parametric resonance when a group of waves is encountered and all characteristics of these waves lay within a certain range.

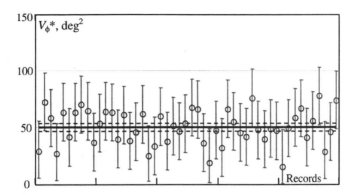

Fig. 7.14 Estimates of variance values of records and ensemble

As can be seen from Fig. 7.3, once started, parametric roll continues for relatively long time. It can be explained by a possible widening of the frequency range once the amplitudes of roll motion become large enough. This effect was found, explained, and demonstrated by Spyrou [12], Neves and Rodrigues [9], and others.

The relatively long duration of parametric roll occurrences leads to slow decay of autocorrelation function, as frequency of parametric roll response is close to natural roll frequency. As a result, the spectrum is narrow and the distribution is leptokurtic.

As was already noted, the random character of the occurrences of parametric roll and their durations lead to significant variability of all characteristics of the records: spectral estimates Fig. 7.4, autocorrelation estimates (Fig. 7.6), mean value estimates (Fig. 7.13), and variance estimates (Fig. 7.14).

As a result, a single record contains relatively small amount of independent statistical data. This leads to large confidence bands of record variance estimates and represents a manifestation of the practical nonergodicity, meaning that several independent records must be used in order to devise any judgment on statistical characteristics of the parametric roll response.

7.8 Number of Records Required for Analysis

The practical nonergodicity of the parametric roll response raises a question of the number of records required for an analysis of the results of numerical simulation or model tests.

Theoretically, increased length of the record could lead to convergence. However, this approach is hardly practical. Available model basin length effectively limits duration of a record with forward speed. Self-repeating effect makes a long record prohibitively expensive for any numerical simulation using Fourier series for presentation of wave elevation. The auto-regression model of irregular waves

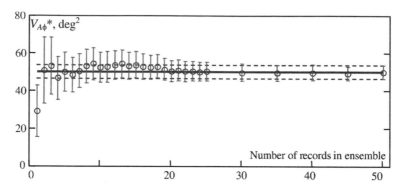

Fig. 7.15 Convergence of ensemble estimate of variance

(Degtayrev [6]), in principle, can be considered as an alternative, but a practical algorithm of pressure calculations has not been developed yet.

Figure 7.15 shows a series of calculations of the ensemble estimate of variance performed for ensembles with varying number of records.

As it can be seen from Fig. 7.15, there is a very small change in the estimate and its confidence interval after 20 records. Significant decrease of the confidence interval is observed after 4 records. However, some steps of the calculation may go unstable. In particular, smoothing of averaged spectrum and cutting autocorrelation function may be problematic. Based on the considered example, calculations of autocorrelation are stabilized after 12 records. However, it may be possible to get a reasonable estimate of autocorrelation function out of 5 records by using statistical procedure only or, possibly, by applying an alternative spectral smoothing.

These numbers were obtained from a rather limited number of calculations, so they can be considered only as a first and very approximate assessment.

7.9   Conclusions

Statistical and spectral post-processing of numerical simulation of parametric roll of C11 class container carrier in head seas allowed some insight into the probabilistic properties of this phenomenon.

Parametric roll in irregular waves "comes and goes". However, the occurrences of parametric roll seem to be rather long. This can be possibly explained by alteration of the frequency range caused by nonlinearity of restoring.

It takes a long time for the autocorrelation function to decay (7–11 min versus less than 1 min for wave elevations) and the spectrum is narrow (width parameter is about 0.4).

The probability distribution is leptokurtic (estimate of kurtosis is about 4). This means a "fat tail" and a significant time spent with large roll angle. It is a symptom of a dangerous roll motions mode.

There is significant variability between different records expressed in spectral estimates, variance estimates, and estimates of autocorrelation function. Decay time of autocorrelation function also varies significantly.

Preliminary convergence study shows that 5 records (25 min each) may be minimally sufficient for express analysis. For more detailed analysis, the data set should include 12–20 such records.

Practical nonergodicity of parametric roll is one of the governing factors in statistical post-processing of parametric roll records from numerical simulation or model tests.

Acknowledgements This work was supported by Office of Naval Research (ONR) under Dr. L. Patrick Purtell. The development of the LAMP System has been supported by the US Navy, the Defence Advanced Research Projects Agency (DARPA), the US Coast Guard, ABS, and SAIC.

References

1. Belenky, V.: On risk evaluation at extreme seas. Proc of 7th International Ship Stability Workshop, Shanghai, China (2004)
2. Belenky, V.: On long numerical simulations at extreme Seas. Proc. of 8th International Ship Stability Workshop, Istanbul, Tyrkey (2005)
3. Belenky, V., Weems, K. M., Lin, W. M., Paulling, J. R.: Probabilistic analysis of roll parametric resonance in head seas, Proc. of 8th International Conference on Stability of Ships and Ocean Vehicles, Madrid, Spain (2003)
4. Belenky, V. L., Yu, H. C., Weems, K.: Numerical Procedures and practical experience of assessment of parametric roll of container carriers. Proc. of the 9th International Conference on Stability of Ships and Ocean Vehicle (STAB'06), Vol. 1, Rio de Janeiro, Brazil, pp. 119–130 (2006)
5. Belenky, V., Weems, K. M.: Procedure for probabilistic evaluation of large amplitude roll motions. Proc. of the Osaka Colloquium on Seakeeping and Stability of Ships, Osaka, Japan (2008)
6. Degtyarev, A. B.: New approach to wave weather scenarios modeling. Proc. of 8th International Ship Stability Workshop, Istanbul, Turkey (2005)
7. France, W. N., Levadou, M., Treakle, T. W., Paulling, J. R., Michel, R. K., Moore, C.: An Investigation of Head-Sea Parametric Rolling and its Influence on Container Lashing Systems. Marine Technology 40(1):1–19 (2003)
8. Hashimoto, H. N., Umeda, N., Matsuda, A., Nakamura, S.: Experimental and numerical studies on parametric roll of a post-panamax container ship in irregular waves, Proc. of the 9th International Conference on the Stability of Ships and Ocean Vehicles, Rio de Janeiro, Brazil, Vol. 1, pp. 181–190 (2006)
9. Neves, M. A. S., Rodrguez, C. A.: Nonlinear aspects of coupled parametric rolling in head seas, Proc. of the 10th Internatinal Symposium on Practical Design of Ships and Other Floating Structures, Houston, TX (2007)
10. Pristley, M. B.: Spectral analysis and time series. Academic Press, London (1981)
11. Shin, Y. S., Belenky, V. L., Paulling, J. R., Weems, K. M., Lin, W. M.: Criteria for parametric roll of large containerships in longitudinal seas, SNAME Trans-112 (2004)
12. Spyrou, K. J.: On the parametric rolling of ships in a following sea under simultaneous nonlinear periodic surging, Phil. Trans. R. Soc. Lond. A, Vol. 358, pp. 1813–1834 (2000)

Chapter 8
Experience from Parametric Rolling of Ships

Anders Rosén, Mikael Huss, and Mikael Palmquist

8.1 Introduction

Parametric rolling of merchant ships are rare events but probably less rare than most shipping companies and officers consider them to be. The cases with large amplitude rolling that result in shift of cargo or other critical consequences are of course noted. Many of these incidents have, however, due to lack of detailed records and understanding, been categorized as "normal" heavy weather damage even though they have developed due to parametric excitation. The vast majority of less critical parametric rolling events are probably not even noted and most certainly not documented.

Historically, ship safety standards have developed as reactions on serious casualties. Famous examples are the loss of Titanic which initiated the development of the SOLAS convention, and the loss of Estonia which forced numerous amendments to the convention. In a similar manner, parametric rolling incidents with a container vessel reported in [6] and with a Ro–Ro Pure Car and Truck Carrier (PCTC) reported in [13] have influenced the ongoing process of development of new intact stability standards, e.g., [14, 15].

Many initiatives at IMO nowadays aim at turning from the traditional reactive development of regulations toward a more proactive regime for the future. In order

A. Rosén (✉)
KTH Royal Institute of Technology, Centre for Naval Architecture,
Teknikringen 8, SE-10044, Stockholm, Sweden
e-mail: aro@kth.se

M. Huss
Wallenius Marine AB, PO Box 17086, S-104 62, Stockholm, Sweden
e-mail: mikael.huss@walleniusmarine.com

M. Palmquist
Seaware AB, PO Box 1244, SE-131 28, Nacka Strand, Sweden
e-mail: mikael.palmquist@seaware.se

T.I. Fossen and H. Nijmeijer (eds.), *Parametric Resonance in Dynamical Systems*,
DOI 10.1007/978-1-4614-1043-0_8, © Springer Science+Business Media, LLC 2012

to be able to identify potential hazards before they develop into large-scale losses, it is of utmost importance to analyze also incidents and near critical events that are far more common than the recognized accidents.

From the early works by Grim [8] until the present day, a large number of studies on parametric rolling based on analytical and numerical models and model scale experiments have been published. The availability of data from real full-scale events is, however, very scarce. This chapter reviews three recent full-scale events with parametric rolling. The objective is to make the experiences from these events available to the research community and in light of these events discuss on-board operational guidance for assisting crews in avoiding parametric rolling.

8.2   Parametric Rolling of Ships

The primary mechanism in parametric rolling of ships is the time variation of the restoring moment created by the varying relation between the geometry of the ship hull and the wave surface as the ship travels through the waves. The prerequisites for parametric rolling to develop can be summarized as follows:

1. Sufficiently large relative variation of stability is generated when the ship travels through waves. This in turn requires a combination of:

    (a) flared hull form with large beam/draft ratio,
    (b) sufficiently high waves, and
    (c) relatively low average stability ($\overline{GM_0}$).

2. Resonance, meaning

    (a) encounter period about half of (or equal to) the roll natural period, and
    (b) resonance condition according to 2(a) being kept for a sufficient number of wave encountering cycles (regularity).

3. Sufficiently low roll damping.

Passenger cruise ships and fishing vessels are examples of ship types that might experience stability variations that can be critical regarding parametric roll excitation, e.g., [3, 20]. Also, ships optimized for large volumes of low weight cargo, such as Lo–Lo container vessels and Ro–Ro ships, can experience very large stability variations and therewith related sensitivity to parametric roll excitation, e.g., [6, 13]. At the far end on this scale are modern Ro–Ro PCTCs. The standard ocean going PCTC with a length of about 200 m and panamax breadth has developed from a rather traditional hull form of the 1970s into today's highly stability optimized hull form which is able to carry significantly more cars by increased cargo hold height.

During the last decade, a new class of larger car and truck carriers (LCTC) with 230 m length and a capacity of about 8,000 cars have been introduced. An example of such a vessel is given in Fig. 8.1, typical main particulars are given in Table 8.1, and typical stability variations in regular waves are exemplified in Fig. 8.2.

Fig. 8.1   A modern LCTC

Table 8.1   Main particulars for a typical LCTC

Length over all	227.8	m
Beam, moulded	32.26	m
Height to upper deck	34.7	m
Draft, design/max	9.5/11.3	m
Deadweight at max draft	30,137	t
Number of car decks	13	–

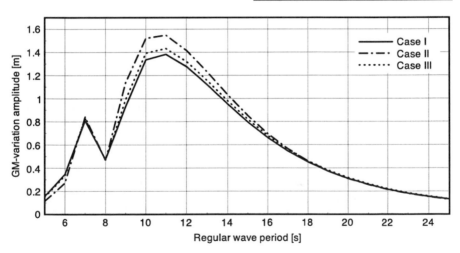

Fig. 8.2   Quasi-static stability variation in regular head/following waves with a wave height of 4 m for the three presented cases

8.3   Service Experience

As a forerunner in the development of modern PCTCs and LCTCs, the Swedish shipping company Wallenius had an early awareness of the potential problems with stability variations in waves for this type of ship. Early in the 1990s, the company supported the research project at the KTH Royal Institute of Technology in Stockholm [11] which eventually resulted in the advanced Seaware EnRoute Live onboard decision support system for seakeeping. Wallenius was then the first shipping company to make the Seaware EnRoute Live system a standard within its fleet. The cooperation with Seaware has extended to include also follow-up analysis of incidents and proactive analysis of dynamic stability properties of new ships at the design stage. In 2004, Wallenius and Seaware prepared a common report of a head sea parametric roll incident that was submitted by Sweden to the IMO SLF sub-committee [13]. The report describes probably the first time ever when full 6-DOF motions have been recorded from an actual parametric rolling event in irregular seas. In 2009, Wallenius, Seaware, and KTH joined a cooperative research program to further develop knowledge and tools in this area.

In addition to making the technical decision support system a standard on board the ships, Wallenius has also at various occasions informed their ship officers about underlying causes of parametric rolling, the specific character of the ships, and recommended actions to avoid critical situations. The shared knowledge and understanding has most likely prevented a number of critical situations but has also made it possible to identify situations where parametric roll actually occurred and enabled collection of important data for further analysis. We will here discuss three such events that represent three principally different modes of parametric rolling:

Case I: *Principal parametric resonance where the period of encounter is half of the roll natural period in following seas*

Case II: *Principal parametric resonance where the period of encounter is half of the roll natural period in head seas*

Case III: *Fundamental parametric resonance where the period of encounter coincides with the roll natural period in following seas*

The events have occurred within the last two years and with the same ship design but with two different ship individuals.

8.3.1   Case I – Principal 2:1 Parametric Resonance in Following Seas

In this case, the ship had been idling for some hours in head sea at about 5 knots outside a port that was closed due to bad weather. Some 20 min after the ship had slowly turned back and increased speed to about 10 knots, a sudden heavy rolling developed. Time series of roll, pitch, speed, and heading are given in Fig. 8.3. As seen the rolling developed very fast, in the first sequence from moderate 2–4° up to 20° in just four roll cycles. In the second sequence, a rolling amplitude of 30° was reached. A clear 2:1 relation between the roll and pitch periods can be observed during the critical sequences. The rolling was stopped by the Master changing over to hand steering and turning the vessel back toward the waves.

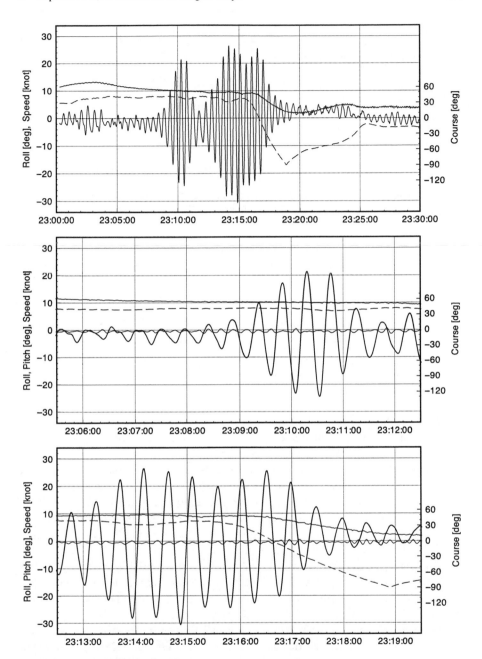

Fig. 8.3 Recorded data for Case I

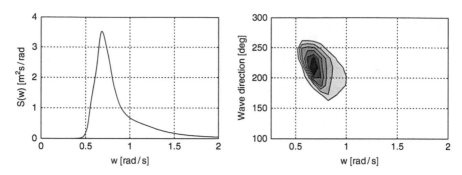

Fig. 8.4 Unidirectional (*left*) and directional (*right*) reanalysis wave spectrum for the area in question 3 h before the incident. (ECMWF Mediterranean Wave Analysis)

Table 8.2 Wave spectral parameters from re-analysis data for three hours before and after the event

Time [h]	H_s [m]	T_p [s]	T_z [s]
−3	4.3	9.2	6.9
+3	3.8	9.2	7.0

Table 8.3 Conditions summary for Case I

$\overline{GM}_0$ [m]	T_0 [s]	H_s [m]	T_p [s]	λ_p/L [-]	V [kn]	β [deg]	T_{Ep} [s]	T_0/T_{Ep} [-]
1.2	28	4.1	9.2	0.55	10	0	14.3	1.95

With the purpose of establishing as good as possible depiction of the sea state at the time and position of interest, re-analysis wave data in terms of directional wave spectra for a 0.5×0.5 degree grid has been obtained from ECMWF. This data covers from three hours before to three hours after the incident. The wave spectra consisted mainly of wind waves with limited directional spread and a narrow frequency peak. The wind data from the same re-analysis data source was concluded to be in compliance with the on-board wind observations. The directional and the corresponding unidirectional wave spectra assembled from the ECMWF data three hours before the incident are displayed in Fig. 8.4.

The wave spectrum parameters calculated from the available spectra 3 h before and after the incident displayed in Table 8.2 indicates that the sea state was rather stationary. Through interpolation, the significant wave height and the spectrum peak period at the incident are determined to 4.1 m and 9.2 s, respectively.

Table 8.3 presents a summary of characteristic parameters for this case. The ship was loaded close to its design draft and had a $\overline{GM}$ of 1.2 m. The sea state parameters are here based on the reanalysis of the wave spectra discussed above and the assumption that the average period within high amplitude wave groups tends to approach the peak period. The characteristic wave length λ_p and the encounter period T_{Ep} are here hence determined for a harmonic wave whose period equals the

Table 8.4 Conditions summary for Case II

$\overline{GM}_0$ [m]	T_0 [s]	H_s [m]	T_z [s]	T_p [s]	λ_p/L [-]	V [kn]	β [deg]	T_{Ep} [s]	T_0/T_{Ep} [-]
2.3	21	5–6	8–9	11–13.5	0.9–1.3	9–14	140–190	8–11.5	1.8–2.6

Table 8.5 Conditions summary for Case III

$\overline{GM}_0$ [m]	T_0 [s]	H_s [m]	T_p [s]	λ_p/L [-]	V [kn]	β [deg]	T_{Ep} [s]	T_0/T_{Ep} [-]
1.3	23	4.1	10.5	0.8	18–20	20–50	16–26	0.7–1.2

spectral peak period of the irregular sea state (the same comes for Tables 8.4 and 8.5). It should be noted that this gives a simplified picture of the irregular seaway and that conclusions should be drawn with this in mind.

To the officers on board at this occasion, the parametric rolling occurred "out of nowhere" in conditions that were far from being perceived as potentially dangerous. The significant wave height was only about 4 m and, as seen in Fig. 8.3, the motions were very small prior to the onset with practically no pitch and only a few degrees rolling. However, the fact that the ship was idling with reduced speed put her into a 2:1 parametric resonance that she would not encounter for regular service speeds in the same sea state. In addition, the relatively low $\overline{GM}$, although well above regulatory minimum stability, contributed to the sensitivity.

8.3.2   Case II – Principal 2:1 Parametric Resonance in Head Seas

In the second case, the ship was running in head sea with a speed of about 11 knots on route southward close outside the coast of New South Wales in Australia. During a period of about four hours, the ship encountered clear parametric roll excitation a number of times with amplitudes up to about 20°. Time series of roll, pitch, speed, and heading are given in Fig. 8.5. As for Case I, a clear 2:1 relation between the roll and pitch periods can be observed during the critical sequences. The pitch amplitude is here however significantly larger due to the head sea condition.

For this case, wave data are available from the nearby Eden wave buoy, managed and operated by the Manly Hydraulics Laboratory, Sydney. This buoy is a nondirectional Waverider type buoy manufactured by the Dutch company Datawell BV. In Fig. 8.6, the position of the buoy is shown on a map (marked by triangle), along with the sailed track and positions where parametric rolling occurred (marked by dots). The distance between the buoy and the position of the last parametric rolling event, closest to the buoy, is about 9 nautical miles. Measured wave spectra from the Eden wave buoy, and corresponding spectral parameters, are given in Figs. 8.7 and 8.8.

The conditions for this case, in terms of characteristic parameters, are summarized in Table 8.4.

The ship was in light loaded condition and had a $\overline{GM}$ of about 2.3 m. Sea state and operational parameters are here given in terms of intervals to reflect the fact that

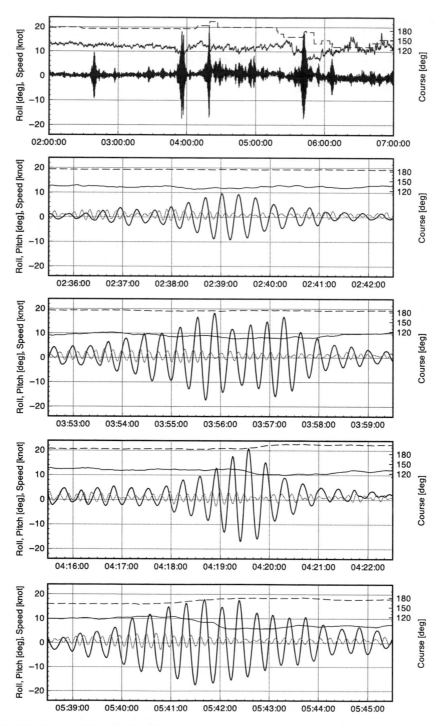

Fig. 8.5 Recorded data for Case II

Fig. 8.6 Map image of the New South Wales coast in Australia showing the sailed track of the ship heading south. The dots indicates positions where parametric rolling occurred. The triangle shows the position of the Eden wave buoy

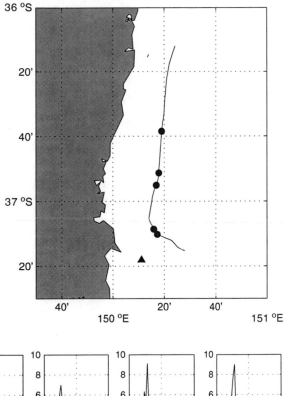

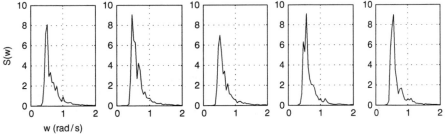

S(w)

w (rad/s)

Fig. 8.7 Measured wave spectra from Eden wave buoy between 02:00 (*leftmost*) and 06:00 (*rightmost*) local time, with one hour increment

parametric rolling sequences occurred over a period of about four hours, over which neither the sea state nor the operational condition may be considered stationary. The forecast available on board at the time showed a rapidly increasing wave height away east of the coast but underestimated the wave height close to the coast. This was due to the relatively low spatial resolution in the forecast which was based on the global ECMWF wave model.

In this case, the officers on board became aware that they were in conditions with potential parametric resonance, but had during the first hours limited options to choose a more favorable heading or speed. Eventually, they made a course deviation of about 50° which led into somewhat more severe seas but enabled the ship to get out of resonant conditions.

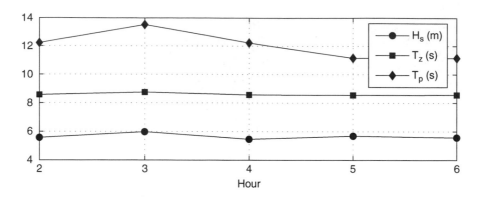

Fig. 8.8 Spectral parameters derived from the wave spectra in Fig. 8.7

8.3.3 Case III – Fundamental 1:1 Parametric Resonance in Following Seas

In the third case, the ship was running at full service speed in quartering seas with a relative heading of 20–40°. The master notified the office about a strange rolling behavior characterized as "neither synchronous nor parametric, perhaps more like loss of stability." On a number of occasions during a couple of hours, the officers turned to manual steering in order to interrupt increasing roll with amplitudes up to 10–15°, but the situation was not considered critical enough to make any general changes in course and speed.

Time series of roll, pitch, speed, and heading are given in Fig. 8.9. During the critical sequences a clear asymmetric development of the roll motion can be seen, which indicates a one-to-one relation between stability variation and rolling periods. The time sequences also clearly show the active steering on board to interrupt the resonance.

In this case with quartering sea, the rolling becomes a complex mixture of direct wave induced synchronous excitation, wind heeling and asymmetric parametric excitation, and less experienced and educated officers would most likely not even have noted the special behavior under these conditions.

For this case, sea state data was obtained from the high-resolution wave model run on the 25th of September 2010, 00 UTC. This means that the wave spectrum presented here represents the +12 h forecast (12 UTC). In relation to using reanalysis data, the present data source is considered by meteorologists to provide a better estimate of the sea state due to the higher spatial resolution in the operational high-resolution wave model, in relation to reanalysis data that is run using lower spatial resolution. Figure 8.10 shows the forecast wave spectrum for a position close to the ship. The significant wave height of this spectrum is 4.1 m, the peak period is 10.5 s and the directional spreading is relatively low.

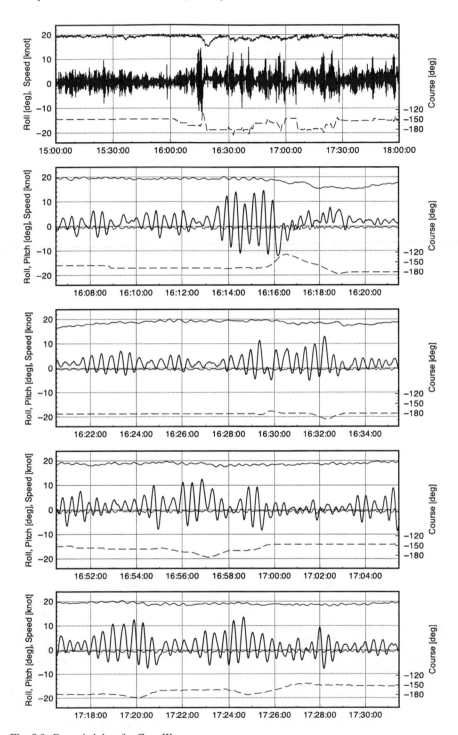

Fig. 8.9 Recorded data for Case III

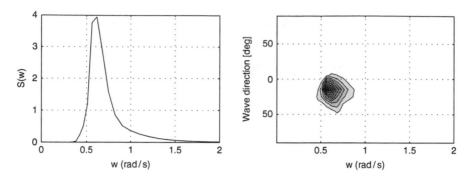

Fig. 8.10 Unidirectional (*left*) and directional (*right*) forecast wave spectrum for 12:00 UTC at a position close to the ship position

The ship was medium loaded and had a $\overline{GM}$ of about 1.3 m. As for the previously presented cases, a summary of characteristic parameters is given in Table 8.5, showing T_0/T_{Ep} ratios in the 1:1 region.

8.4 Operational Guidance

The concept of operational guidance is about providing information on how a specific ship, in a specific loading condition, is expected to behave in a certain environmental condition as a function of ship speed and heading, thus providing assistance to the master on how to operate a vessel with respect to ship dynamics. This can be applied both strategically as part of weather forecast-based route planning procedures, in order to avoid potentially critical situations in the first place, but also and equally important, tactically in terms of real-time guidance based on actual conditions. For nonlinear and rare events such as parametric rolling, the importance of operational guidance is specifically highlighted since it is very hard, even for an experienced seaman, to sense the criticality in advance, as demonstrated particularly in Cases I and II earlier.

There are many different approaches to supplying operational guidance on board ships, ranging from the [14] circulars, through more advanced precalculated polar plot documentation, to on-board computer systems for real-time or forecast-based evaluation of all possible risks related to ship dynamics. The concept based on precalculated polar plots allows for using very sophisticated dynamic models, e.g., nonlinear multi-degree-of-freedom models, but faces problems in terms of recognizing real conditions, e.g., wind waves and swell from different directions and varying wave spectral shapes and bandwidths. The other approach using real-time calculations, will generally have to use more simplified models for CPU efficiency, but can more easily handle the complexity of real wave conditions. The fact that there are such a large number of crucial parameters involved, depending on loading and operating conditions as well as on environmental conditions in terms of wave spectra, speaks in favor for computer-based systems capable of measuring and computing real-time risk indicators that can trig alerts.

Implementation of operational guidance for parametric roll prevention constitutes an integration of many different topics, some frequently addressed within the research community, and some that are of more practical nature. One topic is the mechanical modeling of the dynamic problem. Another is the definition of the input to the dynamic model, which in this case is constituted by the stochastic seaway. Further, the dynamic model include parameters related to the ship, the loading condition and the operating condition, which in practice are not always as easy to define as one might anticipate.

8.4.1 Dynamic Modeling

A simple dynamic model of parametric rolling in heading or following seas can be formulated in terms of a linear single DOF equation of motion:

$$\ddot{\phi} + 2\delta\dot{\phi} + \omega_0^2 \frac{\overline{GM}(t)}{\overline{GM}_0}\phi = 0, \tag{8.1}$$

where δ is the roll damping, ω_0 is the undamped natural frequency, and $\overline{GM}_0$ is the metacentric height in calm water. $\overline{GM}(t)$ is the time variation of the metacentric height which in regular waves with an encounter frequency of ω_e can be expressed as:

$$\overline{GM}(t) = \overline{GM}_m + \overline{GM}_a \cos(\omega_e t), \tag{8.2}$$

where $\overline{GM}_m$ and $\overline{GM}_a$ are the mean and amplitude of the $\overline{GM}$-variation in the regular wave in question, e.g., [21]. Figure 8.11 shows solutions to (8.1) and (8.2) as critical (shaded) areas where combinations of relative stability variations and relative periods result in growing parametric rolling (more than 10° degrees roll amplitude growth in less than ten cycles). The roll damping used has been determined from model tests and is considered representative for the presented cases. For illustration purposes the relative stability variations and relative periods for the cases have been indicated in Fig. 8.11 based on the data in Fig. 8.2 and Tables 8.3–8.5 taking the irregular sea state parameters T_p (peak period) and H_s (significant wave height) as representatives of the regular wave period and wave height. One could consider operational guidance based on a similar approach, e.g., as suggested in [23]. However, depending on rather drastic simplifications of the stochastic properties of the seaway such an approach is less feasible for quantification of actual amplitudes or risk levels.

The stability variations for the three cases in Fig. 8.2 and Fig. 8.11 have been determined by quasi-static balancing in heave and pitch. More advanced

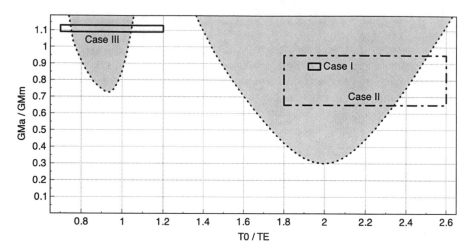

Fig. 8.11 Simplified interpretations of the conditions for the three presented cases in relation to the solutions of (8.1) and (8.2) in terms of critical (shaded) areas where combinations of relative stability variations and relative periods result in growing parametric rolling

single-degree-of-freedom modeling could for example include quasi-dynamic balancing in heave and pitch, nonlinear damping, nonlinear $\overline{GZ}$ variation, and irregular waves, e.g., [7].

The next level of complexity is multidegree-of-freedom models. The state of the art in multidegree-of-freedom parametric-roll modeling has recently been evaluated in a large benchmark study where different simulation methods were compared to model test data [22]. The overall efficiency of the benchmarked methods was concluded to be low, but this was attributed to the wide spread of individual methods both regarding modeling approaches and performances. A group of leading simulation methods was, however, detected, for which the mean probability to successfully detect the parametric roll resonance in relation to the model experiments was estimated to be around 80%.

All three real cases presented here have been realized numerically using such a multidegree-of-freedom simulation model. The used model is based on work of [9] and participated in the above-mentioned benchmark study. For Cases I and III, both in following sea with focused encountering wave spectra, the simulations were able to give consistent and stable results similar to what was measured. However, for Case II which was in head sea, most realizations did not develop parametric roll at all. Figure 8.12 shows an example from one of the rare sequences where similar rolling developed as was measured on board. The simulations for this case were also very sensitive to small variations in conditions and sea state and would not have been conclusive as basis for guidance on board.

Operational guidance based on multidegree-of-freedom modeling of the ship dynamics is being considered as part of the new generation intact stability criteria, which should comprise the design phase as well as the operational phase [15]. The principle for such approach could, for example, be as discussed in [18] where

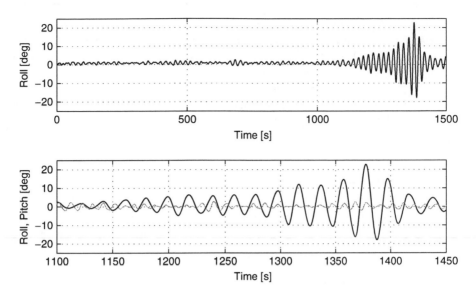

Fig. 8.12 Example of simulated parametric rolling for Case II using a multidegree-of-freedom model

multidegree-of-freedom simulations are used to generate polar plots displaying dangerous combinations of speed and relative heading regarding parametric rolling for a large number of different ship conditions and sea states. With appropriate presentation the precalculated polar plots could then be used for on-board guidance.

8.4.2 Definition of Ship Characteristics and Operational Conditions

Regardless of using single or multidegree-of-freedom dynamic modeling, proper consideration the present sea state and ship condition is of outmost importance in assisting the crew with operational guidance to avoid dangerous situations regarding parametric rolling. In principle, both $\overline{GM}$ and the moment of inertia that together are decisive for the natural roll period T_0, should be determinable based on the data in the loading computer or trim and stability booklet for well-defined loading conditions. However, experience shows that there can be significant uncertainties involved already at this stage, e.g., [17]. A further uncertainty is the effect of free surfaces in liquid tanks. For static stability, this can be equalized with a reduced $\overline{GM}$. In dynamic rolling situations, the effect might, however, be different due to internal flow resistance in the tanks. Internal liquid movements might also have other effects on the roll motions, especially for ships with active or passive roll stabilizing tanks. The most precise method for determining T_0 is probably to measure the roll period in calm water after a sharp turn or in mild beam seas. Under well-controlled conditions possibly also the damping could be evaluated from forced roll tests, but it is hardly feasible for every ship to exercise this under normal service. Due to differences between $\overline{GM}$ in calm water and the average $\overline{GM}$ in waves, one could also expect a difference between the roll natural period in calm water and in waves. For the here presented ship type and cases, this difference is, however, calculated to be rather small and can hardly be noted in the time series for roll. For other ship types and conditions, this effect might, however, be more pronounced.

8.4.3   Definition of Sea State Parameters

Different approaches are used by crews to make visual estimates of the sea state. In the Revised guidelines to the master for avoiding dangerous situations in adverse weather and sea conditions [14] it is, for example, suggested that the wave period should be measured by means of a stop watch as the time span between the generation of a foam patch by a breaking wave and its reappearance after passing the wave trough. However, besides the simple fact that it is dark half of the time, the precision in such manual estimates is all too low to serve as input in high quality operational guidance. Another source of sea state estimation is weather forecast data in terms of wave parameters or complete 2-D wave spectra. Here, the accuracy will depend heavily on the forecast model and its resolution in relation to the ship position. Case II may here serve as an example. The ship was fairly close to the shore while the forecast was a normal resolution forecast from a global wave model incapable of catching the local effects, and the forecast available on board underestimated the wave heights significantly. Using high resolution data from a local wave model would probably have provided a much better sea state estimate. The next step in sea state estimation is ship based in-situ estimation. One such approach is to estimate the sea state from the ship motions. This approach is adopted in the Seaware EnRoute Live system, using a further refined version of a variational method originally described in [10]. A comparison of a ship motion estimation of the wave spectra to the spectra measured by the Eden wave buoy in Case II is shown in Fig. 8.13. Another kind of in-situ estimation based on image processing techniques on raw X-band radar video signals, may provide reliable frequency and directional spectral information.

8.4.4   The Stochastic Problem

One of the main challenges in quantitative prediction of parametric rolling lies in the fact that it is a highly nonlinear dynamic problem subjected to complex stochastic excitation. This means that parametric rolling in ocean waves generally are rare events, thus requiring attention in terms of methodologies that enable CPU-efficient computation of probability quantities or stability boundaries. The

Fig. 8.13 Comparison of wave spectra for Case II. The solid spectrum is estimated in real-time based on ship motions. The dotted spectrum was measured by the Eden wave buoy, approximately 10 nautical miles west of the ship position

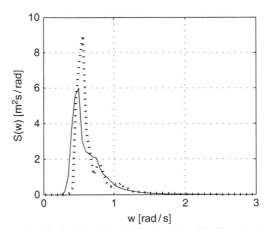

most straightforward approach is to perform massive Monte Carlo simulations in terms of a sufficiently large number of realizations of sufficiently long periods of real time. At present, this is hardly a feasible approach for operational guidance due to the CPU power required. It is therefore necessary to investigate other possibilities. Since in practice, parametric rolling in irregular waves will only occur during encountered wave sequences that are "fairly high and regular," one such possibility is to confine the time-consuming nonlinear simulation to such critical wave sequences. Here, wave group theory, split time methods, and first order reliability based methods play important roles, e.g., [1, 16, 19]. If the objective is to find stability boundaries rather than probabilistic quantities for roll amplitudes, there are a few conceptually different possible approaches by which the stochastic problem can be considered. Bulian [2], for example, describes a method to calculate stochastic stability boundaries based on a simplified analytical approach. Another such approach is presented in [4,5] which is based on the concept of $\overline{GM}$ spectra by considering $\overline{GM}(t)$ as a linear stochastic process in irregular waves.

8.5 Discussions

The ongoing work under the IMO SLF subcommittee developing vulnerability criteria for ships regarding parametric roll and other critical events in waves is welcomed. Eventually, this will make it possible to settle a refined common standard of minimum stability robustness to be used in design of new ships. In addition to this, we believe further developed operational guidance applicable to both existing and new sensitive ship types will be a natural complement to assure safe service under all conditions. Both design criteria and operational guidance limit values should also preferably relate to relevant ship and operation specific design loads or limits, e.g., for cargo lashings, hence putting the criteria and operational guidance into a more risk-based context.

There are many challenges still to be faced in this development. The risk models should account for both the probability of parametric excitation and the severity of the resulting motions and be based on the specific ship characteristics, operational condition and sea state including inherent uncertainties. We believe full scale service data and analysis from incidents and early warnings are essential in this development as they expose complexities and considerations that are normally not included in well-defined model tests or numerical simulations.

Acknowledgment This research has been financially supported by the Swedish Mercantile Marine Foundation (Stiftelsen Sveriges Sjömanshus) and the Swedish Maritime Administration (Sjöfartsverket) which are both gratefully acknowledged. The wave buoy data presented for Case II belongs to the New South Wales Department of Environment, Climate Change and Water, and was collected and kindly provided by the Manly Hydraulics Laboratory, Sydney.

References

1. Belenky, V. L., Weems, K. M., Lin, W. M.: Numerical Procedure for Evaluation of Capsizing Probability with Split Time Method. Proc. 27th Symp. Naval Hydrodynamics, Seoul, Korea (2008)
2. Bulian, G.: Development of Analytical Nonlinear Models for Parametric Roll and Hydrostatic Restoring Variations in Regular and Irregular Waves. Dissertation, University of Trieste, Italy (2006)
3. Dallinga, R., Blok, J., Luth, H.: Excessive Rolling of Cruise Ships in Head and Following Waves. International Conference on Ship Motions & Manoeuvrability, RINA, London, UK (1998)
4. Dunwoody, A. B.: Roll of a Ship in Astern Seas – Metacentric Height Spectra, Journal of Ship Research 33(3)221–228 (1989)
5. Dunwoody, A. B.: Roll of a Ship in Astern Seas – Response to GM Fluctuations, Journal of Ship Research 33(4):284–290 (1989)
6. France, W. N., Levadou, M., Treakle, T. W., Paulling, J. R., Michel, R. K., Moore, C.: An Investigation of Head-Sea Parametric Rolling and its Influence on Container Lashing Systems. Marine Technology 40(1):1–19 (2003)
7. Francescutto, A., Bulian, G., Lugni, C.: Nonlinear and Stochastic Aspects of Parametric Rolling Modeling. Marine Technology 41(2):74–81 (2004)
8. Grim, O.: Das Schiff in von Achtern Mitlaufender See. Transactions STG Hamburg (1951)
9. Hua, J., Palmquist, M. A.: Description of SMS – A Computer Code for Ship Motion Simulation. Report 9502, KTH Royal Institute of Technology, Stockholm, Sweden (1995)
10. Hua, J., Palmquist, M.: Wave Estimation through Ship Motion Measurements – A Practical Approach. Intl. Conf. on Seakeeping and Weather, RINA, London, UK (1995)
11. Huss, M., Olander, A.: Theoretical Seakeeping Predictions On-Board Ships – A System for Operational Guidance and Real Time Surveillance. ISSN 1103-470X, ISRN KTH/FKT/SKP/FR–94/50—SE, KTH Royal Institute of Technology, Stockholm, Sweden (1994)
12. IMO: Guidance to the Master for Avoiding Dangerous Situations in Following and Quartering Seas. MSC/Circ.707 (1995)
13. IMO: Review of the Intact Stability Code – Recordings of head sea Parametric Rolling on a PCTC. SLF 47/INF.5 (2004)
14. IMO: Revised Guidance to the Master for Avoiding Dangerous Situations in Adverse Weather and Sea Conditions. MSC.1/Circ.1228 (2007)
15. IMO: Development of New Generation Intact Stability Criteria. SLF 52/WP.1 (2010)

16. Jensen, J. J.: Efficient Estimation of Extreme Non-linear Roll Motions Using the First-Order Reliability Method (FORM). Journal of Marine Science and Technology 12(4):191–202 (2007)
17. Kaufmann, J.: Fatal Accident on Board the CMV Chicago Express During Typhoon "Hugupit" on September 24 2008 off the Coast of Honk Kong. Bundesstelle für Seeunfalluntersuchung, Investigation Report 510/08 (2009)
18. Levadou, M., Gaillarde, G.: Operational Guidance to Avoid Parametric Roll. Design and Operation of Containerships pp. 75–86, RINA, London, UK (2003)
19. Longuet-Higgins, M. S.: Statistical Properties of Wave Groups in a Random Sea State. Phil. Trans. R. Soc. London, A312(1521):219–250 (1984)
20. Neves, M. A. S., Perez, N. A., Lorca, O. M.: Experimental Analysis on Parametric Resonance for two Fishing Vessels in Head Seas. Proc. 6th International Ship Stability Workshop, Webb Institute, New York, USA (2002)
21. Shin, Y. S., Belenky, V. L., Paulling, J. R., Weems, K., Lin, W. M.: Criteria for Parametric Roll of Large Containerships in Longitudinal Seas, ABS Technical Paper (2004)
22. Spanos, D., Papanikolaou, A.: SAFEDOR International Benchmark Study on Numerical Simulation Methods for the Prediction of Parametric Rolling of Ships in Wave, Revision 1.0, NTUA-SDL Report, School of Naval Architecture and Marine Engineering, National Technical University of Athens, Greece (2009)
23. Thomas, G., Duffy, J., Lilienthal, T., Watts, R., Gehling, R.: On the Avoidance of Parametric Roll in Head Seas. Ships and Offshore Structures, 5(4):295–306 (2010)

Chapter 9
Ship Model for Parametric Roll Incorporating the Effects of Time-Varying Speed

Dominik A. Breu, Christian Holden, and Thor I. Fossen

9.1 Introduction

In this work, we derive a ship model for roll parametric resonance valid for time-varying velocity. In recent years, several ship models of different complexity for parametric roll have been developed [1, 8, 11, 13, 14, 16, 17]. A prevalent, simple model to describe parametrically excited roll motion is the Mathieu equation:

$$m_{44}\ddot{\phi} + d_{44}\dot{\phi} + \left[k_{44} + k_{\phi t}\cos(\omega_e t + \alpha_\phi)\right]\phi = 0 .$$

Here, m_{44} is the sum of the moment of inertia and the added moment of inertia in roll, d_{44} the linear hydrodynamic damping coefficient, and k_{44} the linear restoring moment coefficient. The amplitude of the change in the linear restoring coefficient is $k_{\phi t}$, ω_e is the encounter frequency, and α_ϕ is a phase angle. All the parameters are considered constant.

A system described by the Mathieu equation parametrically resonates at $\omega_e \approx 2\sqrt{k_{44}/m_{44}}$ [12]. The encounter frequency ω_e is the Doppler-shifted frequency of the waves as seen from the ship. Under time-varying velocity, however, the encounter frequency varies, and we show that Mathieu-type equations are not valid to describe the roll dynamics for non-constant ω_e.

D.A. Breu (✉) • C. Holden
Centre for Ships and Ocean Structures, Norwegian University of Science and Technology,
NO-7491 Trondheim, Norway
e-mail: breu@itk.ntnu.no; c.holden@ieee.org

T.I. Fossen
Department of Engineering Cybernetics, Norwegian University of Science and Technology,
NO-7491 Trondheim, Norway

Centre for Ships and Ocean Structures, Norwegian University of Science and Technology,
NO-7491 Trondheim, Norway
e-mail: fossen@ieee.org

T.I. Fossen and H. Nijmeijer (eds.), *Parametric Resonance in Dynamical Systems*,
DOI 10.1007/978-1-4614-1043-0_9, © Springer Science+Business Media, LLC 2012

We derive a highly accurate 6-degree-of-freedom (6-DOF) model of a container ship, taking into account the external forces and moments induced on the ship by the hydrostatic and hydrodynamic pressure field of the surrounding ocean. For each instant in time, the pressure is integrated over the instantaneous submerged hull, giving the restoring forces.

The model is capable of handling complex sea states with nonsteady ship motion, and wave-induced effects enter as first-order forces via the pressure field. Unfortunately, the model is not analytical, and can only be implemented on a computer. Because of this, the model is not suitable for mathematical analysis, but it is highly suitable for simulation purposes.

We simplify the 6-DOF model to the three most important degrees of freedom for ships in parametric roll resonance, i.e., heave, roll, and pitch [11, 14], assuming steady, planar waves. The ship's forward speed is allowed to be non-constant. By assuming that the ship's speed is slowly time-varying and using a quasi-steady approach to derive explicit time-domain solutions to the heave and pitch motions, we further reduce the 3-DOF model to a 1-DOF roll model. We also show how the simplified model is linked to Mathieu-type equations. This reduced-order model is analytical, and preserves the majority of the accuracy of the 6-DOF model.

To verify the simplified 1-DOF model, we implement the 6-DOF model on a computer and show in simulations that the simplified model qualitatively captures the behavior of the highly accurate 6-DOF model also for a time-varying encounter frequency. We furthermore show why the Mathieu-type equation is incapable of describing the ship's rolling motion in parametric roll resonance when the encounter frequency is not constant.

In Chap. 10, we will use the 1-DOF model to design a controller capable of stopping parametric roll resonance by speed changes.

The rest of the chapter is organized as follows. Section 9.2 lays the theoretical and mathematical framework of the 6-DOF model. In Sect. 9.3, the details of the computer implementation of this model are described. Sections 9.4 and 9.5 derive the simplified 1-DOF roll model, which is verified in Sect. 9.6, and Sect. 9.7 contains the conclusion.

9.2 Equations of Motion

This section introduces the reference frames and presents the equations of motion of the 6-DOF model of the ship.

9.2.1 Reference Frames

In this chapter, we are using two reference frames; one fixed to the surface of the ocean, and one fixed to the ship. The one fixed to the ocean surface is assumed inertial, and referred to as the inertial frame or the *n*-frame. We refer to the reference frame fixed to the ship as the body frame or *b*-frame. It is a moving coordinate

Fig. 9.1 Reference frames

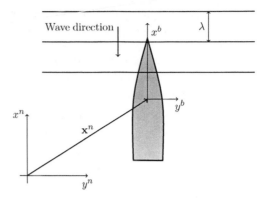

frame which has its origin somewhere along the ship's transversal midline, so that a port–starboard symmetric ship is mirrored about the body frame xz-plane, see Fig. 9.1. The z-axis (not shown in the figure) points towards the center of the Earth (with the gravity field).

Vectors expressed in the inertial reference frame are written in boldface with a superscript n (e.g., $\mathbf{x}^n$), whereas vectors expressed in the body frame are denoted in boldface with a superscript b (e.g., $\mathbf{x}^b$).

The location of the body frame origin relative to the inertial frame, expressed in the inertial frame, is given as $\mathbf{x}^n \in \mathbb{R}^3$, as indicated in Fig. 9.1. The two coordinate systems can be rotated relative to each other. The rotation matrix $\mathbf{R}$ is associated with this rotation so that $\mathbf{x}^n = \mathbf{R}\mathbf{x}^n$. $\mathbf{R}$ satisfies $\mathbf{R}^\top = \mathbf{R}^{-1}$, $\det(\mathbf{R}) = 1$ [3, 7]. The rotation matrix $\mathbf{R}$ belongs to the special orthogonal group of order 3, $\mathrm{SO}(3) \subset \mathbb{R}^{3\times3}$, see [3,7].

The rotation matrix can be fully parametrized with (no less than) three parameters. In this work, we use the so-called roll–pitch–yaw Euler angles. These angles represent simple rotations about the three different body axes; roll (ϕ) about the x-axis, pitch (θ) about the y-axis, and heave (ψ) about the z-axis. See [3,7] for details.

Combining these three angles into the vector $\boldsymbol{\Theta} = [\phi, \theta, \psi]^\top$, we can write $\mathbf{R}$ as:

$$\mathbf{R}(\boldsymbol{\Theta}) = \begin{bmatrix} c\theta c\psi & s\phi s\theta c\psi - c\phi s\psi & c\phi s\theta c\psi + s\phi s\psi \\ c\theta s\psi & s\phi s\theta s\psi + c\phi c\psi & c\phi s\theta s\psi - s\phi c\psi \\ -s\theta & s\phi c\theta & c\phi c\theta \end{bmatrix}, \qquad (9.1)$$

where $c\cdot = \cos(\cdot)$ and $s\cdot = \sin(\cdot)$.

9.2.2 6-DOF Ship Model

To derive the hydrodynamic forces and moments on a ship, we need to make certain assumptions:

A 9.1.   There is no current.

A 9.2.   The hull can be split into triangular or quadrangular panels, where each panel can be parametrized as a two-dimensional surface embedded in $\mathbb{R}^3$.

A 9.3.   The frequency-dependent parameters of the damping, added mass, and Coriolis/centripetal matrices are constant (i.e., maneuvering theory, see for example [7]).

A 9.4.   The ocean is infinitely deep.

A 9.5.   The pressure field in the ocean is unchanged by the passage of the ship (in effect, waves are traveling "through" the ship's hull).

We define the ship's generalized position vector $\mathbf{q}$ as:

$$\mathbf{q} = \left[\mathbf{x}^{n\top}, \boldsymbol{\Theta}^{\top} \right]^{\top} \in \mathbb{R}^6, \tag{9.2}$$

where

$$\mathbf{x}^n = \left[x^n, y^n, z^n \right]^{\top} \in \mathbb{R}^3 \tag{9.3}$$

is the ship's position in the inertial frame and

$$\boldsymbol{\Theta} = \left[\phi, \theta, \psi \right]^{\top} \in \mathbb{R}^3 \tag{9.4}$$

the vector of Euler angles representing the ship's rotation relative to the inertial frame.

We define the ship's generalized velocity vector $\boldsymbol{v}$ as:

$$\boldsymbol{v} = \left[\mathbf{v}^{b\top}, \boldsymbol{\omega}^{b\top} \right]^{\top} \in \mathbb{R}^6, \tag{9.5}$$

where

$$\mathbf{v}^b = \left[v_1^b, v_2^b, v_3^b \right]^{\top} = \mathbf{R}^{\top} \mathbf{v}^n \triangleq \mathbf{R}^{\top} \dot{\mathbf{x}}^n = \mathbf{R}^{\top} \left[v_1^n, v_2^n, v_3^n \right]^{\top} \in \mathbb{R}^3 \tag{9.6}$$

is the ship's linear velocity in the body frame and

$$\boldsymbol{\omega}^b = \left[\omega_1^b, \omega_2^b, \omega_3^b \right]^{\top} \in \mathbb{R}^3 \tag{9.7}$$

the ship's angular velocity relative to the inertial frame, expressed in the body frame.

We let $\boldsymbol{\tau}_c \in \mathbb{R}^6$ be the generalized forces generated by the actuators and $\boldsymbol{\tau}_e \in \mathbb{R}^6$ the environmental disturbances and unmodeled generalized forces.

From [7], we get the 6-DOF ship model

$$\dot{\mathbf{q}} = \mathbf{P}(\boldsymbol{\Theta})\boldsymbol{v} \tag{9.8}$$

$$\mathbf{M}\dot{\boldsymbol{v}} + \mathbf{D}(\boldsymbol{v})\boldsymbol{v} + \mathbf{C}(\boldsymbol{v})\boldsymbol{v} + \mathbf{k}(\mathbf{q},t) = \boldsymbol{\tau}_c + \boldsymbol{\tau}_e, \tag{9.9}$$

where

$$\mathbf{P}(\boldsymbol{\Theta}) = \begin{bmatrix} \mathbf{R}(\boldsymbol{\Theta}) & \mathbf{0}_{3\times3} \\ \mathbf{0}_{3\times3} & \mathbf{G}(\boldsymbol{\Theta}) \end{bmatrix} \in \mathbb{R}^{6\times6} \tag{9.10}$$

with $\mathbf{R}$ as in (9.1) and

$$
\mathbf{G}(\boldsymbol{\Theta}) = \begin{bmatrix} 1 & \sin(\phi)\tan(\theta) & \cos(\phi)\tan(\theta) \\ 0 & \cos(\phi) & -\sin(\phi) \\ 0 & \frac{\sin(\phi)}{\cos(\theta)} & \frac{\cos(\phi)}{\cos(\theta)} \end{bmatrix}, \quad \cos(\theta) \neq 0, \tag{9.11}
$$

and $\mathbf{M} \triangleq \mathbf{M}_{RB} + \mathbf{M}_A = \mathbf{M}^\top > 0 \in \mathbb{R}^{6\times 6}$ is the sum of the rigid-body inertia and added mass, $\mathbf{D}(\mathbf{v}) \in \mathbb{R}^{6\times 6}$ the damping matrix, and $\mathbf{C}(\mathbf{v}) \in \mathbb{R}^{6\times 6}$ the Coriolis/centripetal matrix; $\mathbf{k} = \mathbf{k}_p + \mathbf{k}_g$ where $\mathbf{k}_p$ is the generalized pressure forces, and $\mathbf{k}_g$ the gravity forces. $\mathbf{D}$ satisfies $\mathbf{y}^\top \mathbf{D}(\mathbf{v})\mathbf{y} \geq 0 \, \forall \, \mathbf{y}, \mathbf{v} \in \mathbb{R}^6$.

If

$$
\mathbf{M} \triangleq \begin{bmatrix} \mathbf{M}_{11} & \mathbf{M}_{12} \\ \mathbf{M}_{21} & \mathbf{M}_{22} \end{bmatrix} \in \mathbb{R}^{6\times 6}, \quad \mathbf{M}_{11}, \mathbf{M}_{12}, \mathbf{M}_{21}, \mathbf{M}_{22} \in \mathbb{R}^{3\times 3}
$$

and

$$
\mathbf{S}(\mathbf{x}) \triangleq \begin{bmatrix} 0 & -x_3 & x_2 \\ x_3 & 0 & -x_1 \\ -x_2 & x_1 & 0 \end{bmatrix} \in SS(3) \subset \mathbb{R}^{3\times 3} \, \forall \, \mathbf{x} = [x_1, x_2, x_3]^\top \in \mathbb{R}^3
$$

then

$$
\mathbf{C}(\mathbf{v}) = \begin{bmatrix} \mathbf{0}_{3\times 3} & -\mathbf{S}(\mathbf{M}_{11}\mathbf{v}^b + \mathbf{M}_{12}\boldsymbol{\omega}^b) \\ -\mathbf{S}(\mathbf{M}_{11}\mathbf{v}^b + \mathbf{M}_{12}\boldsymbol{\omega}^b) & -\mathbf{S}(\mathbf{M}_{21}\mathbf{v}^b + \mathbf{M}_{22}\boldsymbol{\omega}^b) \end{bmatrix} = -\mathbf{C}^\top(\mathbf{v}) \in \mathbb{R}^{6\times 6}.
$$

Here, $SS(3)$ is the group of all skew-symmetric matrices of order 3, see [3,7].

The gravity forces $\mathbf{k}_g$ are given by [7]:

$$
\mathbf{k}_g(\mathbf{q}) = -mg \begin{bmatrix} \mathbf{R}^\top(\boldsymbol{\Theta})\mathbf{e}_z \\ (\mathbf{R}^\top(\boldsymbol{\Theta})\mathbf{e}_z) \times \mathbf{r}_g^b \end{bmatrix}, \tag{9.12}
$$

where $\mathbf{r}_g^b$ is the ship's center of gravity in the the body frame, $\mathbf{e}_z = [0,0,1]^\top$, m the ship's mass, and g the acceleration of gravity.

By knowing the pressure field of the surrounding ocean, $\mathbf{k}_p$ can be found. At any given point $\mathbf{r}^n$ in the ocean (in the inertial frame), there will be a local pressure field $\Psi \approx \Psi(\mathbf{r}^n, t) \in \mathbb{R}$ [15, 18].[1] We assume (Assumption A 9.2) that each section of the ship's hull can be parametrized with parameters u and v, so that the vector $\mathbf{r}_i^b(u, v)$ gives the position of a point on the surface of panel i, in the body frame. Defining

$$
\Psi_i(u, v) \triangleq \Psi(\mathbf{R}\mathbf{r}_i^b(u, v) + \mathbf{x}^n, t), \tag{9.13}
$$

---

[1] Technically, the pressure field would also be a function of the ship's state, except for Assumption A 9.5.

we can then take the generalized pressure force on the ship as [2, 15, 18]:

$$
\mathbf{k}_p(\mathbf{q},t) = \sum_i \left[\begin{array}{c} \int\int_{S_{w,i}} \Psi_i(u,v)\frac{\partial \mathbf{r}_i^b}{\partial u}(u,v) \times \frac{\partial \mathbf{r}_i^b}{\partial v}(u,v) \, du \, dv \\ \int\int_{S_{w,i}} \Psi_i(u,v)\mathbf{r}_i^b(u,v) \times \left(\frac{\partial \mathbf{r}_i^b}{\partial u}(u,v) \times \frac{\partial \mathbf{r}_i^b}{\partial v}(u,v) \right) \, du \, dv \end{array} \right], \quad (9.14)
$$

where $S_{w,i}$ is the wetted (submerged) part of panel i and the ship is parametrized so that the normal vector $(\partial \mathbf{r}_i^b / \partial u) \times (\partial \mathbf{r}_i^b / \partial v)^2$ points out of the hull. The effects of current and waves can all be accounted for in the force $\mathbf{k}_p$ [4, 15].

This model itself is nothing new; the computer implementation of the 6-DOF model (9.8)–(9.9) is a novel contribution of this work.

9.3 Computer Implementation of the 6-DOF Model

To implement a computer version of the 6-DOF model (9.8)–(9.9), we use data from a specific, 281 m long container ship. This is the same ship as used in [9–11].

Whereas the model can be used for any sea state in any condition, we have chosen to create an implementation suitable for parametric roll. As such, we assume that the waves are planar and sinusoidal.

9.3.1 Inertia and Damping Forces

We computed the parameters for inertia $\mathbf{M}_{RB}$, added mass $\mathbf{M}_A$, and damping $\mathbf{D}$ in ShipX (VERES) [6]. We set the unmodeled force vector $\boldsymbol{\tau}_e$ to zero and the control force $\boldsymbol{\tau}_c$ as:

$$
\boldsymbol{\tau}_c = - \left[\begin{array}{c} k_p \left(v_1^b - v_{1,d}^b \right) + k_i \int_{t_0}^t \left(v_1^b(T) - v_{1,d}^b(T) \right) \, dT \\ 0 \\ 0 \\ 0 \\ 0 \\ \kappa_p(\psi - \psi_d) + \kappa_d(\dot{\psi} - \dot{\psi}_d) + \kappa_i \int_{t_0}^t (\psi(T) - \psi_d(T)) \, dT \end{array} \right] \quad (9.15)
$$

with $v_{1,d}^b$ the desired surge speed and ψ_d the desired heading. The rudimentary PID controllers in surge and yaw are there to keep the ship on course in the presence

---

[2]See [2] for proof that this is a normal vector.

of the other forces. Without these controllers, the simulated ship tends to drift quite heavily off course. The parameters used are

$$\mathbf{M}_{RB} = \begin{bmatrix} m\mathbf{I}_3 & -m\mathbf{S}(\mathbf{r}_g^b) \\ m\mathbf{S}(\mathbf{r}_g^b) & \mathbf{J}_b \end{bmatrix}, \quad m = 7.7358E7$$

$$\mathbf{r}_g^b = [-3.7486, 0, -1.120]^\top$$

$$\mathbf{J}_b = \begin{bmatrix} 1.41E10 & 0 & 0 \\ 0 & 3.70E11 & 0 \\ 0 & 0 & 3.70E11 \end{bmatrix}$$

$$\mathbf{M}_A = \begin{bmatrix} 0 & 0 & 0 & 0 & 0 & 0 \\ 0 & 7.59E7 & 0 & 6.43E7 & 0 & -1.04E9 \\ 0 & 0 & 7.80E7 & 0 & -7.83E8 & 0 \\ 0 & 6.43E7 & 0 & 2.20E9 & 0 & -9.08E9 \\ 0 & 0 & -7.83E8 & 0 & 3.39E11 & 0 \\ 0 & -1.04E9 & 0 & -9.08E9 & 0 & 4.48E11 \end{bmatrix}$$

$$\mathbf{D}(v) = \begin{bmatrix} 5.66E3 & 0 & 0 & 0 & 0 & 0 \\ 0 & 3.31E7 & 0 & 1.50E7 & 0 & 1.22E8 \\ 0 & 0 & 4.66E7 & 0 & -1.05E9 & 0 \\ 0 & 1.50E7 & 0 & 2.48E8 & 0 & -2.27E9 \\ 0 & 0 & -4.03E8 & 0 & 2.73E11 & 0 \\ 0 & -5.03E8 & 0 & -2.79E9 & 0 & 1.35E11 \end{bmatrix}$$

$$+ \begin{bmatrix} 2.83E4|v_1| & \mathbf{0}_{1\times5} \\ \mathbf{0}_{5\times1} & \mathbf{0}_{5\times5} \end{bmatrix}$$

$$k_p = 7.7358E7, \quad k_i = 7.7358E6, \quad \kappa_p = 8.18E11, \quad \kappa_d = 0, \quad \kappa_i = 0$$

with all values in base SI units (kg–m–s). With these numbers, all forces and moments of the 6-DOF model (9.8)–(9.9) except for $\mathbf{k}_p$ can be computed.

9.3.2 Pressure Forces

The final force that needs to be computed is the pressure spring term $\mathbf{k}_p(\mathbf{q},t)$. This is computed directly from (9.14) with some further simplifications:

A 9.6.   The waves are a simple, planar, and standing sinusoid.
A 9.7.   The "hydrostatic" part of the pressure extends from the instantaneous ocean surface and down.
A 9.8.   The "dynamic" part of the pressure extends from the average ocean surface and down.

By [5], these approximations and the ones used in deriving the 6-DOF model (9.8)–
(9.9) give a first-order approximation of the wave effects.

Pressure Field

From [5], we have that the ocean surface under these conditions is given by:

$$\zeta(t, \mathbf{r}^n) = \zeta_0 \cos(\omega_0 t - k_w r_1^n + \alpha_\zeta) \tag{9.16}$$

in the inertial frame, with ζ giving the wave height at a position $\mathbf{r}^n = [r_1^n, r_2^n, r_3^n]^\top$.
The instantaneous ocean surface is then at $[r_1^n, r_2^n, \zeta(t, \mathbf{r}^n)]^\top$. The constant parameters are the wave amplitude ζ_0, the frequency of the waves as seen by an observer
stationary in the inertial frame ω_0, and the wave number k_w. For waves traveling in
negative x-direction, $k_w < 0$ and vice versa.
    By [5], the pressure field Ψ is given by:

$$\Psi(\mathbf{r}^n, t) = g\rho\zeta_0 e^{-k_w \max(r_3^n, 0)} \cos(\omega_0 t - k_w r_1^n + \alpha_\zeta) + g\rho r_3^n . \tag{9.17}$$

The parameters are the acceleration of gravity g and the density of sea water ρ. The
term $g\rho r_3^n$ is the "hydrostatic" pressure, and $g\rho\zeta_0 e^{-k_w \max(r_3^n, 0)} \cos(\omega_0 t - k_w r_1^n + \alpha_\zeta)$
the "dynamic" pressure.

The Submerged Part of the Ship

The ship's hull, as previously mentioned, is split into panels, each forming a triangle
or quadrangle. In the body frame, panel i has corners $\mathbf{p}_{i,j}^b$, $j \in \{1,2,3,4\}$ for
quadrangles and $j \in \{1,2,3\}$ for triangles. In the inertial frame,

$$\mathbf{p}_{i,j}^n = \mathbf{R}\mathbf{p}_{i,j}^b + \mathbf{x}^n . \tag{9.18}$$

As (9.16) gives the explicit wave surface (in the inertial frame) one can compute
which points, at any given time and physical location, are above or below the wave
surface by solving the equation

$$[0, 0, 1]\mathbf{p}_{i,j}^n = \zeta(t, \mathbf{p}_{i,j}^n) .$$

    Rather than solving the equation explicitly, an approximation is used. First, the
points are transformed using

$$\bar{\mathbf{p}}_{i,j}^n = \mathbf{p}_{i,j}^n - [0, 0, \zeta(t, \mathbf{p}_{i,j}^n)]^\top . \tag{9.19}$$

Any point with a positive z-value is submerged, see Fig. 9.3.

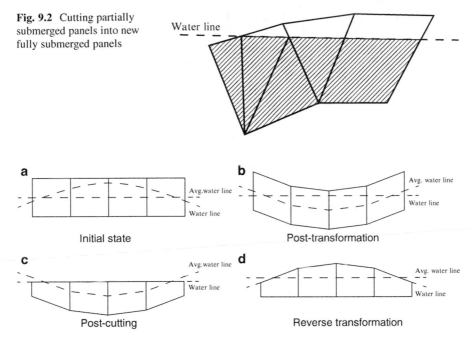

Fig. 9.2 Cutting partially submerged panels into new fully submerged panels

Fig. 9.3 Transforming panels

Each panel is individually parametrized with a bilinear interpolation, so that, for each panel i,

$$\bar{\mathbf{p}}_i^n = \bar{\mathbf{k}}_0 + \bar{\mathbf{k}}_u u + \bar{\mathbf{k}}_v v + \bar{\mathbf{k}}_{uv} uv \tag{9.20}$$

with $u, v \in [0, 1)$ defines all points on panel i.

For each partially submerged panel (one with at least one point underwater and at least one point above water), the parametrization is used to find where the edges of the panel intersect the water line, and to compute the coordinates of these points. The submerged points and the points in the water line then make up (one or more) new panel(s). Note that the panels whose submerged part forms a pentagon are split into three triangular panels, whereas panels whose submerged part forms a triangle or a quadrangle are kept as such, see Fig. 9.2.

The partially submerged panels and the panels wholly above the water line are discarded. The fully submerged panels based on the partially submerged panels, and the original fully submerged panels, are kept. The entire transformation–cutting–reverse transformation process can be seen in Fig. 9.3.

The approximation to find the true intersection of the hull and the ocean surface is good if the average size of the panels is small relative to the wave length.

Generalized Forces

Before computing the forces and moments, the transformation (9.19) needs to be reversed such that all points of the panels are expressed in the inertial frame. We therefore take

$$\mathbf{p}_{i,j}^n = \bar{\mathbf{p}}_{i,j}^n + [0,0,\zeta(t,\mathbf{p}_{i,j}^n)]^\top, \tag{9.21}$$

and parameterize these panels bilinearly so that

$$\mathbf{p}_i^n = \mathbf{k}_0 + \mathbf{k}_u u + \mathbf{k}_v v + \mathbf{k}_{uv} uv \tag{9.22}$$

with $u,v \in [0,1)$ define all points on panel i, in the inertial frame. The partial derivatives of $\mathbf{p}_i^n$ with respect to u and v can be explicitly found as:

$$\frac{\partial \mathbf{p}_i^n}{\partial u} = \mathbf{k}_u + \mathbf{k}_{uv} v, \quad \frac{\partial \mathbf{p}_i^n}{\partial v} = \mathbf{k}_v + \mathbf{k}_{uv} u. \tag{9.23}$$

It is worth noting that for triangular panels, $\mathbf{k}_{uv} = 0$.

For each panel, we can compute the pressure force in the inertial frame as:

$$\mathbf{f}_i^n = \begin{cases} \int_0^1 \int_0^1 \Psi(\mathbf{p}_i^n(u,v),t)[\mathbf{k}_u + \mathbf{k}_{uv} v] \times [\mathbf{k}_v + \mathbf{k}_{uv} u]\, du\, dv & \text{quadrangles} \\ \int_0^1 \int_0^{1-v} \Psi(\mathbf{p}_i^n(u,v),t)\mathbf{k}_u \times \mathbf{k}_v\, du\, dv & \text{triangles} \end{cases}.$$

The integration must be done numerically. Although more points could be used, for increased computational speed, only corner points are used in the calculation of the pressure forces. Thus,

$$\mathbf{f}_i^n \approx \begin{cases} \frac{1}{4}\sum_{j=1}^4 \Psi(\mathbf{p}_i^n(\bar{u}_j,\bar{v}_j),t)[\mathbf{k}_u + \mathbf{k}_{uv}\bar{v}_j] \times [\mathbf{k}_v + \mathbf{k}_{uv}\bar{u}_j] & \text{quadrangles} \\ \frac{1}{6}\sum_{j=1}^3 \Psi(\mathbf{p}_i^n(\bar{u}_j,\bar{v}_j),t)\mathbf{k}_u \times \mathbf{k}_v & \text{triangles} \end{cases},$$

where

$$\bar{u}_1 = 0, \quad \bar{v}_1 = 0,$$
$$\bar{u}_2 = 0, \quad \bar{v}_2 = 1,$$
$$\bar{u}_3 = 1, \quad \bar{v}_3 = 0,$$
$$\bar{u}_4 = 1, \quad \bar{v}_4 = 1.$$

The computation for the torque is similar. However, we need the torque relative to the body origin rather than the inertial origin. We therefore get

$$\mathbf{m}_i^n = \begin{cases} \int_0^1 \int_0^1 \Psi(\mathbf{p}_i^n(u,v),t)[\mathbf{p}_i^n(u,v) - \mathbf{x}^n] \\ \qquad \times ([\mathbf{k}_u + \mathbf{k}_{uv} v] \times [\mathbf{k}_v + \mathbf{k}_{uv} u])\, du\, dv & \text{quadrs.} \\ \int_0^1 \int_0^{1-v} \Psi(\mathbf{p}_i^n(u,v),t)[\mathbf{p}_i^n(u,v) - \mathbf{x}^n] \times (\mathbf{k}_u \times \mathbf{k}_v)\, du\, dv & \text{triangles} \end{cases}.$$

This is approximated as:

$$\mathbf{m}_i^n \approx \begin{cases} \frac{1}{4}\sum_{j=1}^{4}\Psi(\mathbf{p}_i^n(\bar{u}_j,\bar{v}_j),t)\,[\mathbf{p}_i^n(\bar{u}_j,\bar{v}_j)-\mathbf{x}^n] \\ \qquad \times\,([\mathbf{k}_u+\mathbf{k}_{uv}\bar{v}_j]\times[\mathbf{k}_v+\mathbf{k}_{uv}\bar{u}_j]) & \text{quadrs.} \\ \frac{1}{6}\sum_{j=1}^{3}\Psi(\mathbf{p}_i^n(\bar{u}_j,\bar{v}_j),t)\,[\mathbf{p}_i^n(\bar{u}_j,\bar{v}_j)-\mathbf{x}^n]\times(\mathbf{k}_u\times\mathbf{k}_v) & \text{triangles} \end{cases}.$$

Note that this torque is still in the inertial frame, but relative to the body origin.

If the set S consists of all i such that panel i is one of the original, fully submerged panels *or* one of the newly created (also fully submerged) panels, we can then take

$$\mathbf{k}_p(\mathbf{q},t) \approx \begin{bmatrix} \mathbf{R}^\top\sum_{i\in S}\mathbf{f}_i^n \\ \mathbf{R}^\top\sum_{i\in S}\mathbf{m}_i^n \end{bmatrix} \qquad (9.24)$$

to get the total pressure force and moment in the body frame.

The system is simulated with a fixed time step, and for each time instant, the outlined procedure for computing $\mathbf{k}_p$ is performed.

Note that the procedure automatically handles such effects as (first-order) wave-induced forces and Doppler shift of these. The first by simple virtue of the pressure field including the dynamic pressure and the latter by including $\mathbf{x}^n$ in (9.18).

9.4 Encounter Frequency

To an observer standing in a fixed location on the ocean surface, waves will appear to have a specific frequency of oscillation (or a range of frequencies if the waves are irregular). Ocean waves can be seen as planar waves [5], and for regular sinusoidal waves, these can be described by (9.16). To a stationary observer, $\mathbf{r}^n$ is constant.

To a moving observer, the waves will appear to behave differently than to the stationary observer due to the Doppler effect. We take $\mathbf{r}^n = \mathbf{x}^n$ to be the location of the observer in the inertial frame (i.e., the body origin) and assume, without loss of generality, that $\mathbf{x}^n(t_0) = 0$. The observer's velocity in the inertial x-direction is then

$$\dot{x}^n \triangleq v_1^n \qquad (9.25)$$

so that

$$\zeta(t,\mathbf{x}^n) = \zeta_0\cos\left(\omega_0 t - k_w\int_{t_0}^{t}v_1^n(\tau)\,\mathrm{d}\tau + \alpha_\zeta\right). \qquad (9.26)$$

Since the velocity is given in the body frame in the 6-DOF model (9.8)–(9.9),

$$v_1^n = \mathbf{e}_x^\top\mathbf{R}\mathbf{v}^b, \qquad (9.27)$$

where $\mathbf{e}_x = [1,0,0]^\top$.

We then define the encounter frequency, the frequency seen by the observer, as:

$$\omega_e \triangleq \frac{d}{dt}\left(\omega_0 t - k_w \int_{t_0}^{t} v_1^n(\tau)\, d\tau\right) \tag{9.28}$$

$$= \omega_0 - k_w \mathbf{e}_x^\top \mathbf{R} \mathbf{v}^b \approx \omega_0 - k_w \mathbf{e}_x^\top \mathbf{R}[v_1^b, 0, 0]^\top$$

$$= \omega_0 - k_w \cos(\theta)\cos(\psi)v_1^b \approx \omega_0 - k_w v_1^b \tag{9.29}$$

and rewrite (9.26) as:

$$\zeta(t, \mathbf{x}^n) = \zeta_0 \cos\left(\int_{t_0}^{t} \omega_e(\tau)\, d\tau + \alpha_\zeta\right). \tag{9.30}$$

We note that if v_1^n is a constant, then so is ω_e, and the above simply becomes $\zeta(t, \mathbf{x}^n) = \zeta_0 \cos\left(\omega_e t + \alpha_\zeta\right)$.

It is an important fact that whereas we cannot change ω_0, we can change ω_e by changing the velocity v_1^n.

9.5 1-DOF Roll Models

The spring term in the full 6-DOF model is analytically unknown. We make the following extra assumption to derive a 1-DOF roll model:

A 9.9.   The ship is traveling directly into the waves.

Note that the ship is still allowed to change its forward speed.

For ships in parametric resonance, it is well-known that the most important degrees of freedom are heave, roll, and pitch [11, 14]. Heave and pitch are already coupled, and during parametric resonance these transfer energy to roll.

Setting all other degrees of freedom to zero, we define

$$\mathbf{q}_{r3} \triangleq \left[z^n, \phi, \theta\right]^\top, \quad \mathbf{v}_{r3} \triangleq \left[v_3^b, \omega_1^b, \omega_2^b\right]^\top \tag{9.31}$$

and note that

$$\dot{\mathbf{q}}_{r3} = \begin{bmatrix} \cos(\phi)\cos(\theta) & 0 & 0 \\ 0 & 1 & \sin(\phi)\tan(\theta) \\ 0 & 0 & \cos(\phi) \end{bmatrix} \mathbf{v}_{r3} \approx \mathbf{v}_{r3}. \tag{9.32}$$

This 3-DOF model can be written as:

$$\mathbf{M}_{r3}\dot{\mathbf{v}}_{r3} + \mathbf{C}_{r3}\left(\mathbf{v}_{r3}\right)\mathbf{v}_{r3} + \mathbf{D}_{r3}\left(\mathbf{v}_{r3}\right)\mathbf{v}_{r3} + \mathbf{k}_{r3}\left(\mathbf{q}_{r3}, t\right) = \boldsymbol{\tau}_{c,r3} + \boldsymbol{\tau}_{e,r3}. \tag{9.33}$$

For simplicity, we will assume that the velocities in heave and pitch are low, and that the only coupling between these two degrees of freedom and roll exists in the spring term $\mathbf{k}_{r_3}$. This allows us to write

$$
\mathbf{M}_{r_3} = \begin{bmatrix} m_{33} & 0 & m_{35} \\ 0 & m_{44} & 0 \\ m_{53} & 0 & m_{55} \end{bmatrix}, \quad \mathbf{C}_{r_3}(\mathbf{v}_{r_3}) = \mathbf{0}_{3\times3}, \quad \mathbf{D}_{r_3}(\mathbf{v}_{r_3}) = \begin{bmatrix} d_{33} & 0 & d_{35} \\ 0 & d_{44} & 0 \\ d_{53} & 0 & d_{55} \end{bmatrix},
$$

where m_{ij} and d_{ij} are the i,jth element of $\mathbf{M}$ and $\mathbf{D}(0)$ from the 6-DOF model (9.8)–(9.9).

Furthermore, following [11], we simplify $\mathbf{k}_{r_3}$ to

$$
\mathbf{k}_{r_3}(\mathbf{q}_{r_3},t) \approx \begin{bmatrix} k_{33} & 0 & k_{35} \\ 0 & k_{44} & 0 \\ k_{53} & 0 & k_{55} \end{bmatrix} \mathbf{q}_{r_3} + \begin{bmatrix} 0 \\ k_{z\phi}z''\phi + k_{\phi\theta}\phi\theta + k_{\phi3}\phi^3 \\ 0 \end{bmatrix} + \bar{\mathbf{k}}_{r_3}(t)
$$

with

$$
\bar{\mathbf{k}}_{r_3}(t) = -\begin{bmatrix} a_z\zeta_0\cos\left(\int_{t_0}^t \omega_e(\tau)\,d\tau + \alpha_z\right) \\ 0 \\ a_\theta\zeta_0\cos\left(\int_{t_0}^t \omega_e(\tau)\,d\tau + \alpha_\theta\right) \end{bmatrix},
$$

where a_z, α_z, a_θ, and α_θ are constant. We note that $\bar{\mathbf{k}}_{r_3}(t)$ is merely ζ of (9.26) phase-shifted and scaled, effectively sent through a linear filter.

Neither heave nor roll nor pitch are likely to be directly actuated, so $\boldsymbol{\tau}_{c,r_3} = 0$. The unmodeled disturbances $\boldsymbol{\tau}_{e,r_3}$ are also assumed zero. We then rewrite (9.33) as

$$
m_{44}\ddot{\phi} + d_{44}\dot{\phi} + k_{44}\phi + k_{\phi3}\phi^3 = -[k_{z\phi}, k_{\phi\theta}]\mathbf{q}_{r_2}\phi \tag{9.34}
$$

$$
\mathbf{M}_{r_2}\ddot{\mathbf{q}}_{r_2} + \mathbf{D}_{r_2}\dot{\mathbf{q}}_{r_2} + \mathbf{K}_{r_2}\mathbf{q}_{r_2} = \boldsymbol{\tau}_{e,r_2} . \tag{9.35}
$$

with

$$
\mathbf{q}_{r_2} = [z'', \theta]^\top, \quad \mathbf{M}_{r_2} = \begin{bmatrix} m_{33} & m_{35} \\ m_{53} & m_{55} \end{bmatrix}, \quad \mathbf{D}_{r_2} = \begin{bmatrix} d_{33} & d_{35} \\ d_{53} & d_{55} \end{bmatrix}
$$

$$
\mathbf{K}_{r_2} = \begin{bmatrix} k_{33} & k_{35} \\ k_{53} & k_{55} \end{bmatrix}, \quad \boldsymbol{\tau}_{e,r_2}(t) = \begin{bmatrix} a_z\zeta_0\cos\left(\int_{t_0}^t \omega_e(\tau)\,d\tau + \alpha_z\right) \\ a_\theta\zeta_0\cos\left(\int_{t_0}^t \omega_e(\tau)\,d\tau + \alpha_\theta\right) \end{bmatrix} .
$$

We note that the $\mathbf{q}_{r_2}$-subsystem (9.35) is completely decoupled from the roll-subsystem (9.34) and is merely a linear ordinary differential equation with constant coefficients and a sinusoidal input. If we assume that ω_e is constant, then the system will have the steady-state solution

$$
\mathbf{q}_{r_2}(t) = \begin{bmatrix} \bar{a}_z \zeta_0 \cos\left(\int_{t_0}^t \omega_e(\tau)\, \mathrm{d}\tau + \bar{\alpha}_z \right) \\ \bar{a}_\theta \zeta_0 \cos\left(\int_{t_0}^t \omega_e(\tau)\, \mathrm{d}\tau + \bar{\alpha}_\theta \right) \end{bmatrix} . \tag{9.36}
$$

The main purpose of this work is to derive a roll model for changing ω_e, thus we consider the case when ω_e is not constant. We revisit the equation for ω_e, and note

$$
\dot{\omega}_e = \frac{\mathrm{d}}{\mathrm{d}t}\left(\omega_0 - k_w \mathbf{e}_x^\top \mathbf{R} \mathbf{v}^b \right) = -k_w \mathbf{e}_x^\top \mathbf{R} \left(\mathbf{S}(\boldsymbol{\omega}^b)\mathbf{v}^b + \dot{\mathbf{v}}^b \right) \tag{9.37}
$$

giving

$$
|\dot{\omega}_e| \leq |k_w| \|\mathbf{e}_x\| \|\mathbf{R}\| \left(\|\mathbf{S}(\boldsymbol{\omega}^b)\mathbf{v}^b\| + \|\dot{\mathbf{v}}^b\| \right) = |k_w| \left(\|\mathbf{S}(\boldsymbol{\omega}^b)\mathbf{v}^b\| + \|\dot{\mathbf{v}}^b\| \right) \tag{9.38}
$$

since $\|\mathbf{e}_x\| = \|\mathbf{R}\| = 1$.

For large ships, neither the acceleration $\|\dot{\mathbf{v}}^b\|$ nor the term $\|\mathbf{S}(\boldsymbol{\omega}^b)\mathbf{v}^b\|$ is likely to be large. To cause parametric resonance, the wave length has to be approximately the same as the length of the ship, and since k_w is inversely proportional to the wave length, k_w is likely to be quite low. Thus $|\dot{\omega}_e| \approx 0$ and a quasi-steady approach can be used. We therefore take the solution to (9.35) to be given by (9.36) even when ω_e is non-constant.

We insert the solution (9.36) into the right-hand side of (9.34) and get

$$
[k_{z\phi}, k_{\phi\theta}]\mathbf{q}_{r_2} = k_{z\phi} \bar{a}_z \zeta_0 \cos\left(\int_{t_0}^t \omega_e(\tau)\mathrm{d}\tau + \bar{\alpha}_z \right) + k_{\phi\theta} \bar{a}_\theta \zeta_0 \cos\left(\int_{t_0}^t \omega_e(\tau)\, \mathrm{d}\tau + \bar{\alpha}_\theta \right)
$$

$$
= k_{\phi t} \cos\left(\int_{t_0}^t \omega_e(\tau)\, \mathrm{d}\tau + \alpha_\phi \right),
$$

where

$$
k_{\phi t}^2 = \zeta_0^2 \left[k_{z\phi}^2 \bar{a}_z^2 + k_{\phi\theta}^2 \bar{a}_\theta^2 + 2 k_{z\phi} k_{\phi\theta} \bar{a}_z \bar{a}_\theta \cos(\alpha_\theta - \alpha_z) \right]
$$

$$
\alpha_\phi = \arctan\left(\frac{k_{z\phi}\bar{a}_z \sin(\alpha_z) + k_{\phi\theta}\bar{a}_\theta \sin(\alpha_\theta)}{k_{z\phi}\bar{a}_z \cos(\alpha_z) + k_{\phi\theta}\bar{a}_\theta \cos(\alpha_\theta)} \right).
$$

1-DOF roll model (time-varying speed) Under the stated assumptions, the roll motion can be described by the 1-DOF parametric roll model:

$$
m_{44}\ddot{\phi} + d_{44}\dot{\phi} + \left[k_{44} + k_{\phi t} \cos\left(\int_{t_0}^t \omega_e(\tau)\, \mathrm{d}\tau + \alpha_\phi \right) \right]\phi + k_{\phi 3}\phi^3 = 0 . \tag{9.39}
$$

1-DOF roll model (constant encounter frequency) If $\dot{\omega}_e = 0$, then the roll motion is described by the *Mathieu equation*:

$$m_{44}\ddot{\phi} + d_{44}\dot{\phi} + \left[k_{44} + k_{\phi t}\cos(\omega_e t + \alpha_\phi)\right]\phi + k_{\phi 3}\phi^3 = 0. \tag{9.40}$$

For both models, the natural roll frequency ω_ϕ is given by $\omega_\phi \triangleq \sqrt{k_{44}/m_{44}}$.

9.6 Model Verification

To verify the 1-DOF simplified roll model (9.39), we simulate it and compare it to simulations of the full 6-DOF model (9.8)–(9.9) presented in Sect. 9.3. Since the Mathieu equation (9.40) is commonly used to describe ships sailing with constant surge speed experiencing parametric roll resonance, we additionally investigate its ability to describe the dynamics of a ship for a non-constant encounter frequency.

In the simulations, we use the ship described in [9–11] and the same parameters for inertia $\mathbf{M}_{RB}$, added mass $\mathbf{M}_A$, and damping $\mathbf{D}$ as in Sect. 9.3.1. We have implemented the kinematics using quaternions, see [3, 7], instead of the Euler angle representation in the simulations. Those two representations can be used interchangeably and the choice of representation is to a certain degree arbitrary [7].

As the main difference between the models is in the spring term, we compare these. Furthermore, the parameters of these are not known *a priori*, and need to be identified. Denoting the roll angle computed based on the 1-DOF model (9.39) and the Mathieu model (9.40) by ϕ_c and ϕ_m, respectively, we define the spring torque for the 1-DOF roll model (9.39) and the Mathieu model (9.40) as

$$k_{\phi,c}(t;s) = \left[k_{44} + k_{\phi t}\cos\left(\int_{t_0}^t \omega_e(\tau)\,\mathrm{d}\tau + \alpha_\phi\right)\right]\phi_c(t) + k_{\phi 3}\phi_c^3(t) \tag{9.41}$$

$$k_{\phi,m}(t;s) = \left[k_{44} + k_{\phi t}\cos\left(\omega_e t + \alpha_\phi\right)\right]\phi_m(t) + k_{\phi 3}\phi_m^3(t). \tag{9.42}$$

To determine the parameters $s = [k_{44}, k_{\phi t}, \alpha_\phi, k_{\phi 3}]$ in (9.41) and (9.42), we use nonlinear least-squares curve fitting:

$$s_c = \arg\min_s \sum_t |k_4(\mathbf{q}(t),t) - k_{\phi,c}(t;s)|^2 \tag{9.43}$$

$$s_m = \arg\min_s \sum_t |k_4(\mathbf{q}(t),t) - k_{\phi,m}(t;s)|^2, \tag{9.44}$$

where k_4 is the fourth element of $\mathbf{k}_p$ in the full 6-DOF model (9.8)–(9.9).

The instantaneous encounter frequency in the simplified roll model (9.39) is calculated from the simulation of the full 6-DOF model by (9.28). However, even when attempting to keep constant speed, the waves cause the ship's speed to oscillate. This is reflected in the 6-DOF model. Using the instantaneous values of ω_e, the Mathieu model (9.40) will not oscillate. Thus, we use a low-pass filtered encounter frequency when simulating the Mathieu model.

Table 9.1 Simulation parameters, constant speed

Quantity	Symbol	Value
Mean forward speed	v_1^b	7.90 m/s
Mean encounter frequency	ω_e	0.645 rad/s
Wave amplitude	ζ_0	2.5 m
Wave length	λ	281 m
Wave number	k_w	−0.0224
Natural roll frequency	ω_ϕ	0.343 rad/s
Modal wave frequency	ω_0	0.4683 rad/s
	k_{44}	1.7646E9 kg m^2/s^2
Model parameters:	$k_{\phi t}$	7.3224E8 kg m^2/s^2
Simplified roll equation	α_ϕ	0.2295 rad
	$k_{\phi 3}$	2.2741E9 kg m^2/s^2
	k_{44}	1.7685E9 kg m^2/s^2
Model parameters:	$k_{\phi t}$	7.3369E8 kg m^2/s^2
Mathieu equation	α_ϕ	0.2118 rad
	$k_{\phi 3}$	2.2692E9 kg m^2/s^2

9.6.1 Constant Forward Speed

To compare the models when ω_e is kept approximately constant, we simulate the three models with constant speed (barring small variations due to wave-induced forces in surge).

In the following, the signals of the 6-DOF model is represented without subscript, while the subscripts c and m denote the simplified roll equation and the Mathieu equation, respectively. The simulation parameters and the model parameters are summarized in Table 9.1.

Figure 9.4 shows the simulation results for all three models. From Fig. 9.4a it is evident that the ship is experiencing parametric roll resonance in this scenario. Figure 9.4c compares the spring torque divided by the roll angle of the the full 6-DOF model to the ones of the simplified roll equation and the Mathieu equation computed by (9.41) and (9.42), i.e., k_4/ϕ versus

$$\frac{k_{\phi,c}}{\phi_c} = k_{44} + k_{\phi t} \cos \left(\int_{t_0}^{t} \omega_e (\tau) \, d\tau + \alpha_\phi \right) + k_{\phi 3} \phi_c^2$$

$$\frac{k_{\phi,m}}{\phi_m} = k_{44} + k_{\phi t} \cos \left(\omega_e t + \alpha_\phi \right) + k_{\phi 3} \phi_m^2 .$$

Once steady-state is reached, there is good agreement between the 6-DOF model (9.8)–(9.9) and the two 1-DOF models (9.39) and (9.40).

In this scenario, the 1-DOF model (9.39) and the Mathieu model (9.40) behave almost identically. This is as expected, since with $\dot{\omega}_e = 0$ the two models are identical. The slight variations in ω_e in this scenario are not enough to cause any significant discrepancy.

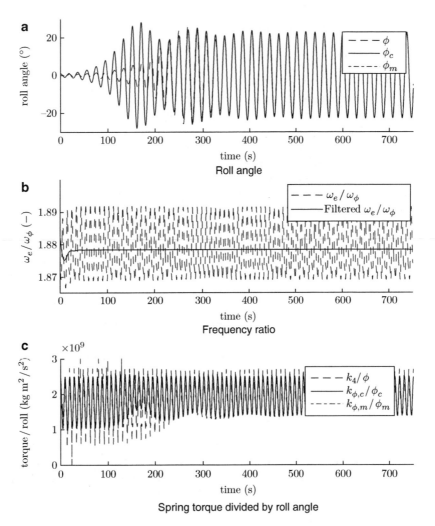

Fig. 9.4 Model comparison. ϕ, ϕ_c, and ϕ_m are roll angles from the 6-DOF model, the simplified roll equation, and the Mathieu equation

Up till about 220 s, there is significant discrepancy between the 6-DOF model (9.8)–(9.9) and the two 1-DOF models (9.39) and (9.40). The two 1-DOF models (9.39) and (9.40) go to the maximum roll angle much faster than the 6-DOF model (9.8)–(9.9). This is because the two 1-DOF models (9.39) and (9.40) are derived under the assumption that heave and pitch are in steady-state. For the first 200 s or so, that is not the case. Once steady-state heave and pitch are achieved, the 6-DOF model (9.8)–(9.9) quickly catches up to the two models (9.39) and (9.40).

9.6.2 Maximum Roll Angle

To compare the models under a wide range of scenarios, we simulate the three models for different (almost constant) forward speeds and different wave amplitudes and compute the maximum roll angle as a function of the encounter frequency and the wave amplitude. The simulation scenarios and parameters are identical for all three cases. The spring torque constants for the simplified 1-DOF roll model (9.39) (s_c of (9.43)) and the Mathieu model (9.40) (s_m of (9.44)) are reestimated for each forward speed and each wave amplitude.

It is well-known that parametric resonance occurs at wave encounter frequencies approximately twice the natural roll frequency [8]. We therefore simulate the models for an initial surge speed from 0.5 to 12.8 m/s, resulting in a frequency ratio ω_e/ω_ϕ from 1.4 to 2.2. The wave amplitude ζ_0 ranges from 0 to 6 m.

Figure 9.5 depicts the maximum roll angle for the models of different complexity as a result of the simulations. The roll amplitude is limited to 90° in the plots for the simplified roll equation (Fig. 9.5b) and the Mathieu equation (Fig. 9.5c) for the sake of presentability. We note that qualitatively the simplified roll model (Fig. 9.5b) is quite close to the 6-DOF model (Fig. 9.5a), at least for low wave amplitudes.

The Mathieu equation (Fig. 9.5c) simulated with the filtered wave encounter frequency also behaves reasonably well and is almost indistinguishable from the 1-DOF model. This is reasonable, as there are only very small variations in ω_e.

9.6.3 Time-Varying Forward Speed

Since the difference between the simplified roll model (9.39) and the Mathieu model (9.40) only becomes apparent when the speed is non-constant, we simulate the system with non-constant forward speed. The scenario tested is a simple speed change, so that the desired forward speed $v_{1,d}^b$ is given by:

$$v_{1,d}^b(t) = \begin{cases} v_{1,0}^b & \forall\, t \in [t_0, t_1] \\ v_{1,0}^b + l(t - t_1) & \forall\, t \in [t_1, t_2] \,, \\ v_{1,1}^b & \forall\, t \in [t_2, \infty) \end{cases}$$

where l is the desired acceleration and $v_{1,1}^b = v_{1,0}^b + l(t_2 - t_1)$. This gives an encounter frequency

$$\omega_e(t) \approx \begin{cases} \omega_{e,0} & \forall\, t \in [t_0, t_1] \\ \omega_{e,0} - k_w l(t - t_1) & \forall\, t \in [t_1, t_2] \\ \omega_{e,1} & \forall\, t \in [t_2, \infty) \end{cases}$$

$$= \begin{cases} \omega_0 - k_w v_{1,0}^b & \forall\, t \in [t_0, t_1] \\ \omega_0 - k_w \left[v_{1,0}^b + l(t - t_1)\right] & \forall\, t \in [t_1, t_2] \,. \\ \omega_0 - k_w v_{1,1}^b & \forall\, t \in [t_2, \infty) \end{cases}$$

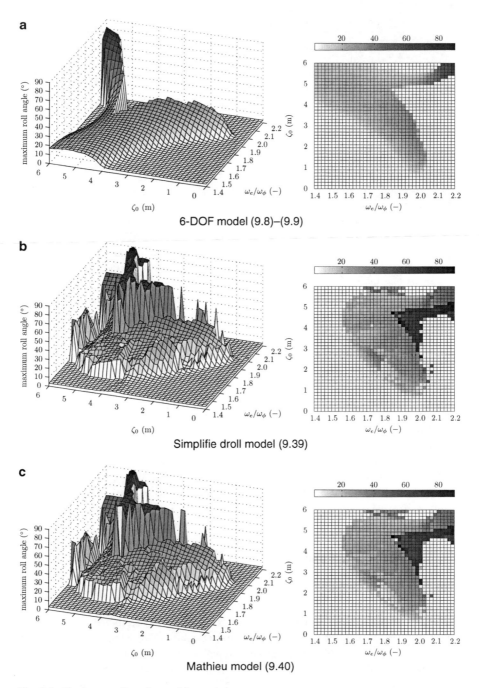

Fig. 9.5 Maximum roll angle, model comparison

Table 9.2 Simulation parameters, time-varying speed

Quantity	Symbol	Value
Initial mean forward speed	$v_{1,0}^b$	7.90 m/s
Desired acceleration	l	0.005 m/s²
Final mean forward speed	$v_{1,1}^b$	9.43 m/s
Initial mean encounter frequency	$\omega_{e,0}$	0.645 rad/s
Final mean encounter frequency	$\omega_{e,1}$	0.680 rad/s
Simulation start time	t_0	0 s
Acceleration start time	t_1	300 s
Acceleration stop time	t_2	607 s
Wave amplitude	ζ_0	2.5 m
Wave length	λ	281 m
Wave number	k_w	−0.0224
Natural roll frequency	ω_ϕ	0.343 rad/s
Modal wave frequency	ω_0	0.4683 rad/s
	k_{44}	1.7646E9 kg m²/s²
Model parameters:	$k_{\phi t}$	7.3224E8 kg m²/s²
Simplified roll equation	α_ϕ	0.2295 rad
	$k_{\phi 3}$	2.2741E9 kg m²/s²
	k_{44}	1.7676E9 kg m²/s²
Model parameters:	$k_{\phi t}$	7.3333E8 kg m²/s²
Mathieu equation	α_ϕ	0.2122 rad
	$k_{\phi 3}$	2.2702E9 kg m²/s²

Due to small oscillations in surge, ω_e does not exactly match the desired value, as seen in Fig. 9.6b. The Mathieu model (9.40) is once again fed the low-pass filtered values of ω_e, while the 1-DOF model (9.39) uses the unfiltered values. The parameters used in the simulation are shown in Table 9.2.

Figure 9.6 depicts the results of the simulation. Again, the ship is in parametric roll resonance, as shown in Fig. 9.6a. The non-constant forward speed results in a non-constant encounter frequency and frequency ratio, respectively, see Fig. 9.6b. The spring torque divided by the roll angle is compared in Fig. 9.6c for the three models. The simplified roll equation is able to estimate the roll motion well even for non-constant speed, whereas it is apparent that the Mathieu equation is not. It gradually becomes out of phase with the roll motion of the full 6-DOF model and never gets back in phase even when steady-state is reached.

We conclude that the simulations indicate that the model based on the simplified roll equation is adequate to describe the ship's dynamics in parametric roll resonance when the wave encounter frequency is non-constant. The Mathieu equation, on the other hand, is not able to capture the dynamics to a sufficient extent unless the encounter frequency is very close to constant and only if the low-pass filtered value of ω_e is used.

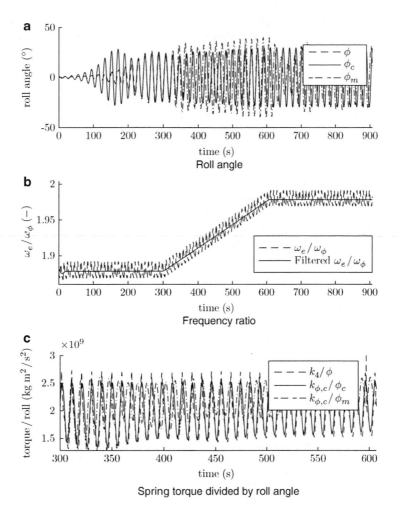

Fig. 9.6 Model comparison. ϕ, ϕ_c, and ϕ_m are roll angles from the 6-DOF model, the simplified roll equation, and the Mathieu equation

9.7 Conclusions

In this work, we have developed a ship model for parametric roll resonance that, unlike most models in literature, is valid for both constant and non-constant ship velocity. The resulting model is a complex and accurate 6-DOF model which considers the induced external forces and moments due to the hydrostatic and hydrodynamic pressure field of the surrounding ocean, which includes the effect of waves. Gravity and viscous damping are also accounted for.

The model assumes that the pressure field is unchanged by the passage of the ship, and wave-induced effects are in practice limited to first-order approximations.

The model is not analytical, and is therefore only suitable for simulations. We implemented the model in Matlab/Simulink using data from a specific, 281 m container ship.

Assuming that the forward speed is slowly time-varying and using a quasi-steady approach to derive explicit time-domain solutions to the heave and pitch motions (which are the modes most tightly coupled to roll), we have also derived a simplified 1-DOF (roll) model which can readily be used for control purposes. For constant wave encounter frequency, we have shown that the 1-DOF model is identical to a Mathieu-type equation, which is commonly used to describe ships in parametric roll resonance. However, unlike the Mathieu equation, the model is suitable also for non-constant velocity.

We have verified the proposed 1-DOF model against the complex 6-DOF model in simulations with constant and non-constant encounter frequencies, and we have shown that it is able to qualitatively match the results of the full 6-DOF model in a wide range of conditions. Furthermore, we have shown that Mathieu-type equations are not capable of capturing the roll dynamics when the encounter frequency is time-varying.

Acknowledgements We would like to thank our colleague Øyvind Ygre Rogne for his assistance with the code for finding the submerged part of the hull. This work was funded by the Centre for Ships and Ocean Structures (CeSOS), NTNU, Norway and the Norwegian Research Council.

References

1. Bulian, G., Francescutto, A., Lugni, C.: On the nonlinear modeling of parametric rolling in regular and irregular waves. International shipbuilding progress **51**(2), 173–203 (2004)
2. Edwards, C.H., Penney, D.E.: Calculus with Analytic Geometry, 5th edn. Prentice Hall, Upper Saddle River, New Jersey, USA (1998)
3. Egeland, O., Gravdahl, J.T.: Modeling and Simulation for Automatic Control. Marine Cybernetics, Trondheim, Norway (2002)
4. Faltinsen, O.M.: Sea Loads on Ships and Offshore Structures. Ocean Technology. Cambridge University Press, Cambridge, UK (1998)
5. Faltinsen, O.M., Timokha, A.N.: Sloshing. Cambridge University Press, New York, USA (2009)
6. Fathi, D.: ShipX Vessel Responses (VERES). SINTEF Marintek AS, Trondheim, Norway (2004)
7. Fossen, T.I.: Handbook of Marine Craft Hydrodynamics and Motion Control. John Wiley & Sons, Ltd., Chichester, UK (2011)
8. France, W.N., Levadou, M., Treakle, T.W., Paulling, J.R., Michel, R.K., Moore, C.: An investigation of head-sea parametric rolling and its influence on container lashing systems. SNAME Annual Meeting, Orlando, Florida, USA (2001)
9. Galeazzi, R., Holden, C., Blanke, M., Fossen, T.I.: Stabilization of parametric roll resonance by combined speed and fin stabilizer control. In: Proceedings of European Control Conference. Budapest, Hungary (2009)
10. Holden, C., Galeazzi, R., Fossen, T.I., Perez, T.: Stabilization of parametric roll resonance with active u-tanks via lyapunov control design. In: Proceedings of European Control Conference. Budapest, Hungary (2009)

11. Holden, C., Galeazzi, R., Rodríguez, C.A., Perez, T., Fossen, T.I., Blanke, M., Neves, M.A.S.: Nonlinear container ship model for the study of parametric roll resonance. Modeling, Identification and Control **28**(4), 87–103 (2007)
12. Nayfeh, A.H., Mook, D.T.: Nonlinear Oscillations. John Wiley & Sons, Inc., Weinheim, Germany (1995)
13. Neves, M.A.S.: On the excitation of combination modes associated with parametric resonance in waves. In: Proceedings of 6th International Ship Stability Workshop. New York, USA (2002)
14. Neves, M.A.S., Rodriguez, C.A.: A coupled third order model of roll parametric resonance. In: Maritime Transportation and Exploitation of Ocean and Coastal Resources: Proceedings of 11th International Congress of the International Maritime Association of the Mediterranean, pp. 243. Taylor & Francis, London, UK (2006)
15. Perez, T.: Ship Motion Control. Advances in Industrial Control. Springer-Verlag, London (2005)
16. Shin, Y., Belenky, V.L., Paulling, J.R., Weems, K.M., Lin, W.M.: Criteria for parametric roll of large container ships in longitudinal seas. Transactions of SNAME **112**, 117–147 (2004)
17. Vidic-Perunovic, J., Juncher Jensen, J.: Parametric roll due to hull instantaneous volumetric changes and speed variations. Ocean Engineering **36**, 891–899 (2009)
18. White, F.M.: Fluid Mechanics. McGraw Hill Text (2002)

Part III
Control of Parametric Resonance in Ships

Chapter 10
Frequency Detuning Control by Doppler Shift

Christian Holden, Dominik A. Breu, and Thor I. Fossen

10.1 Introduction

Control of parametric roll resonance has attracted considerable research in recent years [1, 7–11, 13–15, 18, 19]. The proposed control methods can roughly be categorized as *direct* or *indirect* methods. The direct methods are aimed at directly controlling the roll motion by generating an opposing roll moment, as seen in [11, 19]. Indirect strategies attempt to violate the empirical conditions necessary for the onset of parametric roll resonance, as seen in [1, 14, 15, 18]. A hybrid approach, doing both at the same time, is also possible, as seen in [7–9].

In this work, we consider the indirect approach to control parametrically excited roll motions. A simple model for parametric roll resonance is the *Mathieu equation*:

$$m_{44}\ddot{\phi} + d_{44}\dot{\phi} + \left[k_{44} + k_{\phi t}\cos(\omega_e t + \alpha_\phi)\right]\phi = 0,$$

where m_{44} is the sum of the moment of inertia and the added moment of inertia in roll, d_{44} the linear hydrodynamic damping coefficient, k_{44} the linear restoring moment coefficient and $k_{\phi t}$ the amplitude of its change, ω_e the encounter frequency, and α_ϕ a phase angle. All the parameters are considered constant.

It is known from [17] that such a system parametrically resonates at $\omega_e \approx 2\sqrt{k_{44}/m_{44}}$ (an encounter frequency of twice the natural roll frequency). The

C. Holden (✉) • D.A. Breu
Centre for Ships and Ocean Structures, Norwegian University of Science and Technology,
NO-7491 Trondheim, Norway
e-mail: c.holden@ieee.org; breu@itk.ntnu.no

T.I. Fossen
Department of Engineering Cybernetics, Norwegian University of Science and Technology,
NO-7491 Trondheim, Norway

Centre for Ships and Ocean Structures, Norwegian University of Science and Technology,
NO-7491 Trondheim, Norway
e-mail: fossen@ieee.org

T.I. Fossen and H. Nijmeijer (eds.), *Parametric Resonance in Dynamical Systems*,
DOI 10.1007/978-1-4614-1043-0_10, © Springer Science+Business Media, LLC 2012

encounter frequency ω_e is the Doppler-shifted frequency of the waves as seen from the ship. As the frequency is Doppler-shifted, it can be changed by changing the ship's speed.

The main purpose of this chapter is to show that it is feasible to control parametric roll resonance by changing the encounter frequency to violate the condition $\omega_e \approx 2\sqrt{k_{44}/m_{44}}$. We call this frequency detuning.

As shown in Chap. 9, Mathieu-type equations are not valid for non-constant ω_e. To design and analyze the control system, we therefore use the simplified, 1-DOF model (9.39) developed in Sect. 9.5, allowing the ship's forward speed to change, but only slowly. For ships susceptible to parametric roll – many of which are large [2, 4–6] – this is not an unreasonable assumption.

Based on the 1-DOF roll model (9.39) in Sect. 9.5, we propose a simple controller based on a linear change of the encounter frequency, achieved by variation of the ship's forward speed. We then prove mathematically that the proposed controller is able to drive the ship out of parametric resonance, driving the roll motion to zero. It is worth noting that the controller is in fact simple enough that a human helmsman can perform the necessary control action, rendering a speed controller unnecessary.

The controller is tested with the simplified 1-DOF model (9.39) and the full 6-DOF model presented in Sect. 9.2.2, and is shown to work as expected in both cases.

The rest of this chapter is organized as follows. Section 10.2 lists nomenclature. Section 10.3 briefly summarizes the model used. The controller is derived and its theoretical properties are proven in Sect. 10.4. Its performance in simulation is shown in Sect. 10.5. Section 10.6 contains the conclusion. A short appendix contains the proof of a lemma used in the proof of the controller's performance.

10.2 Nomenclature

In this chapter, the following parameters are used:

$\mathbf{v}^b = [v_1^b, v_2^b, v_3^b]^\top$	The linear velocity of the ship observed in a reference frame attached to the ship.
$\boldsymbol{\omega}^b = [\omega_1^b, \omega_2^b, \omega_3^b]^\top$	The angular velocity of the ship observed in a reference frame attached to the ship.
$\mathbf{v}^n = [v_1^n, v_2^n, v_3^n]^\top$	The linear velocity of the ship observed in an (assumed inertial) reference frame attached to the mean ocean surface.
$\boldsymbol{\omega}^n = [\omega_1^n, \omega_2^n, \omega_3^n]^\top$	The angular velocity of the ship observed in an (assumed inertial) reference frame attached to the mean ocean surface.
$\mathbf{R}$	A rotation matrix rotating vectors from the body-fixed to the inertial frame.
$\boldsymbol{\Theta} = [\phi, \theta, \psi]^\top$	The roll–pitch–yaw Euler angles.
$m_{44} > 0$	The total moment of inertia in roll; the sum of the rigid-body moment of inertia in roll and the added moment of inertia in roll.
$d_{44} > 0$	The linear damping coefficient in roll.

$k_{44} > 0$	The linear restoring coefficient in roll.
$k_{\phi t} > 0$	The amplitude of the time-varying change in the linear restoring coefficient in roll.
$\omega_e(t)$	The time-varying encounter frequency; the Doppler-shifted frequency of the waves as seen from the ship.
α_ϕ	The phase of the time-varying change in the linear restoring coefficient in roll.
$k_{\phi 3} > 0$	The cubic restoring coefficient in roll.
ω_0	The frequency of the waves as seen by an observer in an (assumed inertial) frame attached to the mean ocean surface.
k_w	The wave number as seen by an observer in an (assumed inertial) frame attached to the mean ocean surface.
$u = \dot{\omega}_e$	The control input.
ε	A control parameter.
t	Time.
t_0	The start time.
t_1	The controller is turned on at time $t = t_1$.
t_2	The controller is turned off at time $t = t_2$.
$\omega_{e,0} = \omega_e(t_0)$	The initial value of the encounter frequency.
$\omega_{e,1} = \omega_e(t \geq t_2)$	The final value of the encounter frequency.
$\bar{\alpha}_\phi$	The phase of the time-varying change in the linear restoring coefficient in roll at time $t \geq t_2$.
γ	The normalized damping ratio.
κ	The normalized linear restoring coefficient.
ι	The normalized change in the linear restoring coefficient.
α	The normalized cubic restoring coefficient.
T	Normalized time (fast).
$\bar{t}$	Normalized time (slow).
σ	The detuning parameter.
$a(\bar{t})$	The (slowly) time-varying amplitude of the steady-state roll motion; $\phi \approx a(\bar{t}) \cos(T - \beta(\bar{t})/2)$ for $t \geq t_2$.
$\beta(\bar{t})$	The (slowly) time-varying phase of the steady-state roll motion; $\phi \approx a(\bar{t}) \cos(T - \beta(\bar{t})/2)$ for $t \geq t_2$.

10.3 Roll Model for Non-Constant Speed

As previously mentioned, we use no direct actuation in roll; instead, we are changing ω_e (by changing the forward speed) to detune the encounter frequency and thus violate a necessary condition for the existence of parametric roll resonance.

We make the following assumptions:

A 10.1. The ship is traveling in head or stern seas.

A 10.2. The waves are planar, standing, and sinusoidal, with frequency ω_0 and wave number k_w.

A 10.3. The ship is changing speed only slowly.
A 10.4. The ship's sway and heave velocities are small.
A 10.5. The ship's pitch angle is small.

The ship is traveling with linear and angular velocities

$$\mathbf{v}^b = [v_1^b, v_2^b, v_3^b]^\top \in \mathbb{R}^3 \tag{10.1}$$

$$\boldsymbol{\omega}^b = [\omega_1^b, \omega_2^b, \omega_3^b]^\top \in \mathbb{R}^3 \tag{10.2}$$

as seen from the ship. For an observer standing on the mean ocean surface, the ship
will appear to have linear and angular velocities

$$\mathbf{v}^n = \mathbf{R}\mathbf{v}^b = [v_1^n, v_2^n, v_3^n]^\top \in \mathbb{R}^3 \tag{10.3}$$

$$\boldsymbol{\omega}^n = \mathbf{R}\boldsymbol{\omega}^b = [\omega_1^n, \omega_2^n, \omega_3^n]^\top \in \mathbb{R}^3, \tag{10.4}$$

where $\mathbf{R}$ is a rotation matrix given by:

$$\mathbf{R}(\boldsymbol{\Theta}) = \begin{bmatrix} c\theta c\psi & s\phi s\theta c\psi - c\phi s\psi & c\phi s\theta c\psi + s\phi s\psi \\ c\theta s\psi & s\phi s\theta s\psi + c\phi c\psi & c\phi s\theta s\psi - s\phi c\psi \\ -s\theta & s\phi c\theta & c\phi c\theta \end{bmatrix}, \tag{10.5}$$

where $c\cdot = \cos(\cdot)$, $s\cdot = \sin(\cdot)$ and $\boldsymbol{\Theta} = [\phi, \theta, \psi]^\top$ are the roll, pitch, and yaw angles
as defined in [3], see also Chap. 9.

To analyze the effects of speed changes, we need a model that is valid for time-
varying speed. As discussed in Chap. 9, the commonly used Mathieu equation is not
adequate in this case. We thus use the 1-DOF model (9.39) of Sect. 9.5, given by

$$m_{44}\ddot{\phi} + d_{44}\dot{\phi} + \left[k_{44} + k_{\phi t}\cos\left(\int_{t_0}^t \omega_e(t)\,d\tau + \alpha_\phi\right)\right]\phi + k_{\phi 3}\phi^3 = 0, \tag{10.6}$$

where ϕ is the roll angle, m_{44} the sum of the rigid-body moment of inertia about
the x-axis and the added moment of inertia in roll, d_{44} the linear hydrodynamic
damping, k_{44} the linear restoring moment coefficient, $k_{\phi t}$ the amplitude of its
change, and $k_{\phi 3}$ the cubic restoring force coefficient. These parameters are constant.

We note that the natural frequency of ϕ is $\omega_\phi \triangleq \sqrt{k_{44}/m_{44}}$. From Chap. 9, we
have that the encounter frequency ω_e is given by:

$$\omega_e = \omega_0 - k_w v_1^n = \omega_0 - k_w[1, 0, 0]\mathbf{R}\mathbf{v}^b. \tag{10.7}$$

The encounter frequency is the frequency of the waves as seen from the ship. Due
to the Doppler effect, this is not the same as the frequency of the waves seen by a
stationary observer, ω_0. For a ship traveling at constant velocity, ω_e is constant and
the Mathieu equation can be used to describe the ship's behavior, see Chap. 9.

If the ship is nonrotating (i.e., $\boldsymbol{\omega}^b \equiv \mathbf{0}$), then

$$\dot{\omega}_e = -k_w[1,0,0]\mathbf{R}\dot{v}^b \approx -k_w \mathbf{e}_x^\top \mathbf{R}[\dot{v}_1^b,0,0]^\top$$

$$\approx -k_w \dot{v}_1^b \cos(\theta)\cos(\psi) \approx -k_w \dot{v}_1^b \cos(\psi) \tag{10.8}$$

by the assumption of small sway velocity, yaw rate, and pitch angle. The ship is assumed to be sailing in head or stern seas, that is, $\cos(\psi) = 1$ (head seas) or $\cos(\psi) = -1$ (stern seas).

We can set $\dot{v}_1^b$ directly; this is the forward acceleration and can be changed by increasing or decreasing throttle. It will, however, be limited, so we take it to satisfy $|\dot{v}_1^b| \leq \dot{v}_{1,\max}^b$. Thus, we take $u \triangleq \dot{\omega}_e$ to be the control input, satisfying

$$|u| = |\dot{\omega}_e| \leq u_{\max} = |k_w|\dot{v}_{1,\max}^b . \tag{10.9}$$

Note that the assumption that the forward speed changes only slowly implies that $\dot{v}_{1,\max}^b$ is quite small. The assumption of slow speed change is a necessity to derive the model (10.6), as detailed in Sect. 9.5.

As we can see from the above equation, $u_{\max}$ depends on the size of $\dot{v}_{1,\max}^b$ and k_w. For the type of large, slow vessels that are susceptible to parametric roll, $\dot{v}_{1,\max}^b$ is likely to have quite a low value. For ships to parametrically resonate, the wave length has to be rather long, or $k_{\phi t}$ will be too small [12]. A long wave length implies a small k_w, since $|k_w| = 2\pi/\lambda$ if λ is the wave length. Thus $u_{\max}$ is quite small.

10.4   Control Design

The control objective is to design u such that the origin of the roll system (10.6) is (at least) asymptotically stable. Choosing a $\dot{v}_1^b$ so that $\dot{\omega}_e$ is equal to the desired u is a control allocation problem.[1]

10.4.1   Control Principle

The basic control principle is to (slowly) change the encounter frequency from an undesired value $\omega_{e,0}$ to a desired value $\omega_{e,1}$. We tentatively choose the controller

---

[1] It is also possible to change ω_e by changing course (i.e., changing ψ). This will have the unwanted side effect that the ship will now be directly excited by waves (i.e., there will be an external force on the right-hand side of (10.6) proportional to the wave amplitude), which may also result in relatively large roll amplitude in the type of seas that give rise to parametric resonance. Changing ψ to change the encounter frequency is not investigated in this work.

$$u(t) = \begin{cases} 0 & \forall \ t \in [t_0,t_1] \\ \varepsilon & \forall \ t \in [t_1,t_2] \\ 0 & \forall \ t \in [t_2,\infty) \end{cases} \tag{10.10}$$

for some small constant ε, with $t_2 \geq t_1 \geq t_0$. The initial time is t_0.

If $\omega_e(t_0) = \omega_{e,0}$, then

$$\omega_e(t) = \int_{t_0}^{t} u(\tau)\, d\tau + \omega_{e,0} = \begin{cases} \omega_{e,0} & \forall \ t \in [t_0,t_1] \\ \omega_{e,0} + \varepsilon(t-t_1) & \forall \ t \in [t_1,t_2] \\ \omega_{e,1} & \forall \ t \in [t_2,\infty) \end{cases}, \tag{10.11}$$

where $\omega_{e,1} = \omega_{e,0} + \varepsilon(t_2 - t_1)$. This gives

$$\int_{t_0}^{t} \omega_e(\tau)\, d\tau = \begin{cases} \omega_{e,0}(t-t_0) & \forall \ t \in [t_0,t_1] \\ \omega_{e,0}(t-t_0) + \frac{1}{2}\varepsilon(t-t_1)^2 & \forall \ t \in [t_1,t_2] \\ \omega_{e,1}(t-t_2) + \omega_{e,0}(t_2-t_0) + \frac{1}{2}\varepsilon(t_2-t_1)^2 & \forall \ t \in [t_2,\infty) \end{cases} \tag{10.12}$$

If $\cos(\psi) \equiv \pm 1$ and $v_2^b = v_3^b = \theta = 0$, then $\omega_e(t) = \omega_0 - k_w v_1^b \cos(\psi)$ and the encounter frequency of (10.11) can then be achieved with a surge velocity of

$$v_1^b = \frac{\omega_0 - \omega_e(t)}{k_w \cos(\psi)} = \frac{1}{k_w \cos(\psi)} \begin{cases} \omega_0 - \omega_{e,0} & \forall \ t \in [t_0,t_1] \\ \omega_0 - \omega_{e,0} - \varepsilon(t-t_1) & \forall \ t \in [t_1,t_2] \\ \omega_0 - \omega_{e,1} & \forall \ t \in [t_2,\infty) \end{cases}. \tag{10.13}$$

Proving that the controller (10.10) works is done in two steps: First, ensuring that there exists a (unique finite) solution of (10.6) at $t = t_2$. This step is done in the Appendix. Secondly, we need to prove that if $\omega_e(t) \equiv \omega_{e,1} \ \forall \ t \geq t_2$, then the solution to the initial value problem

$$m_{44}\ddot{\phi} + d_{44}\dot{\phi} + \left[k_{44} + k_{\phi t}\cos\left(\omega_{e,1}t + \bar{\alpha}_\phi\right)\right]\phi + k_{\phi 3}\phi^3 = 0,$$

$$\phi(t_2) = \phi_2, \dot{\phi}(t_2) = \dot{\phi}_2, \tag{10.14}$$

where

$$\bar{\alpha}_\phi \triangleq \alpha_\phi - \omega_{e,1}t_2 + \omega_{e,0}(t_2 - t_0) + \frac{1}{2}\varepsilon(t_2 - t_1)^2$$

is a constant, goes to zero for all $\phi_2, \dot{\phi}_2$.

10.4.2 The System in the Time Interval $t \in [t_2, \infty)$

From the results in the Appendix, we know that there exists a unique finite solution to (10.6), valid at $t = t_2$. From $t \geq t_2$, the trajectories of the system will be the solution to the initial value problem (10.14).

From [17], we know that there are parameter values of $\omega_{e,1}$ which ensure that the trajectories of the system (10.14) go to zero. If we assume that $\omega_{e,0} \approx 2\omega_\phi$ (where parametric resonance of (10.6) is known to occur), we can find theoretical values for the regions of stability from the approximate methods of [17].

Theorem 10.1 (Main result). *Assuming that d_{44} is not very large, the behavior of (10.14) can be categorized into three different categories, depending on the value of $\omega_{e,1}$:*

- *If $0 \leq \omega_{e,1} \leq \underline{\omega}_e$, then the origin of (10.14) is globally attractive.*
- *If $\underline{\omega}_e < \omega_{e,1} \leq \overline{\omega}_e$, then the origin of (10.14) is unstable, and there exists a high-amplitude, stable limit cycle. All trajectories of (10.14) converge to this limit cycle, with the exception of those starting in the origin.*
- *If $\omega_{e,1} > \overline{\omega}_e$, then the origin of (10.14) is locally stable, there exists a high-amplitude, stable limit cycle, and a slightly lower-amplitude, unstable limit cycle.*

$\underline{\omega}_e$ and $\overline{\omega}_e$ are the solutions to the equations

$$\sqrt{1 - \frac{d_{44}^2 \omega_e^2}{k_{\phi t}^2} - \frac{m_{44}\omega_e^2}{k_{\phi t}} \left(2\sqrt{\frac{k_{44}}{m_{44}\omega_e}} - 1 \right)} = 0 \qquad (10.15)$$

$$\sqrt{1 - \frac{d_{44}^2 \overline{\omega}_e^2}{k_{\phi t}^2} + \frac{m_{44}\overline{\omega}_e^2}{k_{\phi t}} \left(2\sqrt{\frac{k_{44}}{m_{44}\overline{\omega}_e}} - 1 \right)} = 0 . \qquad (10.16)$$

If d_{44} is very large, then all solutions to the initial value problem (10.14) go to zero.
In this theorem, asymptotic stability of limit cycles follows the definition of [16, Definition 8.1].

Proof. To simplify the analysis, we define the alternative dimensionless time scale

$$T \triangleq \frac{1}{2}\omega_{e,1}t + \bar{\alpha}_\phi \qquad (10.17)$$

giving

$$\frac{\mathrm{d}}{\mathrm{d}t} = \frac{1}{2}\omega_{e,1}\frac{\mathrm{d}}{\mathrm{d}T}$$

$$\frac{\mathrm{d}^2}{\mathrm{d}t^2} = \frac{1}{4}\omega_{e,1}^2\frac{\mathrm{d}}{\mathrm{d}T^2} .$$

Using primes to indicate derivatives with respect to T, we rewrite the system (10.14) as:

$$\phi'' + 2\imath\gamma\phi' + [\kappa + 2\imath\cos(2T)]\phi + \alpha\imath\phi^3 = 0, \qquad (10.18)$$

where

$$\iota = \frac{2k_{\phi t}}{m_{44}\omega_{e,1}^2},$$

$$\gamma = \frac{d_{44}\omega_{e,1}}{2k_{\phi t}},$$

$$\kappa = \frac{4k_{44}}{m_{44}\omega_{e,1}},$$

$$\alpha = \frac{2k_{\phi^3}}{k_{\phi t}},$$

are all positive dimensionless parameters. It is assumed that ι is small.

Equation (10.18) is known to parametrically resonate if $\kappa \approx 1$ (i.e., $\omega_{e,0} \approx 2\omega_\phi$; the encounter frequency is twice the natural roll frequency).

Using an $O(\iota)$ (big O notation) approximation to the solution of (10.18), [17] derives a solution using the method of multiple scales (see [17]) given by:

$$\phi = a\cos(T - \beta/2) + O(\iota), \tag{10.19}$$

where a and β are slowly time-varying.

Defining an alternative (also dimensionless) time scale

$$\bar{t} = \iota T \tag{10.20}$$

(which is slowly varying) and letting

$$\sqrt{\kappa} = 1 - \iota\sigma \tag{10.21}$$

(with σ representing the nearness of κ to unity, and thus the system to parametric resonance), a and β satisfy the (nonlinear homogenous ordinary) differential equations

$$\frac{\partial a}{\partial \bar{t}} = -\frac{a}{2\sqrt{\kappa}}\sin(\beta) - \gamma a \tag{10.22}$$

$$a\frac{\partial \beta}{\partial \bar{t}} = 2\sigma a - \frac{a}{\sqrt{\kappa}}\cos(\beta) - \frac{3\alpha}{4\sqrt{\kappa}}a^3. \tag{10.23}$$

The a–β system has equilibrium points (corresponding to a steady-state periodic motion of ϕ, i.e., a limit cycle) given by:

$$a = 0, \quad \beta \text{ is arbitrary} \tag{10.24}$$

Fig. 10.1 Stability regions of (10.18), theoretical

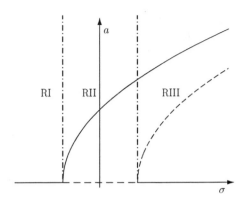

(the trivial solution) and

$$\sin(\beta) = -2\sqrt{\kappa}\gamma, \quad \cos(\beta) = 2\sigma\sqrt{\kappa} - \frac{3\alpha}{4}a^2. \tag{10.25}$$

Since $\sqrt{\kappa} = 1 - \iota\sigma$, the non-trivial steady-state solution of ϕ has the amplitude

$$a^2 = \frac{8\sigma}{3\alpha} \pm \frac{4}{3\alpha}\sqrt{1 - 4\gamma^2}, \tag{10.26}$$

where only the positive root is relevant.

If $2\gamma > 1$, then (10.26) has no real roots and only the trivial steady-state solution exists. As this is equivalent to high damping, if $2\gamma > 1$, parametric resonance will not occur. (The origin of (10.18) is then globally attractive for all $\omega_{e,1}$).

If $2\gamma \le 1$, then there is one real root of (10.26) if $2|\sigma| < \sqrt{1 - 4\gamma^2}$, and two if $2|\sigma| > \sqrt{1 - 4\gamma^2}$. The condition $2\sigma = -\sqrt{1 - 4\gamma^2}$ corresponds to (10.15) (giving $\underline{\omega}_e$) and $2\sigma = \sqrt{1 - 4\gamma^2}$ to (10.16) (giving $\overline{\omega}_e$).

Figure 10.1 illustrates the stability properties of (10.18) for the different cases. Dashed lines represent unstable equilibrium values of a for different values of σ, and solid lines stable equilibrium values.[2]

In Region RI, there is only the trivial solution. From [17], this is globally attractive.

In Region RII (where we have parametric resonance), the trivial solution is unstable, and there exists a large-amplitude steady-state solution, a limit cycle. Apart from the case where $\phi(t_2) = \dot{\phi}(t_2) = 0$, this limit cycle is globally attractive [17].

---

[2]It is worth noting that Fig. 10.1 bears strong similarity to a cross-section with the wave height kept constant of the simulation of the full 6-DOF model (9.8)–(9.9) of Sect. 9.2.2, except that in that chapter there is no evidence of the high-amplitude solutions of (10.18) in Region III. The stability regions indicated from simulating the 6-DOF model are shown in Fig. 10.2.

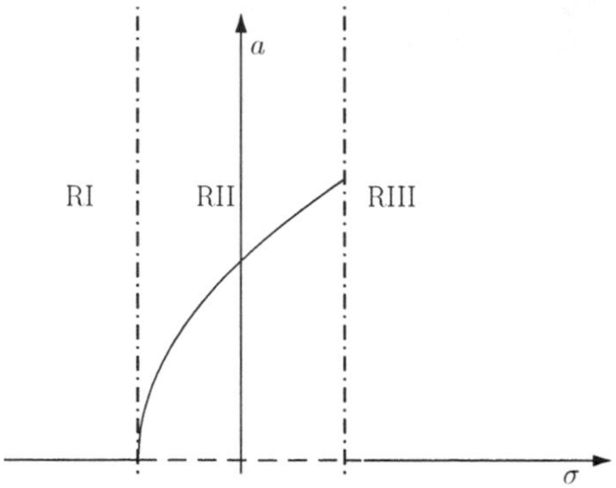

Fig. 10.2 Stability regions of
the 6-DOF model (9.8)–(9.9)
of Sect. 9.2.2, simulation

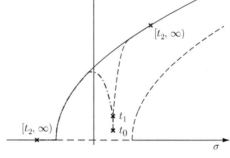

Fig. 10.3 Control of
parametric roll resonance:
Increasing versus decreasing
the encounter frequency

Region RIII has three equilibrium values, and is somewhat more complicated. The value $a = 0$ (equivalent to $\phi = 0$) is (locally) asymptotically stable. However, there exist two limit cycles, one high-amplitude and one low-amplitude. The high-amplitude one is (locally) asymptotically stable, whereas the low-amplitude one is unstable.   □

Based on the proof of Theorem 10.1, we conclude that it is possible that, if one increases ω_e so that $\omega_{e,1} \gg 2\omega_\phi$ (i.e., $\sigma \gg 0$), ϕ does not go to zero but instead to the high-amplitude limit cycle. If one instead decreases ω_e so that $\omega_{e,1} \ll 2\omega_\phi$ (i.e., $\sigma \ll 0$), ϕ will go to zero no matter how large $\phi(t_2)$ is. This is illustrated in Fig. 10.3.

This suggests that reducing the encounter frequency is the most sensible choice, and, in fact, the only option that can be guaranteed to work.

It is, however, worth noting that the analysis is based on a simplification of the ship dynamics. The high-amplitude limit cycle has not been observed in the simulations with the more physically accurate 6-DOF ship model (9.8)–(9.9) of Sect. 9.2.2. [8, 18] came to the opposite conclusion regarding speeding up versus slowing down. But bear in mind that in [8], the conclusion was largely predicated on the need to have sufficient speed for the fins (which were used in addition to speed change) to be effective.

None the less, decreasing the encounter frequency has another benefit: If we assume that σ starts at zero and slowly increases, trajectories will tend to go to a higher-amplitude limit cycle as the steady-state value of a increases with increasing σ in Region RII. However, if we instead decrease σ, trajectories will tend to go to a lower-amplitude limit cycle *even if we are still in parametric resonance*. This phenomenon has been observed in the simulations with the most accurate 6-DOF model (9.8)–(9.9) of Sect. 9.2.2, so there is reason to suspect that this holds true for real-world cases.

10.5 Simulation Results

To test the validity of the controller (10.10), we simulated the closed-loop system using both the simplified model (10.6) and the full 6-DOF model (9.8)–(9.9) of Sect. 9.2.2 in three different simulation scenarios. In all scenarios, we chose the initial conditions such that the ship was experiencing parametric roll resonance.

In accordance with the open-loop simulations in Chap. 9, a reduction of the frequency ratio to $\omega_{e,1}/\omega_\phi < 1.7$ will lead the ship out of the region where the ship is susceptible to parametric roll resonance.

We simulated three different scenarios:

1. *Slow* deceleration. The controller is turned on *after* parametric roll has already fully developed.
2. *Slow* deceleration. The controller is turned on *before* parametric roll has fully developed.
3. *Fast* deceleration. The controller is turned on *before* parametric roll has fully developed.

The simulation parameters (the same as those used in Chap. 9) are listed in Table 10.1. The control parameters are found in Tables 10.2–10.4. The simulation results are summarized in Table 10.5, and can be seen in Figs. 10.4–10.6.

Table 10.1 Simulation parameters

Quantity	Symbol	Value
Wave amplitude	ζ_0	2.5 m
Wave length	λ	281 m
Wave number	k_w	−0.0224 −
Natural roll frequency	ω_ϕ	0.343 rad/s
Modal wave frequency	ω_0	0.4684 rad/s
Simulation start time	t_0	0 s
	k_{44}	1.7533×10^9 kg m^2/s^2
Model parameters	$k_{\phi t}$	7.1373×10^8 kg m^2/s^2
(simplified roll equation)	α_ϕ	0.2741 rad
	$k_{\phi 3}$	2.2627×10^9 kg m^2/s^2

Table 10.2 Control parameters, Scenario #1

Quantity	Symbol	Value
Control action	ε	-1.7889×10^{-4} rad/s^2
Maximum deceleration	$\dot{v}_{1,max}$	0.008 m/s^2
Initial forward speed	$v_1(t_0)$	7.44 m/s
Initial encounter frequency	$\omega_{e,0}$	0.6346 rad/s
Final encounter frequency	$\omega_{e,1}$	0.5831 rad/s
Final forward speed	$v_1(t_2)$	5.14 m/s
Controller turned on	t_1	300 s
Controller turned off	t_2	588 s

Table 10.3 Control parameters, Scenario #2

Quantity	Symbol	Value
Control action	ε	-1.7889×10^{-4} rad/s^2
Maximum deceleration	$\dot{v}_{1,max}$	0.008 m/s^2
Initial forward speed	$v_1(t_0)$	7.44 m/s
Initial encounter frequency	$\omega_{e,0}$	0.6346 rad/s
Final encounter frequency	$\omega_{e,1}$	0.5831 rad/s
Final forward speed	$v_1(t_2)$	5.14 m/s
Controller turned on	t_1	93 s
Controller turned off	t_2	381 s

Table 10.4 Control parameters, Scenario #3

Quantity	Symbol	Value
Control action	ε	-3.5778×10^{-4} rad/s^2
Maximum deceleration	$\dot{v}_{1,max}$	0.016 m/s^2
Initial forward speed	$v_1(t_0)$	6.67 m/s
Initial encounter frequency	$\omega_{e,0}$	0.6174 rad/s
Final encounter frequency	$\omega_{e,1}$	0.5660 rad/s
Final forward speed	$v_1(t_2)$	4.37 m/s
Controller turned on	t_1	55 s
Controller turned off	t_2	199 s

Table 10.5 Simulation results, maximum roll angles

Scenarios	Simplified 1-DOF model			Full 6-DOF model		
	Uncontrolled	Controlled	Reduction	Uncontrolled	Controlled	Reduction
#1	25.34°	25.34°	0%	23.34°	23.34°	0%
#2	25.34°	22.40°	11.6%	23.34°	20.33°	9.0%
#3	23.57°	13.71°	41.8%	17.99°	4.83°	73.2%

Figure 10.4 shows the simulation results for the controlled system in comparison with the uncontrolled system for the first scenario. It is obvious from Fig. 10.4a that the ship is experiencing large roll amplitudes caused by parametric resonance. The frequency ratio is gradually decreased after 300 s (Fig. 10.4c), which causes the expected gradual reduction of the roll motion to zero.

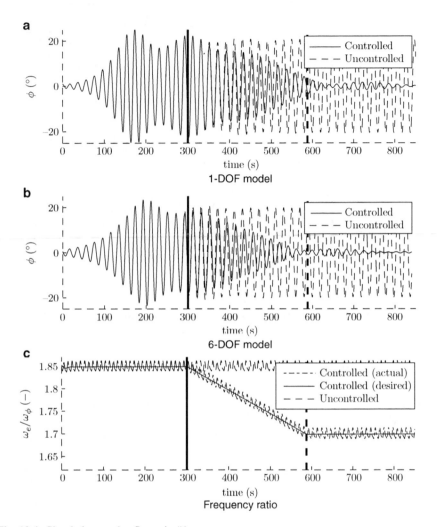

Fig. 10.4 Simulation results, Scenario #1

The simulation results with the full 6-DOF model (9.8)–(9.9) of Sect. 9.2.2 are shown in Fig. 10.4b. The controller works equally well with the more complex model.

Of course, since the controller is turned on only after parametric roll has fully developed, the maximum roll angle in Scenario #1 is the same for the controlled and uncontrolled cases. (The steady-state roll angle is zero as predicted.)

The simulation results of the second scenario are shown in Fig. 10.5. In this scenario, we reduce the encounter frequency when the roll angle is much lower than in the first scenario, early enough that parametric rolling has not yet fully developed (specifically, when the roll angle is about 5°). Figure 10.5 shows that both the simplified 1-DOF model and the full 6-DOF model behave similarly.

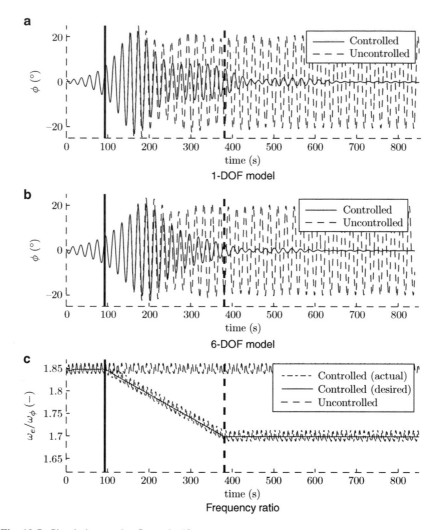

Fig. 10.5 Simulation results, Scenario #2

However, despite the controller being turned on when roll is only at 5°, the maximum roll angle is not greatly reduced compared to the uncontrolled case. This is simply because the ship is moving very slowly out of resonant condition. The steady-state roll angle is none the less zero, as predicted.

To get the ship to move out of resonant condition before the roll angle has reached dangerous levels requires, as it turned out, significantly faster deceleration than in Scenarios #1 and #2, even if the controller was turned on at a lower roll angle.

To this effect, we simulated Scenario #3. The controller is turned on early, at a time when the roll angle is about 2°. The ship is decelerating at twice the rate of Scenarios #1 and #2. Also, both the initial and final encounter frequencies are lower in Scenario #3 than in the two others. The results are plotted in Fig. 10.6.

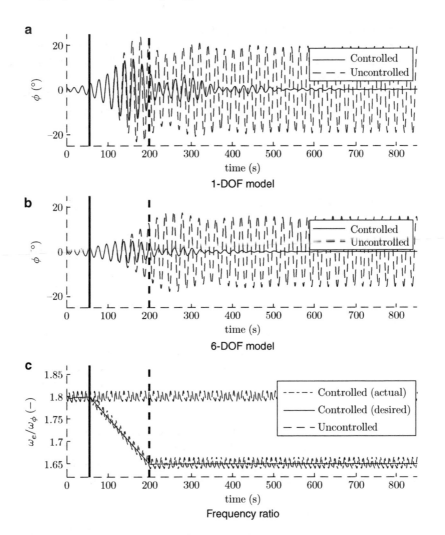

Fig. 10.6 Simulation results, Scenario #3

From Fig. 10.6, we see that the controller is capable of reducing the roll angle sufficiently fast such that the maximum roll angle is only from 1/2 (1-DOF model) to 1/4 (6-DOF model) of the maximum roll angle of the uncontrolled case. Interestingly, from Fig. 10.6 we see that the controller works significantly better for the full 6-DOF model than for the simplified 1-DOF model. In steady-state, the roll angle is zero, as expected.

From the simulation results, we see that the controller is capable of bringing the ship out of parametric resonance and – assuming sufficient deceleration capability – reduce the maximum roll angle significantly. It is also vital to turn on the controller as early as possible. The simulations confirm the theoretical derivations presented in Sect. 10.4.

How practical the controller is in a real-world scenario depends almost entirely on the ability of the captain (or the automated systems) to detect parametric resonance, and the ability of the ship to rapidly decelerate. If these capabilities are present, then the controller could prove useful. In the absence of one or both of these abilities, the practicality of the controller is limited, at least on its own. However, it might be possible to pair it with another control scheme such as fins as done in [7, 9], u-tanks (investigated on their own in Chap. 12), gyro stabilizers, or other active controllers.

10.6 Conclusions

A necessary, but not sufficient, condition for parametric resonance is that the frequency of the parametric excitation has certain values. For ships, this frequency can be changed (due to the Doppler effect) by changing the velocity. In this work, we have derived a controller for parametric roll resonance in ships that takes advantage of this and can drive the roll motion to zero. We call this frequency detuning control.

Based on the simplified 1-DOF roll model (9.39) developed in Sect. 9.5, we proposed a simple controller incorporating a linear change of the wave encounter frequency, accomplished by changing the forward speed of the ship. We showed mathematically and in simulations – using both the simplified model and the 6-DOF model (9.8)–(9.9) of Sect. 9.2.2 – that the proposed controller drives the roll motion to zero. The derived controller is so simple that it can be implemented by a helmsman, even without a speed controller on board.

However, while the controller drives the roll angle to zero, the transient behavior can be problematic. Even if the controller is turned on at a very early stage, the ship will have to be capable of a fairly rapid speed change to prevent high transient roll angles. Frequency detuning does have the advantage that it can easily be paired with direct actuation, such as the use of u-tanks, fins, or gyro stabilizers.

The effectiveness of the frequency detuning controller can be further increased by course changes in addition to speed changes to alter the encounter frequency. However, this can cause regular, directly induced roll excitation to become a problem, and was not investigated in this work.

Frequency detuning can be used for other parametrically resonating systems as long as it is possible to change the frequency of excitation. However, for most systems with the ability to change the frequency of excitation, one presumably also has the ability to change the amplitude of excitation. In that case, it is probably easier to do so. In practice, this limits the applicability of the proposed control scheme to a few systems, most notably those where the parametric resonance is induced by flow past a free-moving body.

Acknowledgements This work was funded by the Centre for Ships and Ocean Structures (CeSOS), NTNU, Norway, and the Norwegian Research Council.

Appendix

In this appendix, we prove the existence and uniqueness properties of (10.6).

From [17], we get the behavior of the system when ω_e is a constant, but not when it is changing. We need to guarantee a unique finite solution of (10.6) also for time-varying ω_e.

To prove the existence (and uniqueness) of the solution to (10.6), we will use the following theorem and lemma, repeated here for convenience:

Theorem 10.2 ([16, Theorem 3.3]). *Let $f(t,x)$ be piecewise continuous in t and locally Lipschitz in x for all $t \geq t_0$ and all x in a domain $D \subset \mathbb{R}^n$. Let W be a compact subset of D, $x_0 \in W$, and suppose it is known that every solution of*

$$\dot{x} = f(t,x), \qquad x(t_0) - x_0$$

lies entirely in W. Then there is a unique solution that is defined for all $t \geq t_0$.

Lemma 10.1 ([16, Lemma 3.2]). *If $f(t,x)$ and $\frac{\partial f}{\partial x}(t,x)$ are continuous on $[a,b] \times D$, for some domain $D \subset \mathbb{R}^n$, then f is locally Lipschitz in x on $[a,b] \times D$.*

If we take $x = [\phi, \dot{\phi}]^\top$, we can rewrite (10.6) as

$$
\dot{x} = \begin{bmatrix} x_2 \\ -\frac{d_{44}}{m_{44}} x_2 - \frac{1}{m_{44}} \left[k_{44} + k_{\phi t} \cos \left(\int_{t_0}^t \omega_e(\tau)\, d\tau + \alpha_\phi \right) \right] x_1 - \frac{k_{\phi 3}}{m_{44}} x_1^3 \end{bmatrix} = f(t,x)
$$

$$
= \begin{bmatrix} 0 & 1 \\ -\frac{k_{44}}{m_{44}} & -\frac{d_{44}}{m_{44}} \end{bmatrix} x + \begin{bmatrix} 0 \\ -\frac{k_{\phi t}}{m_{44}} \cos \left(\int_{t_0}^t \omega_e(\tau)\, d\tau + \alpha_\phi \right) x_1 - \frac{k_{\phi 3}}{m_{44}} x_1^3 \end{bmatrix}
$$

$$
= Ax + g(t,x_1) \tag{10.27}
$$

with

$$
f(t,x) \triangleq \begin{bmatrix} x_2 \\ -\frac{d_{44}}{m_{44}} x_2 - \frac{1}{m_{44}} \left[k_{44} + k_{\phi t} \cos \left(\int_{t_0}^t \omega_e(\tau)\, d\tau + \alpha_\phi \right) \right] x_1 - \frac{k_{\phi 3}}{m_{44}} x_1^3 \end{bmatrix}
$$

$$
A \triangleq \begin{bmatrix} 0 & 1 \\ -\frac{k_{44}}{m_{44}} & -\frac{d_{44}}{m_{44}} \end{bmatrix}
$$

$$
g(t,x_1) \triangleq \begin{bmatrix} 0 \\ -\frac{k_{\phi t}}{m_{44}} \cos \left(\int_{t_0}^t \omega_e(\tau)\, d\tau + \alpha_\phi \right) x_1 - \frac{k_{\phi 3}}{m_{44}} x_1^3 \end{bmatrix}.
$$

Lemma 10.2. *There is a unique solution of (10.27) (and thus (10.6)) defined for all $t \geq t_0$.*

Proof. It is clear that $f(t,x)$ of (10.27) is continuous in x for all $x \in \mathbb{R}^2$. It is also continuous in t for all $t \geq t_0$, as long as $\omega_e(t)$ is piecewise continuous. Our choice of ω_e satisfies this.

The partial derivative of f with respect to x is given by:

$$\frac{\partial f}{\partial x}(t,x) = A - \begin{bmatrix} 0 & 0 \\ \frac{k_{\phi t}}{m_{44}} \cos\left(\int_{t_0}^{t} \omega_e(\tau)\, d\tau + \alpha_\phi\right) + 3\frac{k_{\phi 3}}{m_{44}}x_1^2 & 0 \end{bmatrix} \tag{10.28}$$

which, by the same argument, is continuous in x for all $x \in \mathbb{R}^2$ and $t \geq t_0$. By [16, Lemma 3.2], f is therefore locally Lipschitz in x for all $t \geq t_0$ and all $x \in \mathbb{R}^2$. The first part of [16, Theorem 3.3] is then satisfied.

To prove that the trajectories of the system are bounded, we use the Lyapunov function candidate

$$V = \frac{1}{2}x^\top P x + \frac{1}{4}\left(1 + \frac{m_{44}}{d_{44}}\right)k_{\phi 3}x_1^4 \tag{10.29}$$

with

$$P = \begin{bmatrix} d_{44} + k_{44}\left(1 + \frac{m_{44}}{d_{44}}\right) & m_{44} \\ m_{44} & m_{44}\left(1 + \frac{m_{44}}{d_{44}}\right) \end{bmatrix} = P^\top > 0. \tag{10.30}$$

The time derivative of V along the trajectories of the system (10.27) is given by:

$$\dot{V} = x^\top P\left(Ax + g(t,x)\right) + \left(1 + \frac{m_{44}}{d_{44}}\right)k_{\phi 3}x_1^3 x_2$$

$$= -\left(k_{44} + k_{\phi t}\cos\left(\int_{t_0}^{t}\omega_e(\tau)\, d\tau + \alpha_\phi\right)\right)x_1^2 - d_{44}x_2^2 - k_{\phi 3}x_1^4$$

$$- k_{\phi t}\cos\left(\int_{t_0}^{t}\omega_e(\tau)\, d\tau + \alpha_\phi\right)\left(1 + \frac{m_{44}}{d_{44}}\right)x_1 x_2$$

$$\leq -\left(k_{44} - k_{\phi t}\right)x_1^2 - d_{44}x_2^2 - k_{\phi 3}x_1^4 + k_{\phi t}\left(1 + \frac{m_{44}}{d_{44}}\right)|x_1||x_2|. \tag{10.31}$$

While $k_{44} > k_{\phi t}$, $\dot{V}$ is only negative definite for sufficiently small values of $k_{\phi t}$. If $k_{\phi t}$ *is* sufficiently small, then the origin of the system (10.27) would be globally uniformly exponentially stable, by [16, Theorem 4.10]. A priori we know that this is not the case; in parametric resonance, the origin is, in fact, unstable.

However, V can be used to prove that the trajectories of (10.27) are bounded.

For $|x_1| \geq \mu > 0 \Rightarrow \|x\| \geq \mu$ it holds that

$$\dot{V} \leq -\left(k_{44} - k_{\phi t}\right) x_1^2 - d_{44} x_2^2 - k_{\phi 3} x_1^4 + k_{\phi t} \left(1 + \frac{m_{44}}{d_{44}}\right) |x_1||x_2|$$

$$\leq -d_{44} x_2^2 - k_{\phi 3} \mu^2 x_1^2 + k_{\phi t} \left(1 + \frac{m_{44}}{d_{44}}\right) |x_1||x_2|$$

$$= -(1-\delta) d_{44} x_2^2 - (1-\delta) k_{\phi 3} \mu^2 x_1^2$$

$$+ k_{\phi t} \left(1 + \frac{m_{44}}{d_{44}}\right) |x_1||x_2| - \delta d_{44} x_2^2 - \delta k_{\phi 3} \mu^2 x_1^2 \qquad (10.32)$$

for some $\delta \in (0,1)$. Furthermore, the term

$$k_{\phi t} \left(1 + \frac{m_{44}}{d_{44}}\right) |x_1||x_2| - \delta d_{44} x_2^2 - \delta k_{\phi 3} \mu^2 x_1^2$$

is negative semidefinite if

$$k_{\phi t}^2 \left(1 + \frac{m_{44}}{d_{44}}\right)^2 \leq 4 d_{44} \delta^2 k_{\phi 3} \mu^2 \quad \Rightarrow \quad \mu \geq \frac{1}{2\delta \sqrt{d_{44} k_{\phi 3}}} k_{\phi t} \left(1 + \frac{m_{44}}{d_{44}}\right).$$

$$\qquad (10.33)$$

Therefore, for μ satisfying the above inequality,

$$\dot{V} \leq -(1-\delta) d_{44} x_2^2 - (1-\delta) k_{\phi 3} \mu^2 x_1^2 \qquad (10.34)$$

which is negative definite. By [16, Theorem 4.18] the trajectories of (10.27) are bounded for any initial condition $x(t_0)$.

Therefore, the second condition of [16, Theorem 3.3] is satisfied, and there exists a unique solution of (10.27) (and thus (10.6)) that is defined for all $t \geq t_0$. $\qquad \square$

References

1. Breu, D.A., Fossen, T.I.: Extremum seeking speed and heading control applied to parametric resonance. In: Proceeding of IFAC Conference on Control Applications in Marine Systems. Rostock, Germany (2010)
2. Carmel, S.M.: Study of parametric rolling event on a panamax container vessel. Journal of the Transportation Research Board **1963**, 56–63 (2006)
3. Fossen, T.I.: Handbook of Marine Craft Hydrodynamics and Motion Control. John Wiley & Sons, Ltd., Chichester, UK (2011)

4. France, W.N., Levadou, M., Treakle, T.W., Paulling, J.R., Michel, R.K., Moore, C.: An investigation of head-sea parametric rolling and its influence on container lashing systems. SNAME Annual Meeting, Orlando, Florida, USA (2001)
5. Francescutto, A.: An experimental investigation of parametric rolling in head waves. Journal of Offshore Mechanics and Arctic Engineering **123**, 65–69 (2001)
6. Francescutto, A., Bulian, G.: Nonlinear and stochastic aspects of parametric rolling modelling. In: Proceedings of 6th International Ship Stability Workshop. New York, USA (2002)
7. Galeazzi, R.: Autonomous supervision and control of parametric roll resonance. PhD thesis, Technical University of Denmark (2009)
8. Galeazzi, R., Blanke, M.: On the feasibility of stabilizing parametric roll with active bifurcation control. In: Proceedings of IFAC Conference on Control Applications in Marine Systems. Bol, Croatia (2007)
9. Galeazzi, R., Holden, C., Blanke, M., Fossen, T.I.: Stabilization of parametric roll resonance by combined speed and fin stabilizer control. In: Proceedings of European Control Conference. Budapest, Hungary (2009)
10. Galeazzi, R., Vidic-Perunovic, J., Blanke, M., Jensen, J.J.: Stability Analysis of the Parametric Roll Resonance under Non-constant Ship Speed In: Proceedings of ESDA2008, 9th Biennial ASME Conference on Engineering Systems Design and Analysis. Haifa, Israel (2008)
11. Holden, C., Galeazzi, R., Fossen, T.I., Perez, T.: Stabilization of parametric roll resonance with active u-tanks via lyapunov control design. In: Proceedings of European Control Conference. Budapest, Hungary (2009)
12. Holden, C., Galeazzi, R., Rodríguez, C.A., Perez, T., Fossen, T.I., Blanke, M., Neves, M.A.S.: Nonlinear container ship model for the study of parametric roll resonance. Modeling, Identification and Control **28**(4) (2007)
13. Holden, C., Perez, T., Fossen, T.I.: A lagrangian approach to nonlinear modeling of anti-roll tanks. Ocean Engineering **38**(2–3), 341–359 (2011)
14. Jensen, J.J., Pedersen, P.T., Vidic-Perunovic, J.: Estimation of parametric roll in stachastic seaway. In: Proceedings of IUTAM Symposium on Fluid-Structure Interaction in Ocean Engineering. Hamburg, Germany (2007)
15. Jensen, J.J., Vidic-Perunovic, J., Pedersen, P.T.: Influence of surge motion on the probability of parametric roll in a stationary sea state. In: Proceedings of 10th International Ship Stability Workshop. Hamburg, Germany (2007)
16. Khalil, H.: Nonlinear Systems, 3rd edn. Prentice Hall, Upper Saddle, New Jersey, USA (2002)
17. Nayfeh, A.H., Mook, D.T.: Nonlinear Oscillations. John Wiley & Sons, Inc., Weinheim, Germany (1995)
18. Ribeiro e Silva, S., Santos, T.A., Soares, C.G.: Parametrically excited roll in regular and irregular head seas. International Shipbuilding Progress **51**, 29–56 (2005)
19. Umeda, N., Hashimoto, H., Vassalos, D., Urano, S., Okou, K.: An investigation of different methods for the prevention of parametric rolling. Journal of Marine Science and Technology **13**, 16–23 (2008)

Chapter 11
Optimal Speed and Heading Control for Stabilization of Parametric Oscillations in Ships

Dominik A. Breu, Le Feng, and Thor I. Fossen

11.1 Introduction

Recently, several active control strategies for the stabilization of parametric roll oscillations in ships have been proposed [2,7–10,14,15,17,18,27,29]. The potential of violating one of the conditions for the onset of parametric roll resonance (see [6]) has been effectively shown in [2, 8–10, 17, 18, 27]. Those control strategies, called *frequency detuning control* in Chap. 10, are designed to change the frequency of the parametric excitation for instance via the *Doppler-shift* of the encounter frequency – that is, the frequency of the waves as seen from the ship. The Doppler-shift can be achieved by variations of the ship's speed and heading angle.

Whereas the effectiveness of frequency detuning control to stabilize parametrically excited roll oscillations in ships has been reported, research on how to change the encounter frequency with respect to optimality has been conducted only recently [2]. Since changes of the ship's speed and heading angle result in a shift of the encounter frequency, optimal control methodologies can be used to determine the optimal encounter frequency and the optimal setpoints for the ship's speed and heading angle.

In this work, two optimal control methods for the stabilization of parametric roll resonance are proposed. Based on the results in Breu and Fossen [2], the extremum

D.A. Breu (✉) • L. Feng
Centre for Ships and Ocean Structures, Norwegian University of Science and Technology,
NO-7491 Trondheim, Norway
e-mail: breu@itk.ntnu.no; feng.le@ntnu.no

T.I. Fossen
Department of Engineering Cybernetics, Norwegian University of Science and Technology,
NO-7491 Trondheim, Norway

Centre for Ships and Ocean Structures, Norwegian University of Science and Technology,
NO-7491 Trondheim, Norway
e-mail: fossen@ieee.org

T.I. Fossen and H. Nijmeijer (eds.), *Parametric Resonance in Dynamical Systems*, 213
DOI 10.1007/978-1-4614-1043-0_11, © Springer Science+Business Media, LLC 2012

seeking (ES) methodology is adapted to iteratively determine the optimal setpoint of the encounter frequency. The mapping of the encounter frequency to the two controllable states, the ship's forward speed and heading angle, is a constrained optimization problem which can be posed in a two-step sequential least-squares formulation. By defining an appropriate objective function and designing globally exponential stable speed and heading controllers, it is shown that the proposed ES controller is able to stabilize the roll oscillations caused by parametric excitation effectively.

As a second approach, the application of a model predictive controller (MPC) to ships experiencing parametric roll resonance is proposed. Constraints on inputs and states as well as an objective function aiming to violate one of the empirical conditions for the onset of parametric roll resonance are formulated in the MPC framework. It is illustrated in simulations that the proposed MPC approach is apt to be used for the stabilization of parametrically excited roll oscillations.

11.2   3-DOF Ship Model

A simplified 3-DOF model of the ship describing the coupled motions in *surge*, *roll*, and *yaw* is used to represent the ship dynamics.

11.2.1   Reference Frames

Two reference frames are considered; a geographical reference frame fixed to the ocean surface, and a reference frame fixed to the vessel (body frame). Figure 11.1 depicts the two reference frames in the horizontal plane, that is, the z-axis is not shown in Fig. 11.1.

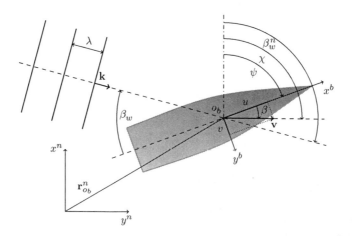

Fig. 11.1 Horizontal plane, reference frames, and angle definitions

The reference frame fixed to the vessel is moving with the vessel and it has its origin at a location o_b midships. The body axes x^b, y^b, and z^b coincide with the principle axes of inertia, as defined in [5]. The vectors decomposed in the body frame are denoted in boldface with a superscript b.

We make the following assumption:

Assumption 11.1. *The geographical reference frame fixed to the ocean surface is inertial.*

The ocean surface reference frame is defined by the *North-East-Down* coordinate system in [5], with the axes accordingly. Boldface and a superscript n denote the vectors expressed in the inertial frame.

11.2.2 Ship Dynamics

The generalized velocity vector expressed in the body frame is denoted by the vector $\mathbf{v} = [u,v,w,p,q,r]^{\top}$. The generalized external forces $\boldsymbol{\tau}_{\mathrm{RB}} = [X,Y,Z,K,M,N]^{\top}$, expressed in the body frame, are the sum of the generalized environmental forces $\boldsymbol{\tau}_{\mathrm{env}}$ and the generalized control forces $\boldsymbol{\tau}$, that is, $\boldsymbol{\tau}_{\mathrm{RB}} = \boldsymbol{\tau}_{\mathrm{env}} + \boldsymbol{\tau}$. The generalized position (position and attitude) vector is denoted as $\boldsymbol{\eta} = [N,E,D,\phi,\theta,\psi]^{\top}$, where the position vector $[N,E,D]^{\top}$ is expressed in the inertial frame and the elements of the attitude vector $[\phi,\theta,\psi]^{\top}$ are the Euler angles.

The relationship between the generalized position and the velocities satisfies [5]

$$\dot{\boldsymbol{\eta}} = \mathbf{J}(\boldsymbol{\eta})\mathbf{v}, \tag{11.1}$$

where $\mathbf{J}(\boldsymbol{\eta})$ is the transformation matrix consisting of the linear and angular velocity transformation matrices as defined by Fossen [5].

The rigid-body kinetics are given by:

$$\mathbf{M}_{\mathrm{RB}}\dot{\mathbf{v}} + \mathbf{C}_{\mathrm{RB}}(\mathbf{v})\mathbf{v} = \boldsymbol{\tau}_{\mathrm{RB}}, \tag{11.2}$$

where $\mathbf{M}_{\mathrm{RB}}$ is the rigid-body inertia matrix, satisfying $\mathbf{M}_{\mathrm{RB}} = \mathbf{M}_{\mathrm{RB}}^{\top} > 0$ and $\dot{\mathbf{M}}_{\mathrm{RB}} = \mathbf{0}$. The rigid-body Coriolis and centripetal matrix $\mathbf{C}_{\mathrm{RB}}(\mathbf{v}) = -\mathbf{C}_{\mathrm{RB}}^{\top}(\mathbf{v})$ is due to the rotation of the body frame about the inertial frame. By the superscript $\{1,4,6\}$, we indicate that only the motions in surge, roll, and yaw – the first, fourth, and sixth rows and columns of the 6-DOF model are considered.

Assumption 11.2. *The mass is distributed homogeneously and the ship has xz-plane symmetry.*

Since the origin of the body frame is in the centerline of the ship and the body axes coincide with the principle axes of inertia, the rigid-body inertia matrix takes the following form:

$$\mathbf{M}_{RB}^{\{1,4,6\}} = \begin{bmatrix} m & 0 & 0 \\ 0 & I_x & 0 \\ 0 & 0 & I_z \end{bmatrix}, \qquad (11.3)$$

where m denotes the ship mass, whereas I_x and I_z are the moments of inertia about the x_b- and the z_b-axis, respectively.

The rigid-body Coriolis and centripetal matrix can expressed by:

$$\mathbf{C}_{RB}^{1,4,6}\left(\mathbf{v}^{\{1,4,6\}}\right) = \begin{bmatrix} 0 & mz_g r & -mx_g r \\ -mz_g r & 0 & 0 \\ mx_g r & 0 & 0 \end{bmatrix} = 0, \qquad (11.4)$$

where $\mathbf{r}_g^b = [x_g, y_g, z_g]^\top$ denotes the vector from the body origin to the center of gravity (CG) of the ship, expressed in the body frame. Next the following assumptions are made.

Assumption 11.3. *The CG and the origin of the body frame coincide, that is,* $\mathbf{r}_g^b = 0$.

Assumption 11.4. *For a maneuvering ship in a seaway, the surge and yaw motions are approximated by the zero-frequency potential coefficients while added mass and damping in roll is approximated at the natural roll frequency ω_ϕ. Furthermore, the fluid memory effects are neglected.*

By Assumption 11.4, it follows that

$$\mathbf{M}_A^{\{1,4,6\}} = \begin{bmatrix} A_{11}(0) & 0 & 0 \\ 0 & A_{44}(\omega_\phi) & 0 \\ 0 & 0 & A_{66}(0) \end{bmatrix}. \qquad (11.5)$$

From (11.5) it follows that the hydrodynamic Coriolis and centripetal matrix is

$$\mathbf{C}_A^{\{1,4,6\}}\left(\mathbf{v}^{\{1,4,6\}}\right) = 0. \qquad (11.6)$$

The linear damping is the sum of the potential and viscous damping and becomes

$$\mathbf{D}_l^{\{1,4,6\}} = \mathbf{D}_p^{\{1,4,6\}} + \mathbf{D}_V^{\{1,4,6\}} := -\begin{bmatrix} X_u & 0 & 0 \\ 0 & K_p & 0 \\ 0 & 0 & N_r \end{bmatrix}, \qquad (11.7)$$

where the viscous damping matrix $\mathbf{D}_V^{\{1,4,6\}}$ is approximated by a diagonal matrix and the couplings between the roll and yaw motions are neglected. The nonlinear damping is

$$\mathbf{D}_n^{\{1,4,6\}}\left(\mathbf{v}^{\{1,4,6\}}\right) := -\begin{bmatrix} X_{|u|u}|u| & 0 & 0 \\ 0 & K_{|p|p}|p| & 0 \\ 0 & 0 & N_{|r|r}|r| \end{bmatrix}. \tag{11.8}$$

The quadratic damping coefficient in surge may be modeled as, see Fossen [5],

$$X_{|u|u} = -\frac{1}{2}\rho S(1+k_{\mathrm{f}})\frac{0.075}{\left(\log_{10}R_n - 2\right)^2}, \qquad R_n = \frac{uL_{\mathrm{pp}}}{v_k}.$$

The water density is denoted by ρ, the wetted surface of the hull by S, and the form factor yielding a viscous correction by k_{f}. The Reynolds number R_n depends on the length between perpendiculars L_{pp} and the kinematic viscosity of the fluid v_k.

Motivated by the results presented in Galeazzi et al. [9], Neves and Rodríguez [24], Shin et al. [28] and based on the model introduced in Chap. 9, the restoring forces for the surface vessel are approximated as:

$$\mathbf{g}^{\{1,4,6\}}\left(\boldsymbol{\eta}^{\{1,4,6\}}\right) \approx \begin{bmatrix} 0 \\ \rho g \nabla \overline{\mathrm{GM}}_T \phi - K_{\phi 3}\phi^3 \\ 0 \end{bmatrix}, \tag{11.9}$$

where g is the acceleration of gravity, ∇ the displaced water volume, and $\overline{\mathrm{GM}}_T$ the transverse metacentric height given by:

$$\overline{\mathrm{GM}}_T = \overline{\mathrm{GM}}_m + \overline{\mathrm{GM}}_a \cos\left(\int_0^t \omega_e(\tau)\,d\tau\right). \tag{11.10}$$

Here, $\overline{\mathrm{GM}}_m$ is the mean metacentric height, $\overline{\mathrm{GM}}_a$ the amplitude of the metacentric height change in waves, and ω_e the encounter frequency. This model takes into account velocity changes since ω_e is allowed to vary with time.

The ship dynamics can be written as:

$$\dot{\boldsymbol{\eta}}^{\{1,4,6\}} = \mathbf{J}^{\{1,4,6\}}\left(\boldsymbol{\eta}^{\{1,4,6\}}\right)\mathbf{v}^{\{1,4,6\}} \tag{11.11}$$

$$\mathbf{M}\dot{\mathbf{v}}^{\{1,4,6\}} + \mathbf{C}\left(\mathbf{v}^{\{1,4,6\}}\right)\mathbf{v}^{\{1,4,6\}} + \mathbf{D}\left(\mathbf{v}^{\{1,4,6\}}\right)\mathbf{v}^{\{1,4,6\}} + \mathbf{g}\left(\boldsymbol{\eta}^{\{1,4,6\}}\right) = \boldsymbol{\tau}^{\{1,4,6\}} \tag{11.12}$$

where

$$\mathbf{J}^{\{1,4,6\}}\left(\boldsymbol{\eta}^{\{1,4,6\}}\right) = \begin{bmatrix} \cos(\psi) & 0 & 0 \\ 0 & 1 & 0 \\ 0 & 0 & \cos(\phi) \end{bmatrix}$$

$$\mathbf{M} = \mathbf{M}_{RB}^{\{1,4,6\}} + \mathbf{M}_A^{\{1,4,6\}}$$

$$= \begin{bmatrix} m + A_{11}(0) & 0 & 0 \\ 0 & I_x + A_{44}(\omega_\phi) & 0 \\ 0 & 0 & I_z + A_{66}(0) \end{bmatrix}$$

$$\mathbf{C}\left(\mathbf{v}^{\{1,4,6\}}\right) = \mathbf{C}_{RB}^{\{1,4,6\}}\left(\mathbf{v}^{\{1,4,6\}}\right) + \mathbf{C}_A^{\{1,4,6\}}\left(\mathbf{v}^{\{1,4,6\}}\right) = \mathbf{0}$$

$$\mathbf{D}\left(\mathbf{v}^{\{1,4,6\}}\right) = \mathbf{D}_l^{\{1,4,6\}} + \mathbf{D}_n^{\{1,4,6\}}\left(\mathbf{v}^{\{1,4,6\}}\right)$$

$$= -\begin{bmatrix} X_u + X_{|u|u}|u| & 0 & 0 \\ 0 & K_p + K_{|p|p}|p| & 0 \\ 0 & 0 & N_r + N_{|r|r}|r| \end{bmatrix}$$

$$\mathbf{g}\left(\boldsymbol{\eta}^{\{1,4,6\}}\right) = \mathbf{g}^{\{1,4,6\}}\left(\boldsymbol{\eta}^{\{1,4,6\}}\right)$$

$$= \begin{bmatrix} 0 \\ \rho g \nabla \left[\overline{\mathrm{GM}}_m + \overline{\mathrm{GM}}_a \cos\left(\int_0^t \omega_e(\tau)d\tau\right)\right]\phi - K_{\phi 3}\phi^3 \\ 0 \end{bmatrix}.$$

For simplicity, it is assumed that the ship is controlled by a single rudder such that

$$\tau^{\{6\}} = -N_\delta \delta, \tag{11.13}$$

where δ denotes the rudder angle. Then, the yaw subsystem can be approximated by a first-order *Nomoto model* with time and gain constants T and K, respectively (Fossen [5]):

$$T\dot{r} + r = K\delta. \tag{11.14}$$

The Nomoto constants T and K can be related to the hydrodynamic ship coefficients such as the acceleration derivatives and the velocity derivatives. These coefficients may be approximated by considering the geometrical dimensions of the ship, that is, the length between the perpendiculars and the draft of the ship, as stated by Clarke et al. [4].

The control inputs to the 3-DOF ship model (11.11) and (11.12) are the control forces in roll τ_ϕ and surge τ_u, as well as the rudder deflection δ. Measured outputs of the system are the roll angle ϕ, the roll rate $\dot{\phi}$, the surge speed u, the heading angle ψ and the heading rate $\dot{\psi}$.

11.2.3 Encounter Frequency Model

In this chapter, the following assumption is made:

Assumption 11.5. *The waves are planar and regular sinusoidal.*

Consequently, the waves can be described by:

$$\zeta\left(\mathbf{r},t\right) = \bar{\zeta}\cos\left(\omega_0 t - \mathbf{k}^\top \mathbf{r}^n + \phi_\zeta\right), \tag{11.15}$$

where $\zeta\left(\mathbf{r},t\right)$ is the sea surface elevation at a location $\mathbf{r}^n$ at a time t. The vector $\mathbf{r}^n$ is expressed in the inertial frame. The amplitude of the sinusoid is $\bar{\zeta}$, the modal wave frequency ω_0, and the initial phase shift ϕ_ζ. The wave vector $\mathbf{k}$ implicitly defines the wave number k:

$$\mathbf{k} = k\mathbf{e}$$

where $\mathbf{e}$ is the propagation vector, satisfying $\|\mathbf{e}\| = 1$. The wave length for a planar wave is

$$\lambda = \frac{2\pi}{\|\mathbf{k}\|} = \frac{2\pi}{k} \tag{11.16}$$

and the phase velocity is

$$c = \frac{\omega_0}{\|\mathbf{k}\|} = \frac{\lambda}{T_w} \tag{11.17}$$

with T_w the period. We assume that $\bar{\zeta}$, ω_0, and $\mathbf{k}$ are constants for simplicity.

To an observer at a fixed location in the inertial reference frame, the frequency at which the waves encounter the ship equals the modal wave frequency. This is however not true when the observer is moving with the ship at a nonzero velocity. A moving ship causes a shift in the peak frequency of the wave spectrum which can be accounted for by introducing the encounter frequency.

From Fig. 11.1 it is evident that the encounter angle expressed in the inertial frame is given by:

$$\beta_w^n = \beta_w + \psi.$$

We assume without loss of generality that β_w^n is constant; that is, the waves are always coming from the same compass direction. The horizontal velocity of the ship $\mathbf{v} = \mathbf{v}^{\{1,2\}}$ is expressed in the body frame and can readily be expressed in the inertial frame:

$$\mathbf{v}^n = \mathbf{R}\left(\psi\right)\mathbf{v}, \tag{11.18}$$

where $\mathbf{R}\left(\psi\right) \in SO\left(2\right)$ is the rotation matrix about ψ. It satisfies $\mathbf{R}\left(\psi\right)\mathbf{R}^\top\left(\psi\right) = \mathbf{R}^\top\left(\psi\right)\mathbf{R}\left(\psi\right) = \mathbf{I}$ and $\det\mathbf{R}\left(\psi\right) = 1$, that is, it is orthogonal. The rotation matrix is given by:

$$\mathbf{R}\left(\psi\right) = \begin{bmatrix} \cos\left(\psi\right) & -\sin\left(\psi\right) \\ \sin\left(\psi\right) & \cos\left(\psi\right) \end{bmatrix}.$$

The peak frequency shift of the wave spectrum is due to the Doppler shift. The projection of the ship velocity $\mathbf{v}^n$ on the wave vector $\mathbf{k}$ is

$$\mathbf{v_p} = v_p \mathbf{e} = \|\mathbf{v}^n\| \cos(\beta_w^n - \chi)\mathbf{e}, \qquad (11.19)$$

where the course angle χ is the sum of the heading angle ψ and the sideslip angle β. The encounter frequency, that is the frequency of oscillation of the waves as it appears to an observer on the ship, can then be calculated by considering the Doppler shift and by combining (11.18) and (11.19):

$$\omega_e = \omega_0 \left(1 - v_p/c\right)$$

$$= \omega_0 \left(1 - \frac{k}{\omega_0} \|\mathbf{R}(\psi)\mathbf{v}\| \cos(\beta_w^n - \chi)\right)$$

$$= \omega_0 - k\sqrt{u^2 + v^2}\cos(\beta_w^n - \psi - \beta). \qquad (11.20)$$

Under the assumption of deep water ($h \geq \lambda/2$), where h is the water depth, the dispersion relationship holds:

$$k = \frac{\omega_0^2}{g}. \qquad (11.21)$$

To decouple the surge model from the sway–yaw subsystem, it is assumed that the forward speed of the vessel is slowly time-varying only, which implies:

$$\|\mathbf{v}\| = \sqrt{u^2 + v^2} \approx u \qquad (11.22)$$

and that there is no ocean current present. From Fig. 11.1 it is apparent that the sideslip angle β is

$$\beta = \arcsin\left(\frac{v}{\|\mathbf{v}\|}\right) \approx \frac{v}{\|\mathbf{v}\|} \qquad (11.23)$$

when β is small. Since the sway component of the ship velocity is neglected, the sideslip angle is disregarded as well.

Hence, the encounter frequency can be expressed by:

$$\omega_e(u, \psi, \omega_0, \beta_w^n) = \omega_0 - \frac{\omega_0^2}{g}u\cos(\beta_w^n - \psi). \qquad (11.24)$$

Notice that the encounter frequency (11.24) couples the roll dynamics to the surge and yaw dynamics in (11.12), respectively.

11.3 Extremum Seeking Control

Extremum seeking control is a real-time optimization methodology, popular in both research and industry. It is characterized by the online tuning of the a priori unknown setpoint of a system to achieve an optimal output, with respect to an objective functional for example. ES is not model based – it is applicable also when the model is not perfectly known. In the ES methodology, a perturbation signal is added to the system to find an estimate of the gradient of the objective signal. In Ariyur and Krstić [1], a thorough introduction to ES control, including many applications in various research areas, is presented.

11.3.1 Extremum Seeking Applied to Ships in Parametric Roll Resonance

In this section, the ES method is adapted to the regulation of parametrically excited roll motions in ships as depicted in Fig. 11.2. The proposed, overall scheme consists of the ship – the surge and the yaw motions are coupled to the roll motion by the encounter frequency (11.24) – the control block and a dynamic feedback loop (ES loop).

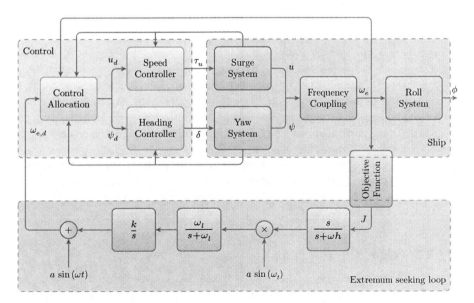

Fig. 11.2 Extremum seeking control applied to ships in parametric roll resonance

The output J of the objective function has an extremum, that is a minimum or a maximum, at $\omega_{e,d}^*$. The ES loop adds a slow perturbation to the best current estimate of $\omega_{e,d}$ in order to iteratively and online tune the parameter $\omega_{e,d}$ to its optimal value $\omega_{e,d}^*$.

By assuming the perturbation signal to be sufficiently slow compared to the open-loop dynamics, the system can be viewed as a static map and its dynamics can be neglected for the ES loop. The high-pass filter $s/(s+\omega_h)$ serves to eliminate the offset of the cost signal J and the second perturbation creates a sinusoidal response of J. Adding a sinusoidal perturbation signal to the best estimate of $\omega_{e,d}$ causes the two sinusoids to be in phase or out of phase depending on whether the best estimate is smaller or larger than its optimal value $\omega_{e,d}^*$. Whereas the low-pass filter $\omega_l/(s+\omega_l)$ is used to extract the offset caused by multiplying the two sinusoids, the integrator in the ES loop gives the approximate gradient update law. ([1])

The proposed ES control schemes requires three time scales in the overall system. Since it is assumed that the map from the reference to the output of the objective function is a static map, the time constant of the plant needs to be the fastest. Furthermore, the perturbation signal must be sufficiently slow compared to the plant, or it would not be fed through the plant properly. The filters give an estimate of the gradient update law, implying that their time constants are required to be the slower than those of both the plant and the perturbation signal.

The encounter frequency (11.24) depends among others on the ship's forward speed and heading angle which are controllable. The best estimate of the optimal encounter frequency is therefore mapped to the desired surge speed u_d and the desired heading angle ψ_d by a (nonlinear) control allocation as depicted by the control block in Fig. 11.2. Speed and heading controllers are then used to compute the required control force in surge τ_u and the rudder deflection δ.

It is noteworthy that the optimal setpoint of the encounter frequency – and as a matter of fact, the ship's speed and heading angle – is a priori not known. There lies the power of the ES control which iteratively tunes the encounter frequency, as the parameter of the feedback loop, to its optimal value which minimizes a defined objective functional. Hence, the vital role of the choice of the objective function is apparent for the performance of the ES control applied to parametric roll resonance.

Treating the encounter frequency as the sole parameter in the proposed ES control scheme seems advantageous compared to the formulation as a multiparameter ES method. In particular, control allocation allows to take into account restrictions on the ship's speed and heading angle as well as on their variation rate.

11.3.2 Objective Function

The objective function is one of the key factors with respect to the performance of the proposed ES control method applied to ships in parametric roll resonance, as depicted in Fig. 11.2. Its choice determines the ability to regulate the roll motion as well as it accounts for mission dependent restrictions.

It is well known that certain ships are prone to experience parametric roll resonance when the encounter frequency is close to double the natural roll frequency of the ship, that is, see Nayfeh and Mook [23]:

$$\omega_e \approx 2\omega_\phi. \tag{11.25}$$

The objective function is constructed as the weighted superposition of two cost functionals, accounting for the frequency condition (11.25) and the deviation from the nominal cruise condition, respectively:

$$J = w_1 J_1 + w_2 J_2, \tag{11.26}$$

where w_1 and w_2 are the weights. The two cost functionals are expressed by:

$$J_1 = c_1 e^{-c_2 \left(\omega_e - 2\omega_\phi\right)^2} \tag{11.27}$$

$$J_2 - c_3 \left(\omega_e - \omega_{e,0}\right)^2, \tag{11.28}$$

where $c_i > 0, i \in \{1,2,3\}$ are constants. Equation (11.27) represents the penalty of the ship not violating the frequency condition (11.25). Equation (11.28), on the other hand, penalizes the deviation of the ship from its nominal cruise condition expressed by the nominal encounter frequency $\omega_{e,0}$, that is, the encounter frequency (11.24) with the nominal setpoints for the ship's surge speed u_0 and heading angle ψ_0. By the choice of the constant parameters $c_i, i \in \{1,2,3\}$ in (11.27) and (11.28), the shape of the cost functionals can be adjusted.

It is apparent from the definition of the objective function (11.26) that in order to avoid parametric roll resonance the objective has to be minimized. Thus, the ES loop is designed such that its parameter $\omega_{e,d}$ is iteratively tuned to the optimal value $\omega_{e,d}^*$, resulting in a minimum of the objective function.

11.3.3 Control Allocation

The control allocation block depicted in Fig. 11.2 maps the parameter of the ES loop $\omega_{e,d}$ – the desired encounter frequency – to the desired trajectory of the control variables, that is the ship's desired surge speed u_d and heading angle ψ_d. Revisiting (11.24), the desired encounter frequency is approximated by a first-order Taylor expansion, taking into account small variations of the ship's forward speed and heading angle:

$$\omega_{e,d}\left(u + \Delta u, \psi + \Delta \psi, \cdot\right) = \omega_0 - \frac{\omega_0^2}{g}\cos\left(\beta_w^n - \psi\right)u - \frac{\omega_0^2}{g}\cos\left(\beta_w^n - \psi\right)\Delta u$$

$$- \frac{\omega_0^2}{g}\sin\left(\beta_w^n - \psi\right)u\Delta \psi. \tag{11.29}$$

Here, it is assumed that the desired encounter frequency can be achieved by a deviation of Δu and $\Delta \psi$ from the ship's forward speed and heading angle, respectively. Equation (11.29) suggest that the virtual control input can be chosen as:

$$
\begin{aligned}
\tau_v &= -\frac{g}{\omega_0^2} \left[\omega_{e,d} \left(u + \Delta u, \psi + \Delta \psi, \cdot \right) - \left(\omega_0 - \frac{\omega_0^2}{g} \cos \left(\beta_w^n - \psi \right) u \right) \right] \\
&= -\frac{g}{\omega_0^2} \left[\omega_{e,d} \left(u + \Delta u, \psi + \Delta \psi, \cdot \right) - \omega_e \left(u, \psi, \cdot \right) \right]
\end{aligned}
\tag{11.30}
$$

The relation between the virtual control input (11.30) and the variations in surge speed and heading angle can be expressed by the constrained linear mapping

$$
\tau_v = \mathbf{B} \left(u, \psi, \beta_w^n \right) \boldsymbol{\zeta},
\tag{11.31}
$$

$$
\boldsymbol{\zeta}_{\min} \leq \boldsymbol{\zeta} \leq \boldsymbol{\zeta}_{\max}
\tag{11.32}
$$

where the control effectiveness matrix $\mathbf{B} \left(u, \psi, \beta_w^n \right)$ and the variation vector $\boldsymbol{\zeta}$ are given by:

$$
\mathbf{B} \left(u, \psi, \beta_w^n \right) = \begin{bmatrix} \cos \left(\beta_w^n - \psi \right) \\ \sin \left(\beta_w^n - \psi \right) u \end{bmatrix}, \quad \boldsymbol{\zeta} = \begin{bmatrix} \Delta u \\ \Delta \psi \end{bmatrix}.
\tag{11.33}
$$

The constraints are expressed in (11.32) where $\boldsymbol{\zeta}_{\min}$ and $\boldsymbol{\zeta}_{\max}$ denote the lower and upper bounds on $\boldsymbol{\zeta}$, respectively. The desired ship's surge speed and heading angle are then merely

$$
u_d = u + \Delta u
\tag{11.34}
$$

$$
\psi_d = \psi + \Delta \psi.
\tag{11.35}
$$

According to [11, 30], the control allocation problem (11.31) and (11.32) can be split up into a two-step sequential least-squares problem to find the variation of the ship's forward speed and heading angle:

$$
S = \arg \min_{\boldsymbol{\zeta}_{\min} \leq \boldsymbol{\zeta} \leq \boldsymbol{\zeta}_{\max}} \left\| \mathbf{W}_{\tau_v} \left(\mathbf{B} \left(u, \psi, \beta_w^n \right) \boldsymbol{\zeta} - \tau_v \right) \right\|
\tag{11.36}
$$

$$
\boldsymbol{\zeta}_{\text{opt}} = \arg \min_{\boldsymbol{\zeta} \in S} \left\| \mathbf{W}_{\boldsymbol{\zeta}} \left(\boldsymbol{\zeta} - \boldsymbol{\zeta}_d \right) \right\|,
\tag{11.37}
$$

where $\mathbf{W}_{\tau_v}$ and $\mathbf{W}_{\boldsymbol{\zeta}}$ are weight matrices. First, the set of feasible solutions S that minimize $\mathbf{B} \left(u, \psi, \beta_w^n \right) \boldsymbol{\zeta} - \tau_v$ is computed. Then, the best solution – the solution which minimizes $\mathbf{W}_{\boldsymbol{\zeta}} \left(\boldsymbol{\zeta} - \boldsymbol{\zeta}_d \right)$ – is determined. $\boldsymbol{\zeta}_d$ is the vector of desired variations in ship's surge speed and heading angle and is presumably null.

The sequential least-squares problem (11.36) and (11.37) is solved in *Matlab* using the *Quadratic Programming Control Allocation Toolbox* (QCAT) (see Härkegård [12]).

11.3.4   Speed and Heading Controllers

The speed and heading controllers determine the appropriate control force in surge and the rudder deflection from the desired surge speed and heading angle, respectively, see Fig. 11.2.

By assumption, the inner control loop, consisting of the ship and the control block in Fig. 11.2, is designed such that its dynamics can be neglected for the ES loop. Thus, the inner control loop needs to be considerably faster than the overall closed-loop system, yielding that the controllers are required to be fast in comparison to the perturbation signal and the filters of the ES loop.

Speed Controller

The surge dynamics is given by the first row in (11.12). Assuming, that the mass and the damping terms are perfectly known, the speed controller can be designed by using feedback linearization:

$$\tau_u = (m + A_{11}(0)) v_u + \left(-X_u - X_{|u|u}|u| \right) u. \tag{11.38}$$

By taking the virtual control input v_u as an ordinary proportional controller, the closed-loop surge dynamics becomes

$$\dot{u} = v_u = -k_{u,p}(u - u_d), \quad k_{u,p} > 0 \tag{11.39}$$

where $k_{u,p}$ is the controller gain, chosen such that the error dynamics is globally exponentially stable (GES); see Fossen [5] or Khalil [20].

Heading Controller

The yaw dynamics is represented by the first-order Nomoto model (11.14). To design the heading controller, is is assumed that $\dot{\psi} \approx r$ and that the rudder deflection δ is the control input:

$$\delta = -k_{\psi,p}(\psi - \psi_d) - k_{\psi,d}(\dot{\psi} - \dot{\psi}_d), \quad k_{\psi,p}, k_{\psi,d} > 0. \tag{11.40}$$

The desired yaw rate $\dot{\psi}_d$ is generated by using a third-order reference model. The proportional and derivative gains, $k_{\psi,p}$ and k_{ψ_d}, in (11.40) are determined such that the error dynamics of the closed-loop system

$$T\ddot{\psi} + (1 + Kk_{\psi,d})\dot{\psi} + Kk_{\psi,p}\psi = Kk_{\psi,p}\psi_d + Kk_{\psi,d}\dot{\psi}_d \tag{11.41}$$

is GES (Fossen [5] or Khalil [20]).

11.3.5   Stability Considerations

It can be proven that the ES parameter converges to a neighborhood of its optimal value and that the ES algorithm is exponentially stable (see Krstić and Wang [21] and Breu and Fossen [2]). Consider the single-input, single-output nonlinear system:

$$\dot{\mathbf{x}} = \mathbf{f}(\mathbf{x}, u), \tag{11.42}$$

$$y = h(\mathbf{x}), \tag{11.43}$$

where $\mathbf{x} \in \mathbb{R}^n$ is the state vector, $u \in \mathbb{R}$ the input, $y \in \mathbb{R}$ the output; $\mathbf{f} : \mathbb{R}^n \times \mathbb{R} \to \mathbb{R}^n$ and $h : \mathbb{R}^n \to \mathbb{R}$ are smooth. The control law $u = \alpha(\mathbf{x}, \theta)$ is parametrized by θ, and assumed to be smooth. The closed-loop system corresponding to (11.42) and (11.43) then becomes

$$\dot{\mathbf{x}} = \mathbf{f}(\mathbf{x}, \alpha(\mathbf{x}, \theta)) \tag{11.44}$$

and it has equilibria parametrized by θ. For the stability analysis, the following assumptions are made (see Krstić and Wang [21]).

Assumption 11.6. *There exists a smooth function* $\mathbf{l} : \mathbb{R} \to \mathbb{R}^n$ *such that:*

$$\mathbf{f}(\mathbf{x}, \alpha(\mathbf{x}, \theta)) = \mathbf{0} \quad \text{if and only if} \quad \mathbf{x} = \mathbf{l}(\theta). \tag{11.45}$$

Assumption 11.7. *The equilibrium* $\mathbf{x} = \mathbf{l}(\theta)$ *of* (11.44) *is locally exponentially stable (LES) with decay and overshoot constants uniform in* θ *for each* $\theta \in \mathbb{R}$.

Assumption 11.8. *There exists* $\theta^* \in \mathbb{R}$ *such that:*

$$(h \circ l)'(\theta^*) = 0 \tag{11.46a}$$

$$(h \circ l)''(\theta^*) > 0. \tag{11.46b}$$

Assumptions (11.6) and (11.7) guarantee the robustness of the control law with respect to θ, i.e., any equilibria produced by θ can be stabilized by the control law. Assumption 11.8 implies that the output equilibrium map has a minimum when $\theta = \theta^*$.

It was proven by averaging for a static system and by the singular perturbation method for a dynamic system that (11.44) converges to a unique, exponentially stable, periodic solution in a neighborhood of the origin [21]. The perturbation signal and the filters in the ES loop determine the size of this neighborhood.

Due to the three different time scales in the proposed ES control (see Sect. 11.3.1) the plant – the surge and yaw subsystems – can be viewed as a static map. The ES parameter ω_e, determined from the ship's forward speed and heading angle, parametrizes the equilibria of the plant. The speed and heading controllers ensure local exponential stability of the equilibria which may be produced by the ES parameter ω_e, see Sect. 11.3.4, and the objective function defined in Sect. 11.3.2 fulfills locally Assumption 11.8. Thus, the parameter ω_e converges to a neighborhood of its optimal value ω_e^*.

11.4   Model Predictive Control

Model predictive control (MPC) is a rather recent control methodology which is characterized by the usage of an explicit plant model to predict the output of the process. This prediction is consequently used to find an optimal control signal which minimizes a specified objective function. The MPC formulation allows to address the constraints of the states and the input explicitly. MPC has been successfully applied to a wide variety of control problems and the increasing availability of computing power has only added to its popularity in both academia and industry, see, for example, Camacho and Bordons [3]. In ship control, MPC formulation has been applied among others to autopilot control design, roll stabilization, fault-tolerant control of a propulsion system, tracking, and control of ship fin stabilizers, see Kerrigan and Maciejowski [19], Naeem et al. [22], Perez [25], and Perez and Goodwin [26].

11.4.1   Model Predictive Control Applied to Ships in Parametric Roll Resonance

The basic structure of a MPC setup is depicted in Fig. 11.3. The MPC algorithm consists generally of the following elements, see Camacho and Bordons [3]:

- Prediction model
- Objective function
- Optimizer to obtain the control law

The strategy of MPCs can be summarized in a three step loop which is performed at each time instant, see, for example, Camacho and Bordons [3]:

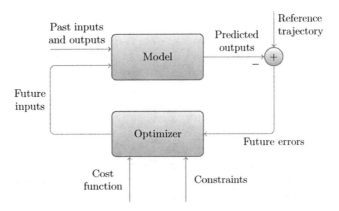

Fig. 11.3 Basic structure of a model predictive controller [3]

1. The future process outputs are predicted for a prediction horizon N, depending on the past inputs and outputs and on the future control signals.
2. An optimization problem is solved to determine the set of future control signals, minimizing the objective function.
3. The first control signal is sent to the process and the steps 1–3 are repeated at the next time instant.

In the context of the control of parametric roll resonance of ships, an approach featuring the MPC formulation to control the ship's forward speed and heading angle simultaneously in order to damp the roll motion is used. Furthermore, it is assumed that no control input affects the roll dynamics, that is, $\tau_\phi = 0$. The ship's surge speed and heading angle can however be changed. This results in a time-varying encounter frequency and the transients due to heading and speed changes must be taken into account. By changing the speed and heading actively it is possible to violate a condition for parametric roll resonance. To that matter, the MPC formulation is adapted to find the optimal surge speed and heading angle to achieve a regulation of the roll motion while taking into account constraints on the inputs as well as on the states.

11.4.2 State-Space Model

Consider the 3-DOF ship model (11.12) in Sect. 11.2.2. The encounter frequency (11.24) couples the roll dynamics to the surge and yaw dynamics, respectively, as derived in Sect. 11.2.3. For convenience the system dynamics is expressed as

$$\dot{\mathbf{x}} = \mathbf{f}(\mathbf{x}, \boldsymbol{\tau}) \qquad\qquad (11.47)$$

$$\mathbf{y} = \mathbf{g}(\mathbf{x}), \qquad\qquad (11.48)$$

where $\mathbf{x} = \left[\boldsymbol{\eta}^{\{1,4,6\}}, \mathbf{v}^{\{1,4,6\}}\right]^{\top}$ and

$$\mathbf{f}(\mathbf{x}, \boldsymbol{\tau}) = \left[\begin{array}{c} \mathbf{J}^{\{1,4,6\}}\left(\boldsymbol{\eta}^{\{1,4,6\}}\right)\mathbf{v}^{\{1,4,6\}} \\ \mathbf{M}^{-1}\left[\boldsymbol{\tau} - \mathbf{C}\left(\mathbf{v}^{\{1,4,6\}}\right)\mathbf{v}^{\{1,4,6\}} - \mathbf{D}\left(\mathbf{v}^{\{1,4,6\}}\right)\mathbf{v}^{\{1,4,6\}} - \mathbf{g}\left(\boldsymbol{\eta}^{\{1,4,6\}}\right)\right] \end{array}\right]$$

$$\mathbf{g}(\mathbf{x}) = \mathbf{x}$$

11.4.3 Objective Function

The MPC objective function is constructed similar to the one in Sect. 11.3.2, that is, as the weighted sum of cost functionals. Following the reasoning in Sect. 11.3.2, the following objective function is proposed:

$$J = w_1 J_1 + w_2 J_2, \qquad\qquad (11.49)$$

where the weights are w_i, $i \in \{1,2\}$ and

$$J_1 = c_1 e^{-c_2 (\omega_e - 2\omega_\phi)^2} \tag{11.50}$$

$$J_2 = c_3 (\omega_e - \omega_{e,0})^2, \tag{11.51}$$

where $c_i > 0$, $i \in \{1,2,3\}$ are constants. As in Sect. 11.3.2, (11.50) represents the penalty of the ship not violating the frequency condition (11.25), and (11.51) penalizes the deviation of the ship from its nominal cruise condition given by $\omega_{e,0}$ – the encounter frequency (11.24) with the nominal setpoints for the ship's surge speed u_0 and heading angle ψ_0.

11.4.4 Obtaining the Control Law

To obtain the control signal, the objective function (11.49) has to be minimized at each time instant. The minimization of (11.49) is subject to equality constraints which, for a state space model as presented in Sect. 11.4.2 and 11.2.2, respectively, are the model constraints given by (see Camacho and Bordons [3]):

$$\mathbf{f}(\mathbf{x}, \boldsymbol{\tau}) = \mathbf{0} \tag{11.52}$$

$$\mathbf{y} - \mathbf{g}(\mathbf{x}) = \mathbf{0}. \tag{11.53}$$

Furthermore, the minimization of (11.49) is as well subject to inequality constraints expressed as

$$\underline{\mathbf{y}} \leq \mathbf{y}(t+j) \leq \overline{\mathbf{y}}, \qquad \forall j = 1, N \tag{11.54}$$

$$\underline{\boldsymbol{\tau}} \leq \boldsymbol{\tau}(t+j) \leq \overline{\boldsymbol{\tau}}, \qquad \forall j = 1, M-1 \tag{11.55}$$

$$\Delta\underline{\boldsymbol{\tau}} \leq \Delta\boldsymbol{\tau}(t+j) \leq \Delta\overline{\boldsymbol{\tau}}, \quad \forall j = 1, M-1, \tag{11.56}$$

where N and M are the prediction horizon and the control horizon, respectively. The solution of the problem to minimize the objective function (11.49) with the model constraints (11.52) and (11.53) and the inequality constraints (11.54)–(11.56) is not a trivial one. It generally involves solving a nonconvex, nonlinear problem.

The nonlinear MPC (NMPC) problem in the form of a general nonlinear programming problem with $\mathbf{w} = \left[\boldsymbol{\tau}^\top, \mathbf{x}^\top, \mathbf{y}^\top\right]^\top$ is, see Camacho and Bordons [3],

$$\min_{\mathbf{w}} \ J(\mathbf{w})$$

$$\text{subject to: } \mathbf{c}(\mathbf{w}) = \mathbf{0}, \ \mathbf{h}(\mathbf{w}) \leq \mathbf{0}. \tag{11.57}$$

Here, c corresponds to the equality constraints (11.52) and (11.53) and h to the inequality constraints (11.54)–(11.56).

The optimization (11.57) is performed by using the *TOMLAB Optimization Environment* (TOMLAB/NPSOL),[1] see Holmström et al. [16].

11.5   Simulation Results

The ship is simulated by applying both the ES control methodology and the nonlinear MPC to the system. The initial values for the simulations are chosen such that the ship is experiencing parametrically excited rolling. The nominal cruise condition is chosen as $u_0 = 7.5 \, \text{m/s}$ and $\psi_0 = 0°$. We assume that the ship is initially in head sea condition, that is, $\beta_w^n = \pi$. Table 11.1 lists the model parameters. In the simulation results, the controlled variables are denoted by the subscript c.

Table 11.1 Model parameters, adopted from [13]

Quantity	Symbol	Value
Moment of inertia, roll	I_x	1.4014×10^{10} kg m^2
Added moment of inertia, roll	A_{44}	2.17×10^9 kg m^2
Nonlinear damping, roll	$K_{|p|p}$	-2.99×10^8 kg m^2
Linear damping, roll	K_p	-3.20×10^8 kg m^2/s
Water density	ρ	1025 kg/m^3
Gravitational acceleration	g	9.81 m/s^2
Water displacement	∇	76468 m^3
Mean meta-centric height	$\overline{GM}_m$	1.91 m
Amplitude of meta-centric height change	$\overline{GM}_a$	0.84 m
Restoring coefficient	$K_{\phi3}$	-2.9740×10^9 kg m^2/s^2
Mass	m	7.6654×10^7 kg
Added mass, surge	A_{11}	7.746×10^6 kg
Linear damping, surge	X_u	-5.66×10^3 kg/s
Wetted surface	S	11800 m^2
Form factor	k_f	0.1 –
Ship length	L_{pp}	281 m
Kinematic viscosity	ν_k	1.519×10^{-6} m^2/s
Nomoto time constant	T	-160.15 s
Nomoto gain constant	K	-0.1986 1/s
Natural roll frequency	ω_ϕ	0.3012 rad/s
Modal wave frequency	ω_0	0.4353 rad/s

---

[1] See http://www.tomopt.com for information about the *TOMLAB Optimization Environment*.

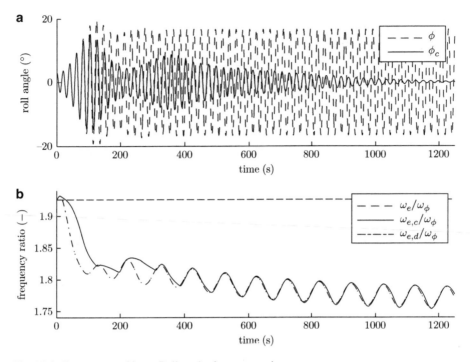

Fig. 11.4 Extremum seeking – Roll angle, frequency ratio

11.5.1 Extremum Seeking

The ES control is initially deactivated but is turned on the time instant when the roll amplitude exceeds $\phi = 3°$ for the first time.

The roll angle and frequency ratio with and without the ES control is shown in Fig. 11.4. Note that, in the uncontrolled scenario, the ship is experiencing parametric roll resonance with high roll amplitudes; see Fig. 11.4a. It is furthermore apparent that, when the ES control is activated, the ship is driven out of the frequency ratio relevant for parametric rolling and consequently the roll motion is reduced significantly.

The frequency ratios $\omega_{e,d}/\omega_\phi$ and $\omega_{e,c}/\omega_\phi$ in Fig. 11.4b denote the desired frequency ratio as output of the ES feedback loop and the actual, controlled, frequency ratio, thus indicating the ability of the controllers to track the desired encounter frequency.

Figure 11.5 shows the ship's surge speed and the control force in surge, whereas Fig. 11.6 depicts the ship's heading angle and the rudder deflection for both the uncontrolled and the controlled scenario.

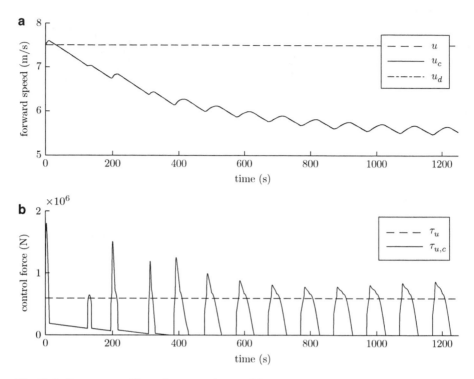

Fig. 11.5 Extremum seeking – Surge speed, control force τ_u

Both the ship's surge speed and heading angle follow the reference trajectory, determined by the control allocation block. Due to the perturbation signal in the ES control, the ship's surge speed and heading angle show an expected oscillatory behavior.

The cost as defined in the objective function is shown in Fig. 11.7 and Fig. 11.8 shows a comparison of the roll angle, when the ES control is activated at different roll angles, that is, at 3°, 5°, and 10°, respectively.

11.5.2   Model Predictive Control

The MPC control, initially turned off, is activated when the roll amplitude exceeds $\phi = 3°$ for the first time. In Fig. 11.9, the roll angle and the frequency ratio is shown for the controlled and the uncontrolled scenario. Figure 11.9a depicts that the ship is experiencing large roll angles due to parametric roll resonance in the uncontrolled scenario. However, the reduction of the frequency ratio reduces the roll angle quickly when the MPC is active.

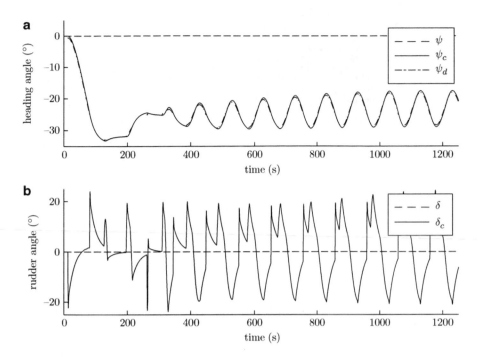

Fig. 11.6 Extremum seeking – Heading angle, rudder deflection

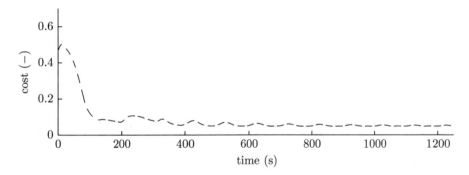

Fig. 11.7 Extremum seeking – Cost

The ship's forward speed and the control force in surge are shown in Fig. 11.10 and the ship's heading angle and the rudder deflection are depicted in Fig. 11.11. Again, the controlled and the uncontrolled scenario is shown.

Figure 11.12 shows the cost defined by the proposed objective function. Finally, in Fig. 11.13, the roll angle is shown, when the MPC is activated at different time instants, corresponding to roll angles of 3°, 5°, and 10°, respectively.

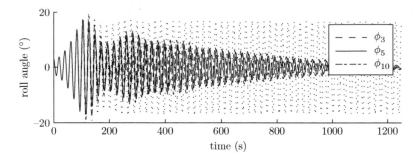

Fig. 11.8 Extremum seeking – Roll angle: Comparison when the controller is activated at 3°, 5°, and 10°, respectively

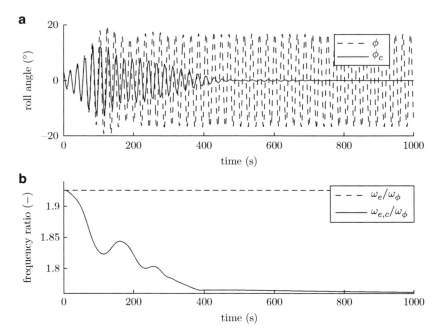

Fig. 11.9 MPC – Roll angle, frequency ratio

11.6 Conclusions

In this chapter, two active control approaches for the stabilization of parametric oscillations in ships by frequency detuning have been proposed. This is done by violating one of the conditions for the onset of parametric roll resonance by varying the ship's forward speed and heading angle simultaneously and thus controlling

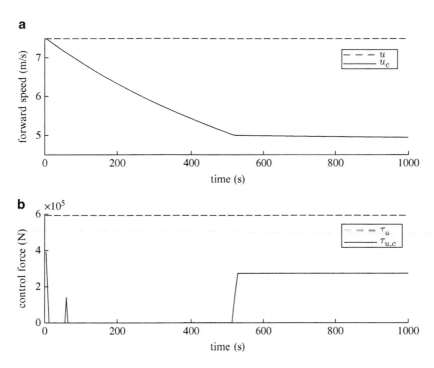

Fig. 11.10 MPC – Surge speed, control force τ_u

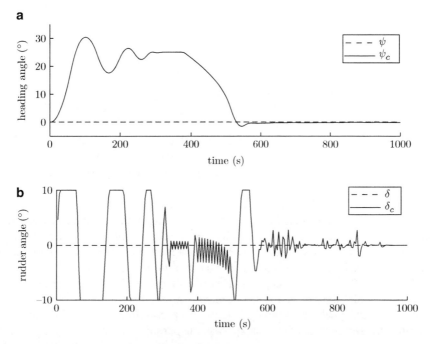

Fig. 11.11 MPC – Heading angle, rudder deflection

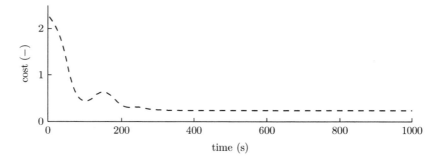

Fig. 11.12 MPC – Cost

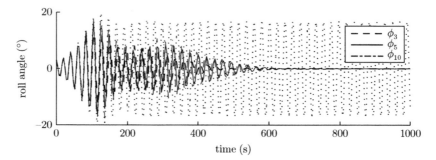

Fig. 11.13 MPC – Roll angle: Comparison when the controller is activated at 3°, 5°, and 10°, respectively

the frequency of encounter. The proposed control strategies feature optimality considerations with respect to the optimal speed and heading changes to stabilize parametrically excited roll motion.

The methodology of ES control has been applied to control ships exhibiting parametric roll resonance. The encounter frequency is tuned in real time to its optimal setpoint by defining an appropriate objective function. The encounter frequency commands are mapped to the ship's forward speed and heading angle by formulating the control allocation problem in the sequential least-squares framework, taking into account constraints on the actuators. The speed and heading controllers guarantee exponentially stable origins of the tracking error dynamics.

Furthermore, MPC is considered as a second approach for the stabilization of parametric roll resonance. By explicitly formulating both constraints on the input and the states as well as an objective function which accounts for the parametric roll resonance condition, a controller has been presented that effectively drives the ship out of parametric resonance and reduces the roll motion significantly.

Both the ES and the model predictive controllers have been successfully verified in computer simulations and it has been shown that the combined variation of the ship's forward speed and heading angle in both control approaches is efficient to stabilize the roll motion of a ship experiencing parametric roll resonance.

Acknowledgements This work was funded by the Centre for Ships and Ocean Structures (CeSOS), NTNU, Norway and the Norwegian Research Council.

References

1. Ariyur, K.B., Krstić, M.: Real-Time Optimization by Extremum-Seeking Control. John Wiley & Sons, Inc., Hoboken, New Jersey, USA (2003)
2. Breu, D.A., Fossen, T.I.: Extremum seeking speed and heading control applied to parametric resonance. In: Proc. IFAC Conference on Control Applications in Marine Systems (CAMS), Rostock, Germany (2010)
3. Camacho, E.F., Bordons, C.: Model Predictive Control, 2nd edition. Advanced textbooks in control and signal processing. Springer-Verlag, London, UK (2004)
4. Clarke, D., Gedling, P., Hine, G.: The application of manoeuvring criteria in hull design using linear theory. In: RINA Transactions and Annual Report 1983, Royal Institution of Naval Architects, London, UK (1983)
5. Fossen, T.I.: Handbook of Marine Craft Hydrodynamics and Motion Control. John Wiley & Sons, Ltd., Chichester, UK (2011)
6. France, W.N., Levadou, M., Treakle, T.W., Paulling, J.R., Michel, R.K., Moore, C.: An investigation of head-sea parametric rolling and its influence on container lashing systems. In: Marine Technology, Vol. 40, No. 1, 2003, Society of Naval Architects and Marine Engineers, New Jersey, USA (2003)
7. Galeazzi, R.: Autonomous supervision and control of parametric roll resonance. PhD thesis, Technical University of Denmark (2009)
8. Galeazzi, R., Blanke, M.: On the feasibility of stabilizing parametric roll with active bifurcation control. In: Proc. IFAC Conference on Control Applications in Marine Systems (CAMS), Bol, Croatia (2007)
9. Galeazzi, R., Vidic-Perunovic, J., Blanke, M., Jensen, J.J.: Stability analysis of the parametric roll resonance under non-constant ship speed. In: Proc. 9th Biennial ASME Conference on Engineering Systems Design and Analysis (ESDA), Haifa, Israel (2008)
10. Galeazzi, R., Holden, C., Blanke, M., Fossen, T.I.: Stabilization of parametric roll resonance by combined speed and fin stabilizer control. In: Proc. European Control Conference (ECC), Budapest, Hungary (2009)
11. Härkegård, O.: Efficient active set algorithms for solving constrained least squares problems in aircraft control allocation. In: Proc. IEEE Conference on Decision and Control (CDC), Las Vegas, Nevada, USA (2002)
12. Härkegård, O.: Dynamic control allocation using constrained quadratic programming. Journal of Guidance, Control, and Dynamics **27**(6), 1028–1034, American Institute of Aeronautics and Astronautics, Inc., (2004)
13. Holden, C., Galeazzi, R., Rodríguez, C., Perez, T., Fossen, T.I., Blanke, M., Neves, M.A.S.: Nonlinear container ship model for the study of parametric roll resonance. Modeling, Identification and Control **28**(4), 87–103 (2007)
14. Holden, C., Galeazzi, R., Fossen, T.I., Perez, T.: Stabilization of parametric roll resonance with active u-tanks via Lyapunov control design. In: Proc. European Control Conference (ECC), Budapest, Hungary (2009)
15. Holden, C., Perez, T., Fossen, T.I.: A Lagrangian approach to nonlinear modeling of anti-roll tanks. Ocean Engineering **38**(2–3), 341–359 (2011)
16. Holmström, K., Göran, A.O., Edvall, M.M.: User's guide for TOMLAB/NPSOL [online] TOMLAB Optimization. Available from: http://tomopt.com/docs/TOMLAB_NPSOL.pdf [Accessed 1 June 2011]

17. Jensen, J.J., Pedersen, P.T., Vidic-Perunovic, J.: Estimation of parametric roll in a stochastic seaway. In: Proc. IUTAM Symposium on Fluid-Structure Interaction in Ocean Engineering, Springer, Hamburg, Germany (2008)
18. Jensen, J.J., Vidic-Perunovic, J., Pedersen, P.T.: Influence of surge motion on the probability of parametric roll in a stationary sea state. In: Proc. International Ship Stability Workshop, Hamburg, Germany (2007)
19. Kerrigan, E.C., Maciejowski, J.M.: Fault-tolerant control of a ship propulsion system using model predictive control. In: Proc. European Control Conference (ECC), Karlsruhe, Germany (1999)
20. Khalil, H.: Nonlinear Systems. 3rd edition. Prentice Hall, New Jersey, USA (2002)
21. Krstić, M., Wang, H.H.: Stability of extremum seeking feedback for general nonlinear dynamic systems. Automatica **36**, 595–601 (2000)
22. Naeem, W., Sutton, R., Ahmad, S.: Pure pursuit guidance and model predictive control of an autonomous underwater vehicle for cable/pipeline tracking. Journal of Marine Science and Environment, 25–35. Citeseer (2004)
23. Nayfeh, A.H., Mook, D.T.: Nonlinear Oscillations. John Wiley & Sons, Inc., Weinheim, Germany (1995)
24. Neves, M., Rodríguez, C.: On unstable ship motions resulting from strong non-linear coupling. Ocean Engineering **33**(14-15), 1853–1883 (2006)
25. Perez, T.: Ship Motion Control: Course Keeping and Roll Stabilisation Using Rudder and Fins. Springer, London, UK (2005)
26. Perez, T., Goodwin, G.: Constrained predictive control of ship fin stabilizers to prevent dynamic stall. Control Engineering Practice **16**(4), 482–494 (2008)
27. Ribeiro e Silva, S., Santos, T.A., Soares, C.G.: Parametrically excited roll in regular and irregular head seas. International Shipbuilding Progress **51**, 29–56 (2005)
28. Shin, Y.S., Belenky, V.L., Paulling, J.R., Weems, K.M., Lin, W.M.: Criteria for parametric roll of large containerships in longitudinal seas. In: Transactions SNAME, Society of Naval Architects and Marine Engineers, New Jersey, USA (2004)
29. Umeda, N., Hashimoto, H., Vassalos, D., Urano, S., Okou, K.: An investigation of different methods for the prevention of parametric rolling. Journal of Marine Science and Technology **13**, 16–23 (2008)
30. Wang, H.D., Yi, J.Q., Fan, G.L.: A dynamic control allocation method for unmanned aerial vehicles with multiple control effectors. Springer-Verlag, Berlin, Germany. Lecture Notes In Artificial Intelligence; Vol. 5314, Proc. Int. Conf. on Intelligent Robotics and Applications (2008)

Chapter 12
A U-Tank Control System for Ships in Parametric Roll Resonance

Christian Holden and Thor I. Fossen

12.1 Introduction

To control parametric resonance, there are two basic approaches; use control force to counter-act the unwanted motion, or ensure that the system's parameters are in such a state that parametric resonance cannot occur. We can call these methods *direct* and *indirect*, respectively. The difference is perhaps best explained by analyzing a differential equation.

A simple model for parametric resonance in ships is the *Mathieu equation*

$$m_{44}\ddot{\phi} + d_{44}\dot{\phi} + \left[k_{44} + k_{\phi t}\cos(\omega_e t + \alpha_\phi)\right]\phi = u_c$$

where ϕ is the roll angle, u_c an externally applied torque and the parameters are constant. The system is known to parametrically resonate when $\omega_e \approx 2\sqrt{k_{44}/m_{44}}$. With indirect control, ω_e is dynamically changed so that ϕ will not parametrically resonate. With direct control, u_c is used to set up a counter-moment to force the system to zero. Direct and indirect methods can be combined, as seen in [8,9].

As shown in Chaps. 9 and 10, it is possible to change the encounter frequency ω_e (which depends on the ship's speed) and thus control the system indirectly. However, in practice, this depends on very early detection and the ability of the ship to rapidly

C. Holden (✉)
Centre for Ships and Ocean Structures, Norwegian University of Science and Technology,
NO-7491 Trondheim, Norway
e-mail: c.holden@ieee.org

T.I. Fossen
Department of Engineering Cybernetics, Norwegian University of Science and Technology,
NO-7491 Trondheim, Norway

Centre for Ships and Ocean Structures, Norwegian University of Science and Technology,
NO-7491 Trondheim, Norway
e-mail: fossen@ieee.org

T.I. Fossen and H. Nijmeijer (eds.), *Parametric Resonance in Dynamical Systems*,
DOI 10.1007/978-1-4614-1043-0_12, © Springer Science+Business Media, LLC 2012

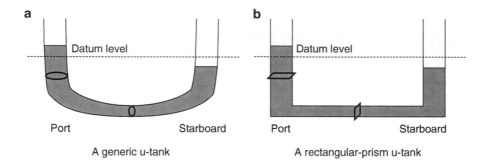

Fig. 12.1 U-tank design

perform a speed change. If the ship has high inertia or is at rest with the engines turned off, it is unlikely that the ship can change its speed fast enough to avoid large roll angles.

There are also some disadvantages associated with direct control. Ships are often not equipped with actuation in roll, as such systems are not necessary for propulsion [5]. Possible actuators include fins, tanks, and gyro stabilizers [21]. In this particular work, we will focus on the use of u-tanks as actuators. These have the advantage that they can be used even if the ship is at rest [18]. As they are internal, they do not increase drag. Unfortunately, they do take up space inside the hull, potentially decreasing the available space for other machinery, cargo, or passengers.

A u-tank (sometimes referred to as u-tube tank or u-shaped anti-roll tank) consists of two reservoirs, one on the starboard side and one on the port side, connected by a duct (see Fig. 12.1a). The basic principle is to use the weight and motion of the fluid to give a direct moment in roll, which can be used to counteract parametric resonance or other unwanted motion.

A disadvantage of u-tanks compared to other potential actuators, is that the roll and tank modes are tightly coupled, and only indirectly give a control moment in roll. Output stabilization (driving roll to zero) tends to not leave the tank in its equilibrium position, as seen in [12].

Most models of u-tanks are derived for tanks shaped like three connected rectangular prisms (see Fig. 12.1b) [14, 15, 17–19, 22], while several actually installed tanks do not match this shape [20–22]. A model for more generic tank shapes is therefore useful. In addition, most models are linear, and technically only valid for small roll angles [6, 11, 15, 17–19]. During parametric resonance, the roll angle can reach 40 to 50° [7, 9, 12, 13].

In this chapter, a novel nonlinear 2-DOF u-tank model is presented for an arbitrarily-shaped u-tank, and a controller that stabilizes parametric roll resonance with the aid of such a tank developed. The model is compared to existing models. The validity of the controller is proved mathematically and tested by simulation.

12.2   Preliminaries

The model will be derived using a combination of Lagrangian (analytical) and Newtonian mechanics. Initially, the tank–ship interaction (which is conservative) will be modeled using Lagrangian mechanics. Forces and moments induced by the surrounding ocean, in addition to friction and other nonconservative forces, will be modeled using Newtonian mechanics and incorporated into the conservative model.

   Only roll and the motion of the tank fluid will be modeled in this chapter. We assume that the ship is not translating relative to the inertial frame.

12.2.1   Coordinate Systems

To use the Lagrangian approach, the dynamics have to be derived in an inertial reference frame [3]. The geometry of the vessel is easier to describe in a reference frame fixed to the body, but as the body is rotating, a body-fixed frame is not inertial. Therefore, we define two coordinate systems: an inertial frame fixed to the surface of the Earth,[1] and a noninertial frame fixed to the body.

   The origin of the inertial reference can be placed arbitrarily. For simplicity, we let the xy-plane coincide with the mean ocean surface and the z-axis point with the gravity field. The body frame is placed at the transversal center of gravity at the calm-water water plane, with the x-axis pointing forwards, the y-axis pointing starboard and the z-axis pointing downwards, see Fig. 12.2. The ship is assumed symmetric around the xz-plane.

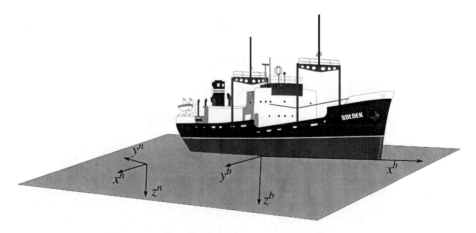

Fig. 12.2 Reference frames used in this chapter

---

[1]As the Earth is not inertial, clearly an Earth-fixed reference frame is not inertial. However, the effects of the non-inertial nature of the Earth's motion are small for many applications [5].

A vector $\mathbf{r}$ is denoted $\mathbf{r}^n$ in the inertial frame and $\mathbf{r}^b$ in the body-fixed frame. These are related by $\mathbf{r}^n = \mathbf{R}\mathbf{r}^b \Leftrightarrow \mathbf{r}^b = \mathbf{R}^\top \mathbf{r}^n$ where $\mathbf{R}$ is a rotation matrix [3].

12.2.2 Modeling Hypothesis

To model the ship–tank system, some assumptions and simplifications have to be made.

12.2.2.1 The Ship

The ship's motion is assumed restricted to a single degree of freedom, namely roll. It can be defined as the number ϕ so that the rotation matrix $\mathbf{R}$ can be written

$$\mathbf{R} = \begin{bmatrix} 1 & 0 & 0 \\ 0 & \cos(\phi) & -\sin(\phi) \\ 0 & \sin(\phi) & \cos(\phi) \end{bmatrix} . \qquad (12.1)$$

This is equivalent to having the body-fixed frame (and the ship with it) rotated an angle ϕ about the inertial x-axis [3]. Note that, with the ship restricted to roll, the inertial and body-fixed x-axes are parallel, and without loss of generality can be assumed to be coinciding.

We note that the ship's angular velocity relative to the inertial frame $\boldsymbol{\omega}$ is then given by

$$\boldsymbol{\omega}^n = \boldsymbol{\omega}^b = [\dot{\phi}, 0, 0]^\top . \qquad (12.2)$$

We also make some assumptions regarding the ship and the ocean:

A12.1.   The ship is port–starboard symmetric (i.e., around the body xz-plane) in mass and geometry.
A12.2.   The ship is not translating relative to the inertial reference frame.
A12.3.   The waves are sinusoidal, planar, and stationary.
A12.4.   The wave length is approximately equal to the ship length.
A12.5.   There is either head or stern seas.

By the first assumption, the ship's center of gravity (excluding the tank fluid) is given by

$$\mathbf{r}_g^b = \left[x_g^b, 0, z_g^b\right]^\top . \qquad (12.3)$$

12.2.2.2   The Tank Fluid

A u-tank is simply two reservoirs of water or another liquid, one on the port side and the other at starboard, with a duct in between to allow the passage of liquid. To be able to model this intrinsically complicated behavior, some assumptions have to be made:

A12.6.   The surface of the fluid in the tank is perpendicular to the centerline of the tank.
A12.7.   The fluid in the tank is incompressible.
A12.8.   The flow of fluid in the tank is one-dimensional.
A12.9.   Tank fluid memory effects are negligible.
A12.10.   The u-tank is placed at the transversal geometrical center of the ship.
A12.11   The tank is symmetrical around the centerline.
A12.12.   There are no air bubbles in the tank.
A12.13.   The centerline of the tank is smooth.
A12.14.   The centerline of the tank runs port–starboard.

Assumption A12.6 is clearly false for a ship in motion; the actual fluid surface in the tank is likely to behave in a complicated and chaotic fashion. Modeling this accurately without resorting to computational fluid dynamics is unfeasible. Assuming the fluid surface to be horizontal would not be much more accurate than Assumption A12.6.

Assumptions A12.6–A12.14 imply that the tank fluid is parameterizable as a tube of varying cross-sectional area. Defining the centerline of the tube of fluid as $\mathbf{r}_t(\sigma)$ with σ as parameter, $\mathbf{r}_t^b(\sigma)$ can be written as

$$\mathbf{r}_t^b(\sigma) = \begin{bmatrix} x_t^b \\ y_t^b(\sigma) \\ z_t^b(\sigma) \end{bmatrix} . \tag{12.4}$$

The parameter σ is defined to have its zero point at the ship centerline and positive in the port direction. The fluid surfaces are located at $\sigma = -\varsigma_s \leq 0$ (starboard side) and $\sigma = \varsigma_p \geq 0$ (port side). Thus, $\sigma \in [-\varsigma_s, \varsigma_p] \subset \mathbb{R}$ defines the fluid-filled part of the tank. When the water level is equal in both the starboard and port side reservoirs, $\varsigma_p = \varsigma_s = \varsigma_0$, and $\sigma \in [-\varsigma_0, \varsigma_0] \subset \mathbb{R}$ defines the fluid-filled part of the tank.

Property 12.1. $\mathbf{r}_t^b$ satisfies the following properties:

- x_t^b is a constant, per Assumption A12.14.
- The functions y_t^b and z_t^b are smooth (specifically, C^1 or greater), per Assumption A12.13.
- y_t^b is odd and lies in the second and fourth quadrants (i.e., $y_t^b(-\sigma) = -y_t^b(\sigma)$, $y_t^b(0) = 0$ and $y_t^b(\sigma) < 0 \,\forall\, \sigma > 0$), per Assumptions A12.10 and A12.11.
- z_t^b is even (i.e., $z_t^b(-\sigma) = z_t^b(\sigma)$), per Assumption A12.11.
- $\max z_t^b = z_t^b(0)$, per Assumptions A12.10 and A12.11.

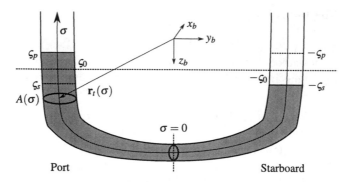

Fig. 12.3 U-tank parameters

To fully describe the tank fluid, the cross-sectional area $A(\sigma)$ is also needed.

Property 12.2. By Assumption A12.12, the fluid fills the entire area $A(\sigma)\forall\sigma\in[-\varsigma_s,\varsigma_p]$. Assumption A12.11 implies that $A(-\sigma)=A(\sigma)>0$.

See Fig. 12.3 for an illustration of the u-tank and its parameters.

The chief physically measurable states of the system are ς_p, ς_s and the volumetric flow of the tank fluid Q (positive to port). ς_p and ς_s are related to the flow rate by

$$\dot{\varsigma}_p = \frac{Q}{A(\varsigma_p)}, \quad \dot{\varsigma}_s = -\frac{Q}{A(\varsigma_s)}.$$

We define the generalized tank coordinate q_t as

$$q_t \triangleq \frac{1}{A_0}\int_{\varsigma_0}^{\varsigma_p} A(\sigma)\,d\sigma, \tag{12.5}$$

where A_0 is an arbitrary constant with unit m^2.

We note that the total fluid volume in the tank, V_t, is constant. Thus,

$$V_t \triangleq \int_{-\varsigma_0}^{\varsigma_0} A(\sigma)\,d\sigma = \int_{-\varsigma_s}^{\varsigma_p} A(\sigma)\,d\sigma = \int_{-\varsigma_s}^{-\varsigma_0} A(\sigma)\,d\sigma + \int_{-\varsigma_0}^{\varsigma_0} A(\sigma)\,d\sigma + \int_{\varsigma_0}^{\varsigma_p} A(\sigma)\,d\sigma$$

$$= \int_{-\varsigma_s}^{-\varsigma_0} A(\sigma)\,d\sigma + V_t + A_0 q_t.$$

This gives

$$q_t = -\frac{1}{A_0}\int_{-\varsigma_s}^{-\varsigma_0} A(\sigma)\,d\sigma. \tag{12.6}$$

The time derivative of q_t is given by

$$\dot{q}_t = \frac{1}{A_0}A(\varsigma_p)\frac{d\varsigma_p}{dt} = \frac{Q}{A_0}. \tag{12.7}$$

By differentiating both sides of (12.5) and (12.6) with respect to q_t, it follows that

$$\frac{d\varsigma_p}{dq_t} = \frac{A_0}{A(\varsigma_p)}, \quad \frac{d\varsigma_s}{dq_t} = -\frac{A_0}{A(\varsigma_s)}. \tag{12.8}$$

The speed of the tank fluid relative to the tank walls (i.e., the ship), at any point σ in the tank, is given by

$$\|\mathbf{v}_{t,r}(\sigma, \dot{q}_t)\| = \frac{Q}{A(\sigma)} = \frac{A_0 \dot{q}_t}{A(\sigma)}.$$

From calculus, we know that velocity is tangential to the path, giving

$$\mathbf{v}_{t,r}(\sigma, \dot{q}_t) = \frac{A_0 \dot{q}_t}{A(\sigma)} \frac{d\bar{\mathbf{r}}_t}{d\sigma}(\sigma), \tag{12.9}$$

where

$$\frac{d\bar{\mathbf{r}}_t}{d\sigma} \triangleq \frac{\frac{d\mathbf{r}_t}{d\sigma}}{\left\|\frac{d\mathbf{r}_t}{d\sigma}\right\|}. \tag{12.10}$$

Noting that $dx_t^b/d\sigma = 0$, we define

$$\frac{d\bar{y}_t^b}{d\sigma} \triangleq [0,1,0] \frac{d\bar{\mathbf{r}}_t^b}{d\sigma} = \frac{\frac{dy_t^b}{d\sigma}}{\sqrt{\left(\frac{dy_t^b}{d\sigma}\right)^2 + \left(\frac{dz_t^b}{d\sigma}\right)^2}}, \tag{12.11}$$

$$\frac{d\bar{z}_t^b}{d\sigma} \triangleq [0,0,1] \frac{d\bar{\mathbf{r}}_t^b}{d\sigma} = \frac{\frac{dz_t^b}{d\sigma}}{\sqrt{\left(\frac{dy_t^b}{d\sigma}\right)^2 + \left(\frac{dz_t^b}{d\sigma}\right)^2}}, \tag{12.12}$$

such that

$$\left(\frac{d\bar{y}_t^b}{d\sigma}\right)^2 + \left(\frac{d\bar{z}_t^b}{d\sigma}\right)^2 \equiv 1. $$

Of course, the ship (and the tank with it) is rotating relative to the inertial frame. Thus, the velocity of the tank fluid relative to the inertial frame, at any point σ in the tank, is given by

$$\mathbf{v}_t(\sigma, \dot{\mathbf{q}}) = \boldsymbol{\omega} \times \mathbf{r}_t(\sigma) + \frac{A_0 \dot{q}_t}{A(\sigma)} \frac{d\bar{\mathbf{r}}_t}{d\sigma}(\sigma). \tag{12.13}$$

12.3   U-Tank Modeling

According to Lagrangian mechanics, it is necessary to derive the system's kinetic and potential energies.

Define the generalized coordinates $\mathbf{q}$ as

$$\mathbf{q} \triangleq [\phi, q_t]^\top \in \mathbb{R}^2. \tag{12.14}$$

Proposition 12.1 (Potential energy). *The potential energy of the roll–tank system is given by*

$$U(\mathbf{q}) = mgz_g^b + 2g\rho_t \int_0^{\varsigma_0} z_t^b(\sigma)A(\sigma)\,d\sigma - g\rho_t \left[\int_{-\varsigma_s(q_t)}^{\varsigma_p(q_t)} y_t^b(\sigma)A(\sigma)\,d\sigma \right] \sin(\phi)$$

$$- \left[mgz_g^b + g\rho_t \int_{-\varsigma_s(q_t)}^{\varsigma_p(q_t)} z_t^b(\sigma)A(\sigma)\,d\sigma \right] \cos(\phi), \tag{12.15}$$

where m is the mass of the ship (excluding tank fluid), g is the acceleration of gravity and ρ_t is the density of the tank fluid. Note that the first integral is a constant, and that $U(\mathbf{0}) = 0$.

Proof. See Appendix 1. □

Proposition 12.2 (Kinetic energy). *The kinetic energy of the roll–tank system is given by*

$$T(q_t, \dot{\mathbf{q}}) = \frac{1}{2}\dot{\mathbf{q}}^\top \mathbf{M}_t(q_t)\dot{\mathbf{q}}, \tag{12.16}$$

where

$$\mathbf{M}_t(q_t) = \begin{bmatrix} J_{11} + J_t(q_t) & m_{4t}(q_t) \\ m_{4t}(q_t) & \bar{m}_t(q_t) \end{bmatrix} \in \mathbb{R}^{2\times2}$$

$$J_t(q_t) = \rho_t \int_{-\varsigma_s(q_t)}^{\varsigma_p(q_t)} A(\sigma)[[y_t^b(\sigma)]^2 + [z_t^b(\sigma)]^2]\,d\sigma$$

$$m_{4t}(q_t) = \rho_t A_0 \int_{-\varsigma_s(q_t)}^{\varsigma_p(q_t)} \left[y_t^b(\sigma)\frac{dz_t^b}{d\sigma}(\sigma) - \frac{dy_t^b}{d\sigma}(\sigma)z_t^b(\sigma) \right] d\sigma$$

$$\bar{m}_t(q_t) = \rho_t A_0^2 \int_{-\varsigma_s(q_t)}^{\varsigma_p(q_t)} \frac{1}{A(\sigma)}\,d\sigma$$

and J_{11} is the ship's moment of inertia around the (body) x-axis (excluding the moment of inertia of the tank fluid).

Proof. See Appendix 2. □

Proposition 12.3 (Lagrangian dynamics). *The (lossless) roll–tank dynamics are given by*

$$\mathbf{M}_t(q_t)\ddot{\mathbf{q}} + \mathbf{C}(q_t,\dot{\mathbf{q}})\dot{\mathbf{q}} + \mathbf{k}_t(\mathbf{q}) = 0, \tag{12.17}$$

where

$$\mathbf{C}(q_t,\dot{\mathbf{q}}) = \frac{\dot{\phi}}{2}\begin{bmatrix} 0 & \frac{\partial J_t}{\partial q_t}(q_t) \\ -\frac{\partial J_t}{\partial q_t}(q_t) & 0 \end{bmatrix} + \frac{\dot{q}_t}{2}\begin{bmatrix} \frac{\partial J_t}{\partial q_t}(q_t) & 2\frac{\partial m_{4t}}{\partial q_t}(q_t) \\ 0 & \frac{\partial \bar{m}_t}{\partial q_t}(q_t) \end{bmatrix}$$

$$\frac{\partial J_t}{\partial q_t}(q_t) = \rho_t A_0 \left[[y_t^b(\varsigma_p(q_t))]^2 - [y_t^b(\varsigma_s(q_t))]^2 + [z_t^b(\varsigma_p(q_t))]^2 - [z_t^b(\varsigma_s(q_t))]^2 \right]$$

$$\frac{\partial m_{4t}}{\partial q_t}(q_t) = \rho_t \frac{A_0^2}{A(\varsigma_p(q_t))} \left[y_t^b(\varsigma_p(q_t))\frac{d\bar{z}_t^b}{d\sigma}(\varsigma_p(q_t)) - \frac{d\bar{y}_t^b}{d\sigma}(\varsigma_p(q_t))z_t^b(\varsigma_p(q_t)) \right]$$

$$\qquad - \rho_t \frac{A_0^2}{A(\varsigma_s(q_t))} \left[y_t^b(\varsigma_s(q_t))\frac{d\bar{z}_t^b}{d\sigma}(\varsigma_s(q_t)) - \frac{d\bar{y}_t^b}{d\sigma}(\varsigma_s(q_t))z_t^b(\varsigma_s(q_t)) \right]$$

$$\frac{\partial \bar{m}_t}{\partial q_t}(q_t) = \rho_t A_0^3 \left[\frac{1}{A^2(\varsigma_p(q_t))} - \frac{1}{A^2(\varsigma_s(q_t))} \right]$$

$$\mathbf{k}_t(\mathbf{q}) = \begin{bmatrix} \left(mgz_g^b + g\rho_t \int_{-\varsigma_s}^{\varsigma_p} z_t^b A \, d\sigma\right)\sin(\phi) - g\rho_t \int_{-\varsigma_s}^{\varsigma_p} y_t^b A \, d\sigma \cos(\phi) \\ -g\rho_t A_0 \left[y_t^b(\varsigma_p) + y_t^b(\varsigma_s) \right]\sin(\phi) - g\rho_t A_0 \left[z_t^b(\varsigma_p) - z_t^b(\varsigma_s) \right]\cos(\phi) \end{bmatrix}$$

Proof. The Lagrangian $\mathcal{L}$ of the roll–tank system is given by

$$\mathcal{L}(\mathbf{q},\dot{\mathbf{q}}) = T(q_t,\mathbf{q}) - U(\mathbf{q})$$

$$= g\rho_t \int_{-\varsigma_s(q_t)}^{\varsigma_p(q_t)} y_t^b(\sigma)A(\sigma)\, d\sigma \sin(\phi) - mgz_g^b - 2g\rho_t \int_0^{\varsigma_0} z_t^b(\sigma)A(\sigma)\, d\sigma$$

$$+ \left[mgz_g^b + g\rho_t \int_{-\varsigma_s(q_t)}^{\varsigma_p(q_t)} z_t^b(\sigma)A(\sigma)\, d\sigma \right]\cos(\phi) + \frac{1}{2}\dot{\mathbf{q}}^\top \mathbf{M}_t(q_t)\dot{\mathbf{q}} \,.$$

$$\tag{12.18}$$

The dynamics of the system are then given by the Euler-Lagrange Equation [10]

$$\frac{d}{dt}\frac{\partial \mathcal{L}}{\partial \dot{\mathbf{q}}} - \frac{\partial \mathcal{L}}{\partial \mathbf{q}} = 0 \,. \tag{12.19}$$

It can be shown that

$$\frac{\partial \mathcal{L}}{\partial \dot{\mathbf{q}}} = \mathbf{M}_t(q_t)\dot{\mathbf{q}}$$

$$\frac{d}{dt}\frac{\partial \mathcal{L}}{\partial \dot{\mathbf{q}}} = \dot{\mathbf{M}}_t(q_t)\dot{\mathbf{q}} + \mathbf{M}_t(q_t)\ddot{\mathbf{q}} = \dot{q}_t\frac{\partial \mathbf{M}_t}{\partial q_t}\dot{\mathbf{q}} + \mathbf{M}_t(q_t)\ddot{\mathbf{q}} = \mathbf{C}(q_t,\dot{\mathbf{q}})\dot{\mathbf{q}} + \mathbf{M}_t(q_t)\ddot{\mathbf{q}}$$

$$\frac{\partial \mathcal{L}}{\partial q_t} = -\mathbf{k}_t(\mathbf{q}) \ .$$

Inserting this into (12.19) gives (12.17).   □

Proposition 12.4 (Two-DOF u-tank model). *The dynamics of the tank–roll system are given by*

$$\mathbf{M}(q_t)\ddot{\mathbf{q}} + \mathbf{C}(q_t,\dot{\mathbf{q}})\dot{\mathbf{q}} + \mathbf{D}(\dot{\mathbf{q}})\dot{\mathbf{q}} + \mathbf{k}(t,\mathbf{q}) = \mathbf{b}u, \qquad (12.20)$$

where

$$\mathbf{M}(q_t) = \begin{bmatrix} m_{a,44} & 0 \\ 0 & 0 \end{bmatrix} + \mathbf{M}_t(q_t) , \quad \mathbf{D}(\dot{\mathbf{q}}) = \mathbf{D}_0 + \mathbf{D}_n(\dot{\mathbf{q}}) ,$$

$$\mathbf{D}_0 = \begin{bmatrix} d_{44} & 0 \\ 0 & d_{tt} \end{bmatrix} , \quad \mathbf{D}_n(\dot{\mathbf{q}}) = \begin{bmatrix} 0 & 0 \\ 0 & d_{tt,n}|\dot{q}_t| \end{bmatrix} , \quad \mathbf{b} = \begin{bmatrix} 0 \\ 1 \end{bmatrix} ,$$

$$\mathbf{k}(t,\mathbf{q}) = \begin{bmatrix} \bar{k}_{44}\phi + k_{\phi t}\cos(\omega_e t + \alpha_\phi)\phi + k_3\phi^3 \\ 0 \end{bmatrix} + \mathbf{k}_t(\mathbf{q})$$

the other matrices are as in Proposition 12.3 and $u \in \mathbb{R}$ is the control force on the tank fluid. All the parameters are positive. $\mathbf{D}$ satisfies $y^\top \mathbf{D}y > 0 \ \forall \ y \neq 0$.

Proof. The forces and moments acting on the roll–tank system that are not captured by the Lagrangian modeling are friction and other dissipative forces, added mass, control forces, and pressure torques in roll.

Since nonviscous damping in roll is quite small [4], we model roll damping linearly. Experiments conducted in [14] indicate that quadratic damping is extremely important for the motion of the tank fluid, so this is included. The generalized damping forces are collected in the term $\mathbf{D}(\dot{\mathbf{q}})\dot{\mathbf{q}}$. [14] experimentally investigated the presence of off-diagonal elements in $\mathbf{D}$, but the influence of such terms was found to be negligible. Such terms have therefore been excluded here.

Added mass is a moment proportional to the acceleration of the ship [5], and is caused by interaction with the surrounding ocean. As the tank fluid is not directly in contact with the ocean, the added mass moment $m_{a,44}\ddot{\phi}$ only (directly) affects roll. In general, $m_{a,44}$ is nonconstant [5], but for simplicity we assume it to be constant in this chapter.

The control force u only affects the tank fluid, which implies that $\mathbf{b} = [0,1]^\top$.

As done in [13], the calm-water pressure moment in roll (hydrostatic and -dynamic buoyancy) is modeled as $\bar{k}_{44}\phi + k_3\phi^3$. The parametric excitation is, as in [9, 12] assumed to take place in the linear roll spring term, giving an additional moment of $k_{\phi t}\cos(\omega_e t + \alpha_\phi)\phi$.

Combining all these forces with the Lagrangian-based dynamics of Proposition 12.3 gives (12.20). $\square$

12.3.1   Analysis

It is prudent to ask how the new model (12.20) compares to existing ones. As pointed out in the introduction, most existing models are made for rectangular-prism u-tanks (Fig. 12.1b) [14, 15, 17–19, 22]. Technically, such models cannot fit into the framework developed here, as the tank centerline functions are only C^0 rather than C^1 as required. The integrals that go into the model can still be computed, but the model will technically be invalid.

However, if we ignore this fact, we can explicitly compute the integrals in (12.20). This renders the model identical to the experimentally validated one of [14]. However, that model requires that the duct is always full of water, a constraint not found in the new model.

The linearization of the model is identical to that of [4] and [17, 18],[2] other than that in these works the models take into account sway and yaw in addition to roll and the tank state, and that in [17, 18] $J_t \equiv 0$.

Excluding the tank moment of inertia J_t is likely to have only a small effect, as it is significantly smaller than the moment of inertia of the ship itself, J_{11}. The coupling to the other degrees of freedom might be significant in general, but in this work the ship is assumed not to be maneuvering.

12.4   Control Design

In [14], experiments with a rectangular-prism tank showed that a linearized model was unsuitable to capture the full dynamics of the roll–tank system. However, the experiments also indicated that the full nonlinear model was needlessly complicated, and suggested an alternative model where the dynamics were linearized, with the exception of the damping. This model was an adequate approximation even for relatively high roll and tank state amplitudes. While the results in [14] were for

---

[2]Note that in [17, 18] the signage is wrong for the tank-induced moment in roll in [17, 18]. [18, (12.54b) p. 266)] reads (neglecting sway and yaw motions) $(I_{44} + a_{44})\ddot{x}_4 + b_{44}\dot{x}_4 + c_{44}x_4 - [a_{4\tau}\ddot{\tau} + c_{4\tau}\tau] = F_{w40}\sin(\omega_e t + \gamma_4)$, but should read $(I_{44} + a_{44})\ddot{x}_4 + b_{44}\dot{x}_4 + c_{44}x_4 + [a_{4\tau}\ddot{\tau} + c_{4\tau}\tau] = F_{w40}\sin(\omega_e t + \gamma_4)$. This error is propagated throughout [17, 18].

a rectangular-prism tank, and not in parametric roll, it seems reasonable that the suggested simplifications would also be applicable in this case. This suggests the model

$$\mathbf{M}_0\ddot{\mathbf{q}} + \mathbf{D}(\dot{\mathbf{q}})\dot{\mathbf{q}} + \mathbf{K}_0(t)\mathbf{q} = \mathbf{b}u, \tag{12.21}$$

where

$$\mathbf{M}_0 \triangleq \mathbf{M}(0)$$

$$\mathbf{K}_0(t) \triangleq \left.\frac{\partial \mathbf{k}}{\partial \mathbf{q}}\right|_{\mathbf{q}=0}$$

$$= \begin{bmatrix} \bar{k}_{44} + mgz_g^b + k_{\phi t}\cos(\omega_e t + \alpha_\phi) + 2g\rho_t \int_0^{\varsigma_0} z_t^b A \,\mathrm{d}\sigma & -2g\rho_t A_0 y_t^b(\varsigma_0) \\ -2g\rho_t A_0 y_t^b(\varsigma_0) & -2g\rho_t \frac{A_0^2}{A(\varsigma_0)} \frac{\mathrm{d}z_t^b}{\mathrm{d}\sigma}(\varsigma_0) \end{bmatrix}.$$

We note that $y_t^b(\varsigma_0) < 0$ and $\frac{\mathrm{d}z_t^b}{\mathrm{d}\sigma}(\varsigma_0) < 0$. We refer to (12.21) as the nominal model.
    We define

$$\mathbf{x} \triangleq [\mathbf{q}^\top, \dot{\mathbf{q}}^\top]^\top \in \mathbb{R}^4 \tag{12.22}$$

and rewrite the dynamics as

$$\dot{\mathbf{x}} = \mathbf{A}\mathbf{x} + \mathbf{B}u + \mathbf{G}(t)\mathbf{x} + \mathbf{g}(\dot{\mathbf{q}}), \tag{12.23}$$

where

$$\mathbf{A} = \begin{bmatrix} \mathbf{0}_{2\times 2} & \mathbf{I}_2 \\ -\mathbf{M}_0^{-1}\mathbf{K}_l & -\mathbf{M}_0^{-1}\mathbf{D}_0 \end{bmatrix} \in \mathbb{R}^{4\times 4}$$

$$\mathbf{K}_l = \begin{bmatrix} \bar{k}_{44} + mgz_g^b + 2g\rho_t \int_0^{\varsigma_0} z_t^b A \,\mathrm{d}\sigma & -2g\rho_t A_0 y_t^b(\varsigma_0) \\ -2g\rho_t A_0 y_t^b(\varsigma_0) & -2g\rho_t \frac{A_0^2}{A(\varsigma_0)} \frac{\mathrm{d}z_t^b}{\mathrm{d}\sigma}(\varsigma_0) \end{bmatrix} \triangleq \begin{bmatrix} k_{44} & k_{4t} \\ k_{4t} & k_{tt} \end{bmatrix} \in \mathbb{R}^{2\times 2}$$

$$\mathbf{B} = \begin{bmatrix} \mathbf{0}_{2\times 1} \\ \mathbf{M}_0^{-1}\mathbf{b} \end{bmatrix} \in \mathbb{R}^{4\times 1}$$

$$\mathbf{G}(t) = \begin{bmatrix} \mathbf{0}_{2\times 2} & \mathbf{0}_{2\times 2} \\ -\mathbf{M}_0^{-1}\begin{bmatrix} k_{\phi t}\cos(\omega_e t + \alpha_\phi) & 0 \\ 0 & 0 \end{bmatrix} & \mathbf{0}_{2\times 2} \end{bmatrix} \in \mathbb{R}^{4\times 4}$$

$$\mathbf{g}(\dot{\mathbf{q}}) = \begin{bmatrix} \mathbf{0}_{2\times 1} \\ -\mathbf{M}_0^{-1}\mathbf{D}_n(\dot{\mathbf{q}})\dot{\mathbf{q}} \end{bmatrix} \in \mathbb{R}^{4\times 1}.$$

Theorem 12.1 (U-tank control). *The origin of the system* (12.23) *is globally (uniformly) exponentially stabilized (following [16, Definition 4.5][3]) by the controller*

$$u = d_{tt,n}|\dot{q}_t|\dot{q}_t - \mathbf{K}_p\mathbf{x}, \tag{12.24}$$

where $\mathbf{K}_p = [K_{p,1}, K_{p,2}, K_{p,3}, K_{p,4}] \in \mathbb{R}^{1\times4}$ *is a matrix such that* $\mathbf{A} - \mathbf{BK}_p$ *is Hurwitz and the eigenvalues of* $\mathbf{A} - \mathbf{BK}_p$ *chosen such that*

$$\lambda_{\max}(\mathbf{P}) < \frac{1}{2k_{\phi t}\|\mathbf{M}_0^{-1}\|_2}, \tag{12.25}$$

where $\mathbf{P}$ *is the solution to the Lyapunov equation*

$$\mathbf{P}(\mathbf{A} - \mathbf{BK}_p) + (\mathbf{A} - \mathbf{BK}_p)^\top\mathbf{P} = -\mathbf{I}_4$$

and $\lambda_{\max}(\mathbf{P})$ *is the maximum eigenvalue of* $\mathbf{P}$. *Moreover, a* $\mathbf{K}_p$ *such that* (12.25) *is satisfied can always be found.*

Proof. See Appendix 3.   □

If we take a closer look at the controller (12.24), it cancels the nonlinear tank damping. This damping is "good" damping; in the absence of the time-varying disturbance (setting $k_{\phi t} = 0$) it is fairly straight-forward to show that the origin of the system (12.23) is GAS by using the energy-like Lyapunov function $\bar{V} = \dot{q}^\top\mathbf{M}_0\dot{q} + q^\top\mathbf{K}_l q$ (via the Krasowskii–LaSalle theorem [16, Theorem 4.4]; this theorem cannot be used for time-varying systems).

It is therefore reasonable to believe that this damping term is also beneficial in the presence of the time-varying disturbance ($k_{\phi t} \neq 0$). However, proving this has shown itself to be difficult, and the controller is therefore canceling this term.

12.5   Simulation Study

We simulated the full system (12.20) both with and without the controller (12.24) to test the validity and the robustness of the controller. For comparison, we also simulated the controlled nominal system (12.23).

---

[3]By this definition, exponential stability is stronger than uniform asymptotic stability, and thus the uniformity is implied.

The tank functions y_t^b, z_t^b, and A were given by

$$
y_t^b(\sigma) = \begin{cases}
\frac{w}{2} & \forall\ \sigma \in (-\infty, -w/2 - \varepsilon] \\
-a_0 - a_1\sigma - a_2\sigma^2 & \forall\ \sigma \in [-w/2 - \varepsilon, -w/2 + \varepsilon] \\
-\sigma & \forall\ \sigma \in [-w/2 + \varepsilon, w/2 - \varepsilon] \\
a_0 - a_1\sigma + a_2\sigma^2 & \forall\ \sigma \in [w/2 - \varepsilon, w/2 + \varepsilon] \\
-\frac{w}{2} & \forall\ \sigma \in [w/2 + \varepsilon, \infty)
\end{cases}
\tag{12.26}
$$

$$
z_t^b(\sigma) = \begin{cases}
r_d + \frac{w}{2} + \sigma & \forall\ \sigma \in (-\infty, -w/2 - \varepsilon] \\
-b_0 - b_1\sigma - b_2\sigma^2 & \forall\ \sigma \in [-w/2 - \varepsilon, -w/2 + \varepsilon] \\
r_d & \forall\ \sigma \in [-w/2 + \varepsilon, w/2 - \varepsilon] \\
-b_0 + b_1\sigma - b_2\sigma^2 & \forall\ \sigma \in [w/2 - \varepsilon, w/2 + \varepsilon] \\
r_d + \frac{w}{2} - \sigma & \forall\ \sigma \in [w/2 + \varepsilon, \infty)
\end{cases}
\tag{12.27}
$$

$$
A(\sigma) = \begin{cases}
A_r & \forall\ \sigma \in (-\infty, -w/2 + \varepsilon] \\
c_0 + c_1\sigma & \forall\ \sigma \in [-w/2 - \varepsilon, -w/2 + \varepsilon] \\
A_d & \forall\ \sigma \in [-w/2 + \varepsilon, w/2 - \varepsilon] \\
c_0 - c_1\sigma & \forall\ \sigma \in [w/2 - \varepsilon, w/2 + \varepsilon] \\
A_r & \forall\ \sigma \in [w/2 + \varepsilon, \infty)
\end{cases}
\tag{12.28}
$$

with $\varepsilon \ll w/2$ and

$$
a_0 = \frac{\left(\varepsilon - \frac{w}{2}\right)^2}{4\varepsilon}, \quad a_1 = \frac{w + 2\varepsilon}{4\varepsilon}, \quad a_2 = b_2 = \frac{1}{4\varepsilon}, \quad b_0 = \frac{\left(\varepsilon - \frac{w}{2}\right)^2}{4\varepsilon} - r_d,
$$

$$
b_1 = \frac{w - 2\varepsilon}{4\varepsilon}, \quad c_0 = \frac{2(A_d + A_r)\varepsilon + w(A_d - A_r)}{4\varepsilon}, \quad c_1 = \frac{A_d - A_r}{2\varepsilon}.
$$

Note that this choice of a_i, b_i, c_i ensures that y_t^b, $z_t^b \in C^1$ and that $A \in C^0$. A tank described by these functions has a centerline function describing half a rounded rectangle. The tank state q_t was limited so that $|q_t| \leq q_{t,\max} = V_t/(2A_0)$ so that there always is tank fluid at the tank center point $\sigma = 0$.

Simulation parameters can be found in Table 12.1. The ship parameters J_{11}, $m_{a,44}$, d_{44}, $\bar{k}_{44}$, $k_{\phi t}$, and k_3 were taken from [13, Experiment 1174]. The tank damping parameters are based on experimental values from [14] and formulas found in [12] and [18]. The encounter frequency ω_e was chosen to be twice the natural roll frequency, when the system is known to parametrically oscillate.

The uncontrolled nominal system (12.23) had eigenvalues

$$
\lambda(\mathbf{A}) \approx [-0.0049 \pm 0.3327i, -0.0051 \pm 0.2558i]
$$

while the controlled system had eigenvalues

$$
\lambda(\mathbf{A} - \mathbf{BK}_p) \approx [-0.0196 \pm 0.3327i, -0.0206 \pm 0.2558i].
$$

With the parameters of Table 12.1, V_t was computed to be $V_t \approx 337.8\ \mathrm{m}^3$. As per the standard rules of u-tank design [18], the tank is dimensioned so that the natural frequency of the tank (here, 0.2978 rad/s) is chosen to be approximately equal to the natural roll frequency (here, 0.2972 rad/s).

Table 12.1 Simulation parameters

Parameter	Value Unit	Parameter	Value Unit
J_{11}	$1.4014\text{E}10\,\text{kg}\,\text{m}^2$	A_r	$30\,\text{m}^2$
$m_{a,44}$	$2.17\text{E}9\,\text{kg}\,\text{m}^2$	A_d	$3.6145\,\text{m}^2$
d_{44}	$3.1951\text{E}8\,\text{kg}\,\text{m}^2/\text{s}$	w	$27\,\text{m}$
d_{tt}	$2.4618\text{E}3\,\text{kg}\,\text{m}/\text{s}$	ε	$1\,\text{m}$
$d_{tt,n}$	$2.2742\text{E}5\,\text{kg}\,\text{m}$	r_d	$2\,\text{m}$
$\bar{k}_{44}$	$2.2742\text{E}9\,\text{kg}\,\text{m}^2/\text{s}^2$	ς_0	$17.5\,\text{m}$
$k_{\phi t}$	$5.0578\text{E}8\,\text{kg}\,\text{m}^2/\text{s}^2$	A_0	$30\,\text{m}^2$
k_3	$2.974\text{E}9\,\text{kg}\,\text{m}^2/\text{s}^2$	$\phi(t_0)$	$5\,^\circ$
m	$7.64688\text{E}7\,\text{kg}$	$\dot{\phi}(t_0)$	$0\,^\circ/\text{s}$
z_g^b	$-1.12\,\text{m}$	$q_t(t_0)$	$0\,\text{m}$
g	$9.81\,\text{m}/\text{s}^2$	$\dot{q}_t(t_0)$	$0\,\text{m}/\text{s}$
ρ_l	$1{,}000\,\text{kg}/\text{m}^3$	$K_{p,1}$	$3.9935\text{E}5\,\text{kg}\,\text{m}/\text{s}^2$
ω_c	$0.594\,\text{rad}/\text{s}$	$K_{p,2}$	$7.2833\text{E}3\,\text{kg}/\text{s}^2$
α_ϕ	$0\,\text{rad}$	$K_{p,3}$	$-4.1664\text{E}5\,\text{kg}\,\text{m}/\text{s}$
$q_{t,\max}$	$5.6307\,\text{m}$	$K_{p,4}$	$3.9916\text{E}5\,\text{kg}/\text{s}$

The results of the simulation study can be seen in Figs. 12.4 and 12.5. We can clearly see that the system trajectory converges to the origin. Note also that the trajectory of the nominal system is almost identical to that of the true system.

From Fig. 12.4, we can also see that a passive (uncontrolled) tank is capable of reducing the roll angle compared to not having a tank at all[4] (a reduction in maximum roll angle of approximately 21° to 7°). However, both roll and the tank fluid will end up in steady-state oscillations.

The issue of correctly tuning the natural frequency of the tank fluid bears some consideration. For a rectangular-prism tank (and a tank like the one used in the simulations), the natural frequency can be changed by adjusting the fluid level ς_0 or the ratio of A_r and A_d (cross-sectional area of the reservoirs and the duct, respectively). The latter can of course only be done when the tank is constructed.

Unfortunately, the natural frequency is quite insensitive to changes in ς_0 [18]. It is therefore almost impossible to change the natural frequency of the tank after it has been built. However, there can be quite some uncertainty in the natural roll frequency, which can also depend on loading conditions [4, 18].

If the natural frequency of the tank is not properly tuned, the effect of a passive tank can be drastically reduced. The more badly tuned it is, the less effective the tank is. However, as proven in Theorem 12.1, an active (controlled) tank will still be able to stabilize the origin of the system.

---

[4]In [12] it was concluded that using a passive tank did not noticeably reduce the roll angle in parametric resonance, but this tank had a badly tuned natural frequency.

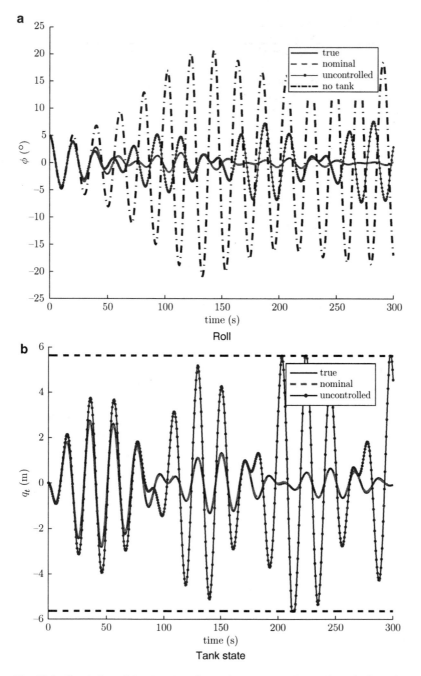

Fig. 12.4 Simulation of the closed- and open-loop system. True and nominal graphs are closed-loop simulations of (12.20) and (12.21), respectively. Uncontrolled is open-loop simulation of (12.20)

The power and energy consumptions of the control system is also worth noting. As can be seen from Fig. 12.5, at peak the controller requires a force of about 250 kN and 210 kW. By integrating the power consumption (over 1,000 s), the total energy use can be found to be approximately 8 MJ.

These numbers require some context. If force on the tank fluid is applied by using high-pressure air in the reservoirs, the pressure difference in the two reservoirs has to be about 8.5 kPa, or 0.085 bar. When considering the maximum power consumption 21 kW, bear in mind that the actuator is moving 337.8 metric tons of fluid, and that the ship itself has a mass of 76,500 metric tons and is likely to have a fairly large power system. The total energy consumption equals about 0.23 liters of gasoline burned in a combustion engine (using 34.8 MJ/liter of gasoline [2]). All in all, the control system if fairly modest in scale.

12.6   Conclusions

This work presents two important contributions: A novel model of u-tanks, suitable for u-tanks of any shape and system response of any magnitude; and a u-tank control system capable of exponentially driving the roll angle to zero and the tank fluid to its equilibrium state during parametric roll resonance.

The proposed model has two degrees of freedom; roll, and one for the motion of the tank fluid. To derive the model, the inherently complex motion of a fluid in a tank was modeled as a one-dimensional flow. While technically not true, the model is only designed to capture the tank fluid's and the ship's mutual effect on each other. Only the macroscopic fluid effects are likely to be relevant in this case.

Unlike most existing u-tank models, which can only be used for tanks of a very specific shape, the new model can describe u-tanks of arbitrary shape. The new model also captures the inherently nonlinear behavior of the system, and is valid for large system motions. Existing models are largely linear and assume small motions.

The control system was developed and its stability properties proven for a simplification of the 2-DOF model. Under the assumption of no constraints on the states or the input, the controller renders the origin of the closed-loop system globally (uniformly) exponentially stable.[5]

The controller was tested in simulation on both the nominal (simplified) system and the full nonlinear system. The responses of these two systems was virtually indistinguishable in simulation, and the origin was stabilized, as shown theoretically. The power and energy consumptions are quite reasonable; total energy required to stabilize parametric roll is equivalent to a quarter of a liter of gasoline, and peak power requirements are quite modest given the size of the ship.

---

[5]In the presence of limitations on the tank state or the input, the origin of the controlled system might only be locally (uniformly) exponentially stable.

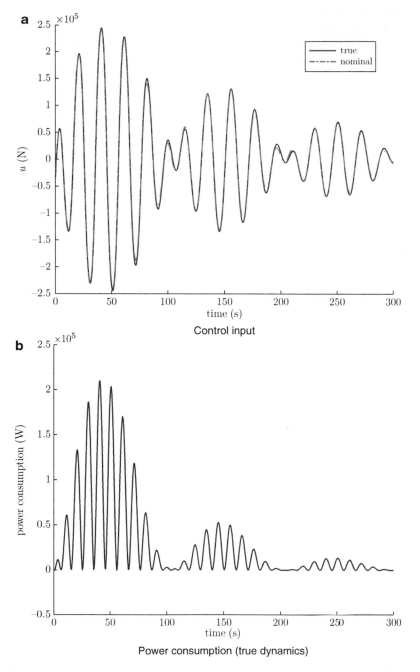

Fig. 12.5 Simulation of the closed-loop system. True and nominal graphs are simulations of (12.20) and (12.21), respectively

Simulations also showed that a passive (uncontrolled) tank would be able to reduce the roll angle significantly ($21°$ to $7°$) in the presence of parametric roll, but not to drive the roll angle to zero. It can also be shown that a passive tank only works if it is correctly tuned (i.e., the natural tank frequency is identical to the natural roll frequency). Changing the natural tank frequency without rebuilding the tank is almost impossible. The natural roll frequency can change depending on loading conditions, sailing conditions, and simple discrepancies between theoretical design and practical implementation makes it, to a certain degree, unknown. This makes correctly tuning the tank difficult.

However, the controlled tank would still be able to drive roll to zero, even with a poorly tuned tank.

Acknowledgements This work was funded by the Centre for Ships and Ocean Structures (CeSOS), Norwegian University of Science and Technology, Norway, and the Norwegian Research Council.

Appendix 1

In this appendix, the potential energy of the ship–tank system is derived, proving Proposition 12.1.

An infinitesimal volume block dV of the tank or ship at a position $\mathbf{r}$ has density $\rho(\mathbf{r})$ given by

$$\rho(\mathbf{r}) = \begin{cases} \rho_t & \text{in the tank} \\ \rho_s(\mathbf{r}) & \text{in the ship} \end{cases} \tag{12.29}$$

and is at a height $h(\mathbf{r})$ above some arbitrary zero point. We note that h is the zero level minus the inertial z-component of $\mathbf{r}$, that is,

$$h(\mathbf{r}) = h_0 - [0,0,1]\mathbf{r}^n = h_0 - [0,0,1]\mathbf{Rr}^b. \tag{12.30}$$

The negative signage is because the z-axis has the same direction as the gravity field.

The potential energy dU of the volume block is then given by

$$dU = g\rho(\mathbf{r})h(\mathbf{r})dV, \tag{12.31}$$

which, in the body frame, can be written

$$dU = g\rho(\mathbf{r}^b)h(\mathbf{r}^b)\,dV = g\rho(\mathbf{r}^b)\left(h_0 - [0,0,1]\mathbf{Rr}^b\right)\,dV. \tag{12.32}$$

The total potential energy U of the ship and the tank fluid is then given by

$$U = \int_{\text{ship and tank}} dU = gm_0 h_0 - g[0,0,1]\mathbf{R}\left[\int_{\text{ship}} \rho_s(\mathbf{r}^b)\mathbf{r}^b \, dV + \rho_t \int_{\text{tank}} \mathbf{r}_t^b \, dV\right]$$

$$= gm_0 h_0 - g[0,0,1]\mathbf{R}\left[m\mathbf{r}_g^b + \rho_t \int_{-\varsigma_s(q_t)}^{\varsigma_p(q_t)} A(\sigma)\mathbf{r}_t^b(\sigma) \, d\sigma\right] \tag{12.33}$$

by the definition of the center of gravity, where m_0 is the combined mass of the ship and tank fluid. From (12.1), $[0,0,1]\mathbf{R} = [0,\sin(\phi),\cos(\phi)]$ and, by assumption, $\mathbf{r}_g^b = [x_g^b, 0, z_g^b]^\top$. This gives

$$U = gm_0 h_0 - mgz_g^b \cos(\phi) - g\rho_t \int_{-\varsigma_s(q_t)}^{\varsigma_p(q_t)} A(\sigma)y_t^b(\sigma) \, d\sigma \sin(\phi)$$

$$- g\rho_t \int_{-\varsigma_s(q_t)}^{\varsigma_p(q_t)} A(\sigma)z_t^b(\sigma) \, d\sigma \cos(\phi). \tag{12.34}$$

A priori, we know that $\mathbf{q} = \mathbf{0}$ is an equilibrium point for the system, so we choose $U(\mathbf{q} = \mathbf{0}) = 0$. This gives

$$U(\mathbf{q} = \mathbf{0}) = gm_0 h_0 - mgz_g^b - 2g\rho_t \int_0^{\varsigma_0} A(\sigma)z_t^b(\sigma) \, d\sigma = 0,$$

since $\varsigma_s(0) = \varsigma_p(0) = \varsigma_0$ and $A(\sigma)z_t^b(\sigma)$ is an even function. We therefore choose

$$gm_0 h_0 = mgz_g^b + 2g\rho_t \int_0^{\varsigma_0} A(\sigma)z_t^b(\sigma) \, d\sigma = 0$$

giving

$$U(\mathbf{q}) = mgz_g^b + 2g\rho_t \int_0^{\varsigma_0} z_t^b(\sigma)A(\sigma) \, d\sigma - g\rho_t \int_{-\varsigma_s(q_t)}^{\varsigma_p(q_t)} y_t^b(\sigma)A(\sigma) \, d\sigma \sin(\phi)$$

$$- \left[mgz_g^b + g\rho_t \int_{-\varsigma_s(q_t)}^{\varsigma_p(q_t)} z_t^b(\sigma)A(\sigma) \, d\sigma\right]\cos(\phi), \tag{12.35}$$

which we recognize as (12.15).

Appendix 2

In this appendix, the kinetic energy of the ship–tank system is derived, proving Proposition 12.2.

An infinitesimal volume block dV of the tank or ship at a position $\mathbf{r}$ in the body frame has density $\rho(\mathbf{r})$ given by (12.29) and velocity $\mathbf{v}(\mathbf{r})$ given by

$$
\mathbf{v}(\mathbf{r}) =
\begin{cases}
\boldsymbol{\omega} \times \mathbf{r} + \frac{A_0 \dot{q}_t}{A(\sigma)} \frac{d\bar{\mathbf{r}}_t}{d\sigma}(\sigma) & \text{in the tank} \\
\boldsymbol{\omega} \times \mathbf{r} & \text{in the ship}
\end{cases},
\tag{12.36}
$$

where $\boldsymbol{\omega}$ is the angular velocity of the ship. The velocity of the tank fluid comes from (12.13).

The volume block has kinetic energy dT given by

$$
dT = \frac{1}{2}\rho(\mathbf{r})\|\mathbf{v}(\mathbf{r})\|_2^2 dV,
\tag{12.37}
$$

which, in the body frame, can be written

$$
dT = \frac{1}{2}\rho(\mathbf{r}^b)\|\mathbf{v}^b(\mathbf{r}^b)\|_2^2 dV.
\tag{12.38}
$$

The total kinetic energy T of the ship and the tank fluid is then given by

$$
\begin{aligned}
T &= \frac{1}{2}\int_{\text{ship and tank}} \rho(\mathbf{r}^b)\|\mathbf{v}^b(\mathbf{r}^b)\|_2^2 \, dV \\
&= \frac{1}{2}\int_{\text{ship}} \rho_s(\mathbf{r}^b)\|\boldsymbol{\omega}^b \times \mathbf{r}^b\|_2^2 \, dV + \frac{\rho_t}{2}\int_{\text{tank}} \left\|\boldsymbol{\omega}^b \times \mathbf{r}^b + \frac{A_0\dot{q}_t}{A(\sigma)}\frac{d\bar{\mathbf{r}}_t^b}{d\sigma}\right\|_2^2 dV \\
&= -\frac{1}{2}\boldsymbol{\omega}^{b\top}\left[\int_{\text{ship}} \rho_s(\mathbf{r}^b)S^2(\mathbf{r}^b)\,dV\right]\boldsymbol{\omega}^b \\
&\quad + \frac{\rho_t}{2}\int_{-\varsigma_s(q_t)}^{\varsigma_p(q_t)} A(\sigma)\left\|\dot{\phi}[0, -z_t^b(\sigma), y_t^b(\sigma)]^\top + \frac{A_0\dot{q}_t}{A(\sigma)}\frac{d\bar{\mathbf{r}}_t^b}{d\sigma}(\sigma)\right\|_2^2 d\sigma.
\end{aligned}
$$

We note that, by definition,

$$
\mathbf{J} = -\int_{\text{ship}} \rho_s(\mathbf{r}^b)S^2(\mathbf{r}^b)\,dV \in \mathbb{R}^{3\times 3}
$$

is the moment of inertia of the ship and assumed a priori known. Furthermore,

$$
\boldsymbol{\omega}^{b\top}\mathbf{J}\boldsymbol{\omega}^b = \dot{\phi}^2[1,0,0]\mathbf{J}[1,0,0]^\top = \dot{\phi}^2 J_{11},
$$

where $J_{11} > 0 \in \mathbb{R}$ is the top left element of J, that is, the moment of inertia about the (body) x-axis. Thus,

$$
\begin{aligned}
T &= \frac{1}{2} J_{11} \dot{\phi}^2 + \frac{\rho_t}{2} \int_{-\varsigma_s(q_t)}^{\varsigma_p(q_t)} A(\sigma) \left\| \dot{\phi}[0, -z_t^b(\sigma), y_t^b(\sigma)]^\top + \frac{A_0 \dot{q}_t}{A(\sigma)} \frac{d\bar{r}_t^b}{d\sigma}(\sigma) \right\|_2^2 d\sigma \\
&= \frac{1}{2} \left[J_{11} + \rho_t \int_{-\varsigma_s(q_t)}^{\varsigma_p(q_t)} A(\sigma) \left[[y_t^b(\sigma)]^2 + [z_t^b(\sigma)]^2 \right] d\sigma \right] \dot{\phi}^2 + \frac{\dot{q}_t^2}{2} \int_{-\varsigma_s(q_t)}^{\varsigma_p(q_t)} \frac{\rho_t A_0^2}{A(\sigma)} d\sigma \\
&\quad + \dot{\phi} \dot{q}_t \rho_t A_0 \int_{-\varsigma_s(q_t)}^{\varsigma_p(q_t)} \left[y_t^b(\sigma) \frac{dz_t^b}{d\sigma}(\sigma) - \frac{dy_t^b}{d\sigma}(\sigma) z_t^b(\sigma) \right] d\sigma
\end{aligned} \tag{12.39}
$$

since $\left\| d\bar{r}_t^b / d\sigma \right\|_2^2 = 1$.

Defining

$$
J_t(q_t) = \rho_t \int_{-\varsigma_s(q_t)}^{\varsigma_p(q_t)} A(\sigma) [[y_t^b(\sigma)]^2 + [z_t^b(\sigma)]^2] d\sigma
$$

$$
m_{4t}(q_t) = \rho_t A_0 \int_{-\varsigma_s(q_t)}^{\varsigma_p(q_t)} \left[y_t^b(\sigma) \frac{dz_t^b}{d\sigma}(\sigma) - \frac{dy_t^b}{d\sigma}(\sigma) z_t^b(\sigma) \right] d\sigma
$$

$$
\bar{m}_t(q_t) = \rho_t A_0^2 \int_{-\varsigma_s(q_t)}^{\varsigma_p(q_t)} \frac{1}{A(\sigma)} d\sigma
$$

$$
\mathbf{M}_t(q_t) = \begin{bmatrix} J_{11} + J_t(q_t) & m_{4t}(q_t) \\ m_{4t}(q_t) & \bar{m}_t(q_t) \end{bmatrix} \in \mathbb{R}^{2 \times 2},
$$

we can rewrite (12.39) as

$$
T(q_t, \dot{q}) = \frac{1}{2} \dot{q}^\top \mathbf{M}_t(q_t) \dot{q}, \tag{12.40}
$$

which we recognize as (12.16).

Appendix 3

This section contains the proof of Theorem 12.1.

We immediately note that by choosing $u = d_{tt,n} |\dot{q}_t| \dot{q}_t + v$, we can transform the dynamics of the system (12.23) into the linear system

$$
\dot{x} = \mathbf{A}x + \mathbf{B}v + \mathbf{G}(t)x. \tag{12.41}
$$

The term $\mathbf{G}(t)\mathbf{x}$ can be viewed as a time-varying disturbance to the system. We cannot directly cancel it, both because the parameters are unlikely to be known and because roll is not directly actuated.

The unperturbed system has the dynamics

$$\dot{\mathbf{x}} = \mathbf{A}\mathbf{x} + \mathbf{B}v. \tag{12.42}$$

This system is controllable if its controllability matrix $\mathscr{C} = \begin{bmatrix} \mathbf{B} & \mathbf{AB} & \mathbf{A}^2\mathbf{B} & \mathbf{A}^3\mathbf{B} \end{bmatrix}$ has full rank (i.e., is nonsingular) [1].

The controllability matrix is given by

$$\mathscr{C} = \begin{bmatrix} \mathbf{M}_0^{-1} & \mathbf{0}_{2\times 2} \\ \mathbf{0}_{2\times 2} & \mathbf{M}_0^{-1} \end{bmatrix} \bar{\mathscr{C}}, \quad \bar{\mathscr{C}} \triangleq \begin{bmatrix} 0 & 0 & a_1 & b_1 \\ 0 & 1 & a_2 & b_2 \\ 0 & a_1 & b_1 & c_1 \\ 1 & a_2 & b_2 & c_2 \end{bmatrix}$$

$$\begin{bmatrix} a_1 \\ a_2 \end{bmatrix} = -\mathbf{D}_0\mathbf{M}_0^{-1}\mathbf{B}$$

$$\begin{bmatrix} b_1 \\ b_2 \end{bmatrix} = (\mathbf{D}_0\mathbf{M}_0^{-1}\mathbf{D}_0 - \mathbf{K}_l)\mathbf{M}_0^{-1}\mathbf{B}$$

$$\begin{bmatrix} c_1 \\ c_2 \end{bmatrix} = (\mathbf{K}_l\mathbf{M}_0^{-1}\mathbf{D}_0 + \mathbf{D}_0\mathbf{M}_0^{-1}\mathbf{K}_l - \mathbf{D}_0\mathbf{M}_0^{-1}\mathbf{D}_0\mathbf{M}_0^{-1}\mathbf{D}_0)\mathbf{M}_0^{-1}\mathbf{B}.$$

From [1], we have that $\mathrm{rank}(\mathscr{C}) = \mathrm{rank}(\bar{\mathscr{C}})$ since the matrix $\mathrm{diag}(\mathbf{M}_0^{-1}, \mathbf{M}_0^{-1}) \in \mathbb{R}^{4\times 4}$ is nonsingular.

$\bar{\mathscr{C}}$ has full rank if its determinant is nonzero [1]. Since

$$\det(\bar{\mathscr{C}}) = -\frac{1}{\det(\mathbf{M}_0)^2}\left([k_{44}m_{12} - k_{4t}m_{11}]^2 + d_1^2 k_{4t}m_{12}\right), \tag{12.43}$$

this gives the condition

$$[k_{44}m_{12} - k_{4t}m_{11}]^2 + d_1^2 k_{4t}m_{12} \neq 0. \tag{12.44}$$

As long as this condition is satisfied, (12.42) is controllable. Since all the parameters in (12.44) are strictly positive, this condition is always satisfied.

As (12.42) is controllable, we can select a $v = -\mathbf{K}_p\mathbf{x}$, $\mathbf{K}_p \in \mathbb{R}^{1\times 4}$, such that $\mathbf{A} - \mathbf{B}\mathbf{K}_p$ is Hurwitz, and can place the poles arbitrarily far into the left half-plane [1]. The closed-loop system is then given by

$$\dot{\mathbf{x}} = (\mathbf{A} - \mathbf{B}\mathbf{K}_p)\mathbf{x} + \mathbf{G}(t)\mathbf{x}. \tag{12.45}$$

From [16], we know that for any positive definite symmetric matrix $\mathbf{N}$, there exists a positive definite symmetric matrix $\mathbf{P}$ such that

$$\mathbf{P}(\mathbf{A} - \mathbf{B}\mathbf{K}_p) + (\mathbf{A} - \mathbf{B}\mathbf{K}_p)^\top\mathbf{P} = -\mathbf{N}. \tag{12.46}$$

We choose Lyapunov function candidate

$$V(\mathbf{x}) = \mathbf{x}^\top \mathbf{P} \mathbf{x} \tag{12.47}$$

with $\mathbf{P}$ as the solution to (12.46). We note that it is positive definite and decrescent. Specifically,

$$\lambda_{\min}(\mathbf{P}) \|\mathbf{x}\|_2^2 \le V(\mathbf{x}) \le \lambda_{\max}(\mathbf{P}) \|\mathbf{x}\|_2^2, \tag{12.48}$$

where $\lambda_{\min}(\mathbf{P})$ and $\lambda_{\max}(\mathbf{P})$ are the minimum and maximum eigenvalues of $\mathbf{P}$.

The time derivative of V along the trajectories of the closed-loop system (12.45) is given by

$$
\begin{aligned}
\dot{V}(\mathbf{x}) &= \mathbf{x}^\top \left[\mathbf{P}(\mathbf{A} - \mathbf{B}\mathbf{K}_p) + (\mathbf{A} - \mathbf{B}\mathbf{K}_p)^\top \mathbf{P} \right] \mathbf{x} + \mathbf{x}^\top \left[\mathbf{P}\mathbf{G}(t) + \mathbf{G}^\top(t)\mathbf{P} \right] \mathbf{x} \\
&= -\mathbf{x}^\top \mathbf{N} \mathbf{x} + \mathbf{x}^\top \left[\mathbf{P}\mathbf{G}(t) + \mathbf{G}^\top(t)\mathbf{P} \right] x \\
&\le -\lambda_{\min}(\mathbf{N}) \|\mathbf{x}\|_2^2 + 2\lambda_{\max}(\mathbf{P}) \max_t \|\mathbf{G}(t)\|_2 \|\mathbf{x}\|_2^2 \\
&\le -\lambda_{\min}(\mathbf{N}) \|\mathbf{x}\|_2^2 + 2k_{\phi t} \lambda_{\max}(\mathbf{P}) \|\mathbf{M}_0^{-1}\|_2 \|\mathbf{x}\|_2^2 \\
&= - \left[\lambda_{\min}(\mathbf{N}) - 2k_{\phi t} \lambda_{\max}(\mathbf{P}) \|\mathbf{M}_0^{-1}\|_2 \right] \|\mathbf{x}\|_2^2, \tag{12.49}
\end{aligned}
$$

where we have used that

$$\max_t \|\mathbf{G}(t)\|_2 = \left\| \begin{bmatrix} \mathbf{0}_{2\times2} & \mathbf{0}_{2\times2} \\ -\mathbf{M}_0^{-1} \begin{bmatrix} k_{\phi t} & 0 \\ 0 & 0 \end{bmatrix} & \mathbf{0}_{2\times2} \end{bmatrix} \right\|_2 \le k_{\phi t} \|\mathbf{M}_0^{-1}\|_2. \tag{12.50}$$

By [16, Theorem 4.10], the origin of the controlled system (12.45) is globally (uniformly) exponentially stable, as long as $\lambda_{\min}(\mathbf{N}) > 2k_{\phi t} \lambda_{\max}(\mathbf{P}) \|\mathbf{M}_0^{-1}\|_2$ or

$$k_{\phi t} < \frac{\lambda_{\min}(\mathbf{N})}{2\lambda_{\max}(P) \|\mathbf{M}_0^{-1}\|_2}.$$

The ratio $\lambda_{\min}(\mathbf{N})/\lambda_{\max}(\mathbf{P})$ is maximized by choosing $\mathbf{N} = \mathbf{I}_4$ [16]. Since we can choose the eigenvalues of $\mathbf{A} - \mathbf{B}\mathbf{K}_p$ arbitrarily far into the left half-plane, we can choose $\lambda_{\max}(\mathbf{P})$ arbitrarily, as this value depends on the eigenvalues of $\mathbf{A} - \mathbf{B}\mathbf{K}_p$ [1].

Thus, for any $k_{\phi t}$, we can find a controller such that the origin of the controlled system (12.45) is globally (uniformly) exponentially stable. $\square$

References

1. Chen, C.T.: Linear System Theory and Design. Oxford University Press (1999), New York, USA
2. Davis, S.C., Diegel, S.W.: Transportation Energy Data Book, 29th edn. Oak Ridge National Laboratory (2010), Oak Ridge, Tennessee, USA
3. Egeland, O., Gravdahl, J.T.: Modeling and Simulation for Automatic Control. Marine Cybernetics (2002), Trondheim, Norway
4. Faltinsen, O.M., Timokha, A.N.: Sloshing. Cambridge University Press (2009), New York, USA
5. Fossen, T.I.: Handbook of Marine Craft Hydrodynamics and Motion Control. John Wiley & Sons, Ltd. (2011), Chichester, UK
6. Frahm, H.: Results of trials of anti-rolling tanks at sea. Trans. of the Institution of Naval Architects **53** (1911)
7. France, W.N., Levadou, M., Treakle, T.W., Paulling, J.R., Michel, R.K., Moore, C.: An investigation of head sea parametric rolling and its influence on container lashing systems. SNAME Annual Meeting (2001), Orlando, Florida, USA
8. Galeazzi, R.: Autonomous supervision and control of parametric roll resonance. Ph.D. thesis, Technical University of Denmark (2009)
9. Galeazzi, R., Holden, C., Blanke, M., Fossen, T.I.: Stabilization of parametric roll resonance with active u-tanks via lyapunov control design. In: Proceedings of the European Control Conference (2009), Budapest, Hungary
10. Goldstein, H., Poole, C., Safko, J.: Classical Mechanics, 3rd edn. Addison Wesley (2002), San Francisco, California, USA
11. Goodrich, G.J.: Development and design of passive roll stabilizers. In: Transactions of the Royal Institution of Naval Architects (1968), London, UK
12. Holden, C., Galeazzi, R., Fossen, T.I., Perez, T.: Stabilization of parametric roll resonance with active u-tanks via lyapunov control design. In: Proceedings of the European Control Conference (2009), Budapest, Hungary
13. Holden, C., Galeazzi, R., Rodríguez, C.A., Perez, T., Fossen, T.I., Blanke, M., Neves, M.A.S.: Nonlinear container ship model for the study of parametric roll resonance. Modeling, Identification and Control **28**(4) (2007)
14. Holden, C., Perez, T., Fossen, T.I.: A Lagrangian approach to nonlinear modeling of anti-roll tanks. Ocean Engineering **38**(2–3), 341–359 (2011)
15. Kagawa, K., Fujita, K., Matsuo, M., Koukawa, H., Zensho, Y.: Development of tuned liquid damper for ship vibrations. In: Transactions of the West Japan Society of Naval Architects, vol. 81, p. 181 (1990)
16. Khalil, H.: Nonlinear Systems, 3rd edn. Prentice Hall (2002), Upper Saddle, New Jersey, USA
17. Lloyd, A.R.: Seakeeping: Ship Behaviour in Rough Weather. Ellis Horwood Series in Marine Technology. Ellis Horwood (1989), Chichester, UK
18. Lloyd, A.R.: Seakeeping: Ship Behaviour in Rough Weather. ARJM Lloyd (1998), Gosport, UK
19. Moaleji, R., Greig, A.R.: On the development of ship anti-roll tanks. Ocean Engineering **34**, 103–121 (2007)
20. Perez, T.: Patrol boat stabilizers. Tech. rep., ADI-Limited, Major Projects Group (2002)
21. Perez, T.: Ship Motion Control. Advances in Industrial Control. Springer-Verlag, London (2005)
22. Sellars, F.H., Martin, J.: Selection and evaluation of ship roll stabilization systems. Marine Technology, SNAME **29**(2), 84–101 (1992)

Part IV
Control of Parametric Resonance in Mechanical Systems

Chapter 13
Parametric and Direct Resonances in a Base-Excited Beam Carrying a Top Mass

Rob H.B. Fey, Niels J. Mallon, C. Stefan Kraaij, and Henk Nijmeijer

13.1 Introduction

The function of many structures in engineering practice is to carry a static load. To minimize costs, it is often desired to reduce the mass of the supporting structure as much as possible, while retaining a high stiffness. Thin-walled structures are often applied for this purpose (e.g. in aerospace and civil engineering), because of their favorable (high) stiffness to mass ratio. It is well-known that thin-walled structures are liable to buckling and static buckling analysis is often carried out to assess their static stability. In many situations, additionally to the static load, a dynamic load is present, e.g. due to motions of the base of the structure. These situations require assessment of the dynamic stability of the structure. Especially excitation frequencies, which bring the structure into resonance, may induce dynamic buckling/instability (large motions and deformations) of the structure. This may lead to damage to the structure or even to total collapse of the structure.

In the last years, there is increasing attention for modeling and analysis to assess the dynamic stability of structures. A well-known phenomenon, which may affect the dynamic stability of a structure, is the occurrence of parametric resonance.

R.H.B. Fey (✉) • H. Nijmeijer
Department of Mechanical Engineering, Eindhoven University of Technology,
PO Box 513, 5600 MB Eindhoven, The Netherlands
e-mail: R.H.B.Fey@tue.nl; H.Nijmeijer@tue.nl

N.J. Mallon
Centre for Mechanical and Maritime Structures, TNO Built Environment and Geosciences,
PO Box 49, 2600 AA Delft, The Netherlands
e-mail: Niels.Mallon@tno.nl

C.S. Kraaij
IHC Lagersmit BV, PO Box 5, 2960 AA Kinderdijk, The Netherlands
e-mail: CS.Kraaij@ihcmerwede.com

T.I. Fossen and H. Nijmeijer (eds.), *Parametric Resonance in Dynamical Systems*,
DOI 10.1007/978-1-4614-1043-0_13, © Springer Science+Business Media, LLC 2012

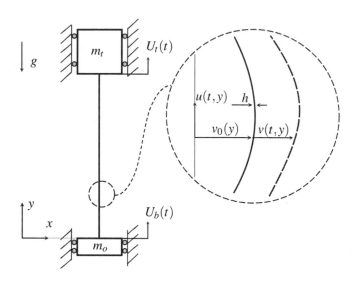

Fig. 13.1 Base-excited slender beam with top mass

Recent textbooks for studying the parametric resonance phenomenon and for dynamic stability assessment of structural elements are among others [1, 9, 17].

In this paper, nonlinear resonances in a coupled shaker-beam-top mass system are investigated both numerically and experimentally. The imperfect, vertical, slender beam carries a top mass and is coupled to and axially excited by the shaker at its base. The top mass can only move in vertical direction. The boundary conditions of the beam are clamped-clamped. Fig. 13.1 shows a schematic overview of the system. This can be seen as an archetype system for studying the dynamic stability of a load carrying, thin-walled structure. The weight of the top mass causes a compressive prestress not large enough to cause static buckling. The harmonic base excitation, however, may induce dynamic buckling due to parametric and/or direct resonance.

In literature, not many papers consider parametric excitation of a beam with a point mass attached to it. In [18, 20], the top mass is completely free in contrast to the situation in the current chapter. In [13], parametric excitation of a horizontal, simply supported elastic beam with a point mass attached to one end is studied.

In the current chapter, a semi-analytical modeling approach is used resulting in a low-dimensional model. Geometric imperfections of the beam and linear as well as quadratic viscous damping are taken into account. The model also includes models of the amplifier and shaker used in the experiments. Advanced numerical tools are used for calculating branches of periodic solutions and their local stability and for bifurcation analysis. In combination with the low-dimensional semi-analytical model, these tools permit fast parameter studies. Single and two mode discretizations for the beam are considered in obtaining steady-state responses. Numerically obtained steady-state response results are validated by experiments.

The outline of this chapter is as follows. In the next section, the semi-analytical amplifier-shaker-structure model will be derived and discussed. In Sect. 13.3, the experimental setup of the base-excited slender beam with top mass will be introduced. The steady-state response results, predicted by the semi-analytical approach and obtained experimentally, will be compared in Sect. 13.4. Finally, in Sect. 13.5, conclusions will be presented.

13.2 Semi-analytical Model

At the experimental setup used, see Sect. 13.3, base-excitation of the slender beam is realized by supplying an amplified harmonic input voltage to an electrodynamic shaker system. The resulting base acceleration will not be purely harmonic, will not have a constant amplitude but will be influenced by the dynamics of the shaker system carrying the slender beam with top mass. Response results for voltage excitation can thus not directly be compared with results for a prescribed harmonic base-acceleration as considered in [6]. To be able to compare the experimental results with the semi-analytical results in a quantitative manner, the equations of motion for the base-excited slender beam with top mass will be coupled with a model of the shaker. The derivation of this coupled model is the topic of this section.

13.2.1 Modeling of the Slender Beam with Top Mass

Figure 13.1 shows the beam under consideration with length L, width b, and thickness h. The beam is very slender, i.e. $h \ll L$. Consequently, the displacements of the beam will be dominated by changes in curvature allowing to assume the beam to be inextensible. In addition, the slender beam is considered to be initially not perfectly straight. In the initial stress free state, the transversal shape of the slender beam is denoted by $v_0(y)$. The axial displacement field relative to the absolute axial base displacement $U_b(t)$ is indicated by $u(t,y)$ and the transversal displacement field relative to $v_0(y)$ by $v(t,y)$.

The length of an infinitesimally small piece of the beam in the initial state is [8]

$$ds^2 = dy^2 + (v_{0,y}\,dy)^2. \tag{13.1}$$

Due to the inextensibility assumption, the length of ds remains constant. In the deformed state this length satisfies [8]

$$ds^2 = (dy + u_{,y}\,dy)^2 + ([v_{0,y}+v_{,y}]\,dy)^2. \tag{13.2}$$

By combining (13.1) and (13.2), the following inextensibility constraint results

$$u_{,y} = \sqrt{1 - 2v_{0,y}\,v_{,y} - v_{,y}^2} - 1. \tag{13.3}$$

In the Cartesian coordinate system x, y, the centerline of the deformed imperfect beam is described by the curve $[X(t, y), Y(t, y)]$, where $X(t, y) = v_0(y) + v(t, y)$ and $Y(t, y) = y + U_b(t) + u(t, y)$. The exact curvature of this curve follows from [4]

$$\kappa = \frac{X(t,y)_{,y}\,Y(t,y)_{,yy} - X(t,y)_{,yy}\,Y(t,y)_{,y}}{\left(X(t,y)_{,y}^2 + Y(t,y)_{,y}^2\right)^{\frac{3}{2}}}, \tag{13.4}$$

and can be evaluated in terms of (derivatives of) $v_0(y)$ and $v(t, y)$ solely, after substitution of (13.3). It is assumed that, depending on the maximum deflection, the constraint (13.3) and the curvature (13.4) can be accurately approximated by their Taylor series expansions in $v_{,y}$ and $v_{0,y}$ up to nth order with n sufficiently high. For example, the 3rd order expansions of (13.3) and (13.4) yield

$$u_{,y} = -v_{0,y}\,v_{,y} - \frac{1}{2}v_{,y}^2, \tag{13.5}$$

$$\kappa = \kappa_0 + v_{,yy} + \frac{1}{2}(v_{0,yy} + v_{,yy})v_{,y}^2 + v_{,yy}\,v_{0,y}\,v_{,y} - \frac{1}{2}v_{,yy}\,v_{0,y}^2, \tag{13.6}$$

where $\kappa_0 = v_{0,yy} - \frac{3}{2}v_{0,yy}\,v_{0,y}^2$ is (in this case) the 3rd order approximation of the initial curvature.

The kinematic boundary conditions for the transversal displacement field of the clamped-clamped beam, see Fig. 13.1, are

$$v(t, 0) = v(t, L) = 0 \text{ and } v(t, 0)_{,y} = v(t, L)_{,y} = 0. \tag{13.7}$$

Each of the following modes a priori obeys these conditions

$$v_i(y) = \cos\left[(i - 1)\pi y/L\right] - \cos\left[(i + 1)\pi y/L\right], \, i = 1, 2, 3, \ldots \tag{13.8}$$

Using these modes, the transversal displacement field is discretized as

$$v(t, y) = \sum_{i=1}^{N} Q_i(t)v_i(y), \tag{13.9}$$

where $Q_i(t)$ [m] are N generalized degrees of freedom (DOFs). In a similar fashion, the initial shape of the beam, i.e. the geometric imperfection, is discretized as

$$v_0(y) = \sum_{i=1}^{N_e} \frac{1}{2}e_i h v_i(y), \tag{13.10}$$

where e_i are *dimensionless* imperfection parameters and $N_e \leq N$. After discretization of $v_0(y)$ and $v(t,y)$, the corresponding axial displacement field $u(t,y)$ can be computed by integrating an nth order expansion of (13.3). Subsequently, the absolute axial displacement of the top mass, see Fig. 13.1, follows from

$$U_t(t) = U_b(t) + u(t,L). \tag{13.11}$$

Note that (in general) U_t depends in a nonlinear fashion on the DOFs Q_i.

The kinetic energy $\mathscr{T}_{\text{beam}}$ and the potential energy $\mathscr{V}_{\text{beam}}$ of the beam with top mass are determined by

$$\mathscr{T}_{\text{beam}} = \frac{1}{2}\rho A \int_0^L \dot{v}^2 \mathrm{d}y + \frac{1}{2}m_t \dot{U}_t^2, \tag{13.12}$$

$$\mathscr{V}_{\text{beam}} = \frac{1}{2}EI \int_0^L (\kappa - \kappa_0)^2 \mathrm{d}y + m_t g U_t, \tag{13.13}$$

where $A = bh$ is the cross-sectional area, $I = bh^3/12$ is the second moment of area, ρ is the mass density, E is the Young's modulus of the beam, g is the acceleration due to gravity, and m_t is the top mass. Note that the axial and rotatory inertia of the beam are neglected, i.e. the case $\rho A l \ll m_t$ and $h/L \ll 1$ (as stated before) is considered. Damping of the beam is modeled by including a linear and a quadratic viscous damping force for each DOF Q_i: $F_d = -c_i \dot{Q}_i - c_{q,i}|\dot{Q}_i|\dot{Q}_i$, where c_i is the linear viscous damping constant and $c_{q,i}$ is the quadratic viscous damping constant for DOF Q_i. With respect to the damping of slender beams, addition of quadratic damping improves the agreement between theoretical and experimental results in many studies [2, 18, 19]. These generalized damping forces result in the following Rayleigh dissipation function

$$\mathscr{R}_{\text{beam}} = \sum_{i=1}^N \left(\frac{1}{2}c_i \dot{Q}_i^2 + \frac{1}{3}c_{q,i}\text{sign}\left(\dot{Q}_i\right)\dot{Q}_i^3 \right). \tag{13.14}$$

Energy and work expressions (13.12)–(13.14) will be used to derive the coupled shaker-structure model in Sect. 13.2.3.

13.2.2 Shaker Model

The dynamics of the electrodynamic shaker are described by the linear ODEs [7]

$$L\dot{I} + RI + \kappa_c \dot{U}_b = E,$$
$$m_b \ddot{U}_b + c_b \dot{U}_b + k_b U_b = \kappa_c I + F_{\text{str}}, \tag{13.15}$$

Table 13.1 Parameters amplifier-shaker model

c_b	278	[kg/s]	L	2.6×10^{-3}	[H]	P_{amp}	-88.3	[-]
m_b	3.0	[kg]	κ_c	11.5	[N/A]	b_{amp}	$1.4 \cdot 10^{-3}$	$[s^{-1}]$
k_b	5.28×10^4	[N/m]	R	0.9	$[\Omega]$			

where I represents the current through the electric circuit of the shaker, R the coil resistance, L the coil inductance, κ_c the current-to-force constant, E the harmonic excitation voltage supplied by a power amplifier, U_b the vertical displacement of the shaker armature mass m_b, k_b the stiffness of the mass suspension, c_b the viscous damping constant of this suspension, and F_{str} the vertical force exerted to the shaker mass by the structure it carries (the beam with top mass). This force in general depends on $\ddot{U}_b$, generalized DOFs Q_i, and their first and second time derivatives. The mass of the shaker armature m_b is a part of the total moving mass of the lower linear sledge m_o, see Fig. 13.1. The latter mass also includes the mass of the bottom clamping of the slender beam, see Sect. 13.3. In the frequency domain, the relation between E and E_0, respectively the harmonic output and (known) harmonic input voltage of the amplifier, is given by

$$E(j\omega) = P_{amp}(j\omega b_{amp} + 1)E_0(j\omega). \tag{13.16}$$

This relation results in a good fit of the shaker-amplifier dynamics for the frequency range of interest (0–300 [Hz]). The time domain version of (13.16), $E(t) = P_{amp}(b_{amp}\dot{E}_0(t) + E_0(t))$, can simply be substituted in (13.15). The parameters of the amplifier-shaker model are identified using frequency domain techniques, see [5] for more details. During the identification procedure of the unknown parameters of the shaker-amplifier system, the bare shaker was used, i.e. $F_{str} = 0$ [N]. The identified parameter values for the amplifier-shaker model are listed in Table 13.1.

13.2.3   The Coupled Shaker-Structure Model

The coupled shaker-structure model will be derived by following a charge displacement formulation of Lagrange's equations [11]. In this formulation, energy and work expressions of the coupled structure are formulated in terms of mechanical DOFs and (in this case) a single additional *charge* coordinate q. The first time derivative of q constitutes the current through the electrical part of the shaker model, i.e. $\dot{q} = I$. The total set of $N + 2$ DOFs is collected in the column

$$\mathbf{Q}^* = [Q_1, \ldots, Q_N, U_b, q]^{\mathrm{T}}, \tag{13.17}$$

where the DOFs Q_i are the N generalized DOFs of the beam, see (13.9).

In the model of the slender beam, the axial motions are defined with respect to an arbitrary base motion U_b. For the coupled shaker/structure system, the energy/work expressions and the Rayleigh dissipation function now become

$$\mathcal{M} = \frac{1}{2}L\dot{q}^2 + \kappa_c \dot{q}U_b, \quad \mathcal{T} = \mathcal{T}_{\text{beam}} + \frac{1}{2}m_0\dot{U}_b^2, \quad \mathcal{V} = \mathcal{V}_{\text{beam}} + \frac{1}{2}k_b U_b^2,$$

$$\mathcal{R} = \mathcal{R}_{\text{beam}} + \frac{1}{2}c_b\dot{U}_b^2 + \frac{1}{2}RI^2, \quad \delta\mathcal{W}_{\text{nc}} = E(t)\delta q, \tag{13.18}$$

where $\mathcal{M}$ is the magnetic energy of the moving coil of the shaker and $\delta\mathcal{W}_{\text{nc}}$ is the virtual work done by the output voltage of the amplifier $E(t)$ [11]. Defining the Lagrangian $\mathcal{L}$ of the complete system by $\mathcal{L} = \mathcal{T} + \mathcal{M} - \mathcal{V}$, the final coupled set of equations of motion can be determined by

$$\frac{\mathrm{d}}{\mathrm{d}t}\mathcal{L}_{,\dot{\mathbf{Q}}^*} - \mathcal{L}_{,\mathbf{Q}^*} + \mathcal{R}_{,\dot{\mathbf{Q}}^*} = \mathbf{b}E(t), \tag{13.19}$$

where $\mathbf{b} = [0,...,0,1]^T$ is an $N+2$ dimensional column vector. Among others, this will lead to an explicit expression for F_{str}, the force exerted to the shaker mass by the slender beam with top mass, which was introduced in (13.15).

To illustrate some key features of the model, the equation of motion of the slender beam structure is given for single-mode expansions of $v(t,y)$ and $v_0(y)$, i.e. $N = N_e = 1$ in (13.9) and (13.10), and using third-order Taylor-series approximations according to (13.5) and (13.6). This results in two ODEs for the shaker, see (13.15), which are coupled to the following single equation of motion for the beam with top mass

$$M(Q_1)\ddot{Q}_1 + G(Q_1,\dot{Q}_1) + C(\dot{Q}_1) + p_1\left[1 - r_0(1 + \ddot{U}_b/g) - p_2 e_1^2\right]Q_1$$
$$+ K(Q_1) = p_3 e_1 r_0\left(1 + \ddot{U}_b/g\right), \tag{13.20}$$

where $r_0 = m_t g/P_c$ is the ratio between the static load due to the weight of the top mass and the static Euler buckling load of the (perfect) beam ($P_c = 4\pi^2 EI/L^2$). In (13.20), the following abbreviations are used

$$M(Q_1) = \left[\frac{3}{2}\rho AL + \frac{m_t\pi^4}{L^2}\left(h^2 e_1^2 + 4he_1 Q_1 + 4Q_1^2\right)\right],$$

$$G(Q_1,\dot{Q}_1) = \frac{2m_t\pi^4}{L^2}\dot{Q}_1^2\left(he_1 + 2Q_1\right), \quad C(\dot{Q}_1) = c_1\dot{Q}_1 + c_{q,1}|\dot{Q}_1|\dot{Q}_1,$$

$$K(Q_1) = \frac{2\pi^6 EI}{L^5}\left(8Q_1^3 + 9he_1 Q_1^2\right), \quad p_1 = \frac{8\pi^4 EI}{L^3}, \quad p_2 = \frac{\pi^2 h^2}{4L^2}, \quad p_3 = \frac{h}{2}p_1. \tag{13.21}$$

As can be noted, due to the nonlinear Eqs. (13.5) and (13.6), (13.20) contains inertia nonlinearities in $M(Q_1)\ddot{Q}_1$ if $m_t > 0$ and stiffness nonlinearities in $K(Q_1)$. For $e_1 = 0$, the inertia nonlinearities are of the softening type (mass increases for increasing $|Q_1|$), whereas the stiffness nonlinearities are of the hardening type (stiffness increases for increasing $|Q_1|$). C contains linear and quadratic dissipative forces. G contains centrifugal and Coriolis forces. Furthermore, Q_1 is excited by $\ddot{U}_b$ in a parametric manner and for $e_1 \neq 0$ also in a direct manner. In the static situation, for $e_1 = 0$, the linear stiffness term becomes negative for $r_0 > 1$, indicating that the trivial static solution $Q_1 = 0$ becomes unstable, if the static Euler buckling load is exceeded; note that this does not depend on the order of the Taylor series expansion.

13.2.4 Discretizations and Parameter Identification

In Sect. 13.4, experimental steady-state response results will be compared to results based on semi-analytical models. For this, two semi-analytical models will be used: a model based on a single mode discretization of v and v_0 ($N = N_e = 1$, see (13.9) and (13.10)) and a model based on a two mode discretization of v and v_0 ($N = N_e = 2$). Later on in this chapter, these two models will be respectively referred to as the 1-MODE model (this beam model coupled to the shaker model actually has five states: $Q_1, \dot{Q}_1, U_b, \dot{U}_b, I$) and the 2-MODE model (this coupled model has two additional states: $Q_2, \dot{Q}_2$). All numerical responses presented in this chapter are based on models using third-order Taylor series expansions of the inextensibility constraint (13.3) and the curvature (13.4), i.e. (13.5) and (13.6). By considering higher order expansions of the exact kinematics and a multi-mode discretization, it is shown in [6], that the third-order single-mode semi-analytical model can (to a large extent) accurately describe the first harmonic resonance and the first (large amplitude) 1/2 subharmonic resonance of the base-excited (initially unbuckled) slender beam. It is noted that for accurate steady-state response prediction of an initially buckled beam, in general higher order approximations of the exact kinematics are required [10].

The semi-analytical models have a number of parameters, i.e. imperfection and damping parameters, which must be experimentally identified. In addition, to cope with small model errors, the Young's modulus E is considered as a parameter to be identified. Consequently, the 1-MODE model has four unknown parameters (i.e. e_1, c_1, $c_{q,1}$, and E) and the 2-MODE model has seven unknown parameters (i.e. e_1, c_1, $c_{q,1}$, e_2, c_2, $c_{q,2}$, and E). The numerical values for these parameters are identified by fitting periodic solutions, calculated using the semi-analytical models, to measured periodic solutions using a weighted least squares method. In this method, harmonic steady-state responses are used for eleven different excitation frequencies (i.e. for 35, 37, 55, 60, 67, 74, 77, 91, 103, 121, and 136 [Hz]) on stable parts of the harmonic branch, see Sect. 13.4. In general, responses around resonances are useful to identify damping parameters, whereas low-amplitude solutions are useful to identify geometric imperfections. The identification results are robust; using different periodic solutions results in minor changes of the identified parameter values. More details on the applied identification procedure can be found in [16].

13.3   Experimental Setup

Both a picture and a schematic overview of the experimental setup are depicted
in Fig. 13.2. The base excitation of the slender, steel beam is realized by using an
electrodynamic shaker system. The slender beam is clamped between two linear
sledges with very low friction in axial direction. The linear sledge at the top side
of the beam is based on air bearings. This sledge with clamping block acts as the
rigid top mass m_t. The top mass can be increased by mounting additional masses
on top of the upper linear sledge. At the bottom side of the beam, a linear sledge is
realized by an elastic support mechanism based on folded leaf springs. The bottom
linear sledge is mounted rigidly on top of the shaker. The moving mass of the lower
linear sledge, including the mass of the bottom clamping block and the mass of
the shaker armature, equals $m_o = 3.2$ [kg]. The beam used for the experiments is

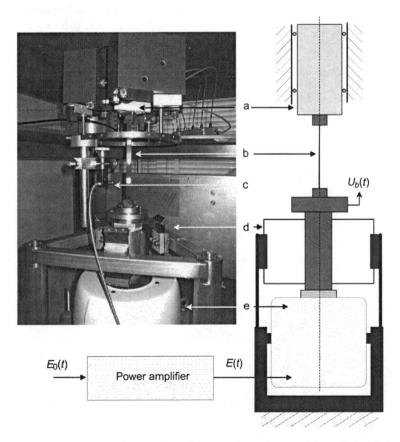

Fig. 13.2 Picture and schematic overview of the experimental setup: (**a**) top linear sledge (top
mass) based on air bearings, (**b**) slender beam, (**c**) laser vibrometer, (**d**) elastic support mechanism
for the base, (**e**) electrodynamic shaker

Table 13.2 Material and geometrical properties of the slender steel beam

E^*	2.0×10^{11}	[N/m^2]
ρ	7,850	[kg/m^3]
L	180	[mm]
b	15	[mm]
h	0.5	[mm]

*This parameter will be slightly modified during the identification procedure with the experimental results

made of spring steel. The material and geometric properties of the beam are listed in Table 13.2. As stated before, note that the value for the Young's modulus E later will be used as a parameter to be identified to account for (small) model errors. This will be discussed in more detail later. Obviously, the identified Young's modulus should not differ too much from its well-known value for steel given in Table 13.2.

At the experimental setup, the base-excitation is introduced by supplying the following harmonic input voltage to the power amplifier

$$E_0(t) = v_d \sin(2\pi f t) \ [\text{V}], \tag{13.22}$$

where v_d is the voltage amplitude and $f = 1/T$ is the excitation frequency. The output voltage of the amplifier $E(t)$, see Fig. 13.2, is supplied to the shaker. The amplifier operates in voltage mode, i.e. the output voltage of the amplifier is kept proportional to its input voltage. No active feedback is used to control the acceleration of the base $\ddot{U}_b$. Consequently, the resulting acceleration of the shaker (and thus the effective axial force on the slender beam with top mass) will not be proportional to the input voltage $E_0(t)$ as given by (13.22), but will be influenced by the dynamics of the shaker system carrying the slender beam with top mass. Due this fact, it is essential to derive a coupled shaker-structure model to be able to compare numerical results with experimental results as has been done in Sect. 13.2.

A laser vibrometer (Ono Sokki LV 1500) is used to measure the transversal velocity ($\dot{v}$) at one point of the beam. In the static equilibrium state obtained for zero input voltage ($E_0 = 0$ [V]), the vibrometer is located at beam height $y = L/4$ (see Fig. 13.1). Note that y is measured relative to the base motion U_b. In the dynamic situation, the vibrometer measures the transversal velocity of the beam at a height $L/4 - U_{b,dyn}(t)$ relative to the static equilibrium position of the base, where $U_{b,dyn}(t)$ is the dynamic part of the base motion caused by a non-zero input voltage E_0. The signal of the laser vibrometer is numerically integrated to obtain measurements in terms of transversal displacements v. To avoid drift during the numerical integration, the measurement signal is filtered using a high pass filter with a cut-off frequency of $f = 1.6$ [Hz]. The data-acquisition and input signal generation is performed using a laptop with Matlab/Simulink in combination with a TUeDACS AQI [15]. A sample frequency of 4 [kHz] is used. Note that $U_{b,dyn}(t) \ll L/4$. Therefore, the experimentally observed transversal velocity and displacement at $L/4 - U_{b,dyn}$ may be compared to the numerically obtained signals $\dot{v}(t, L/4)$ and $v(t, L/4)$.

13.4 Steady-State Response Results

In this section, experimental steady-state response results for the base-excited slender beam with top mass will be compared with semi-analytical results obtained for the 1-MODE model and the 2-MODE model, see Sect. 13.2. The experimental steady-state results are obtained using a stepped sine frequency sweep: harmonic excitation according to (13.22) is applied using a constant voltage amplitude v_d and a stepwise varying excitation frequency $f = 1/T$. Both a sweep-up (the excitation frequency is incrementally increased) and a sweep-down (the excitation frequency is incrementally decreased) are carried out using a step size of $\Delta f = 0.5$ [Hz]. For each discrete value of f, the signals are saved during $N_e = 150$ excitation periods. The data during the first $N_t = 50$ periods is not used to minimize transient effects.

As explained before, the dynamic steady-state response of the beam is experimentally characterized using the measured velocity signal $\dot{V}_{L/4}(t) = \dot{v}(t, L/4 - U_{b,dyn})$ and its corresponding displacement signal $V_{L/4}(t) = v(t, L/4 - U_{b,dyn})$ obtained by filtering and numerical integration of $\dot{V}_{L/4}(t)$. The averaged peak-to-peak amplitude of the steady-state velocity response is determined by

$$\tilde{V}_{L/4} = \frac{1}{N_m} \sum_{k=0}^{N_m-1} \left(\max_{T_m} \dot{v}_{(k)}(t, L/4 - U_{b,dyn}) - \min_{T_m} \dot{v}_{(k)}(t, L/4 - U_{b,dyn}) \right), \quad (13.23)$$

where $T_m = (N_e - N_t)T/N_m$, T is the excitation period, $N_m = 5$ [-], and k refers to the kth record. Averaging over N_m records is applied to cancel measurement noise to some extent. Because $T_m = 20$, the peak-to-peak amplitude of a $1/20$th subharmonic solution still can be estimated. In principle this is not possible for subharmonic solutions with a period time longer than $20T$ and aperiodic solutions.

Later on in this section, numerical steady-state results for the 1-MODE and 2-MODE models are obtained using the software package AUTO97 [3], which is capable of: (1) calculating branches of periodic solutions of a nonlinear dynamic system for a varying system parameter, (2) analyzing the local stability of these branches using Floquet theory, and (3) detecting local bifurcations on these branches. More theoretical background on these topics can be found in [14]. The numerical peak-to-peak amplitude equivalent to the experimental quantity defined in (13.23) is obtained as follows. In AUTO97, the equations of motion given by (13.19) are programmed in first order form. This means that the periodic solutions are available in state space, i.e. in terms of $\mathbf{Q}^*$, see (13.17), and its first time derivative. By substituting the first time derivatives $\dot{Q}_i$ in the first time derivative of (13.9), $\dot{v}(t, L/4)$ is obtained, from which directly the peak-to-peak value can be derived.

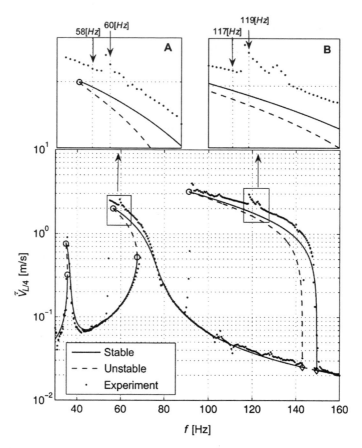

Fig. 13.3 Frequency-amplitude plot with $v_d = 0.03$ [V] and $m_t = 0.51$ [kg] (experimental versus semi-analytical results based on 1-MODE model)

13.4.1 Steady-State Responses for the 1-MODE Model

In Fig. 13.3, the experimentally obtained frequency-amplitude plot in terms of $\tilde{V}_{L/4}$ [m/s] is depicted by means of black dots for $v_d = 0.03$ [V] and $m_t = 0.51$ [kg], i.e. $r_0 = 0.135$. Both frequency sweep-up and frequency sweep-down results are plotted. In this figure, also the numerical steady-state response results are depicted based on the semi-analytic 1-MODE model. Stable branches are indicated by solid lines, unstable branches by dashed lines.

The identified parameter values are listed in the first column of Table 13.3. The identified Young's modulus E is a little bit lower than the theoretical value. The eigenfrequencies and damping ratios of the 1-MODE model linearized around the static equilibrium position are listed in the first column of Table 13.4. The lowest eigenfrequency of the model (f_1) corresponds to the suspension mode of the shaker, i.e. the mode shape is dominated by U_b. This mode is highly damped (although still

Table 13.3 Identified parameter values based on experimental results obtained for $v_d = 0.03$ [V] and $m_t = 0.51$ [kg]

Parameters	1-MODE	2-MODE
e_1 [-]	1.36	1.36
c_1 [Ns/m]	0.0	0.0
$c_{q,1}$ [kg/m]	0.20	0.2
e_2 [-]	–	0.04
c_2 [Ns/m]	–	0.04
$c_{q,2}$ [kg/m]	–	0.0
E [N/m^2]	1.95×10^{11}	1.95×10^{11}

Table 13.4 Eigenfrequencies f_i and damping ratios ξ_i of linearized models with parameters according to Table 13.3 and $m_t = 0.51$ [kg]

	1-MODE	2-MODE
f_1 [Hz]	18.1	18.1
ξ_1 [-]	0.489	0.489
f_2 [Hz]	73.1	73.1
ξ_2 [-]	0.001	0.001
f_3 [Hz]	–	215.8
ξ_3 [-]	–	0.007

undercritically damped). The second eigenfrequency of the model (f_2) corresponds to the first bending mode of the beam, i.e. the mode shape is dominated by Q_1. The linear damping coefficient c_1 is identified zero. Note that this does not result in a zero damping ratio ξ_2 for the first beam mode, see Table 13.4, since this mode has some (linear) coupling with the heavily damped suspension mode of the shaker.

Next, the obtained steady-state responses as depicted in Fig. 13.3 are discussed in detail. The focus is on the dynamics of the beam-top mass system. Therefore, only excitation frequencies above 30 [Hz] are considered. The responses computed with the semi-analytical model show a second superharmonic resonance at $f \approx f_2/2$, a harmonic resonance at $f \approx f_2$, and a 1/2 subharmonic resonance (the latter branch contains periodic solutions with period $2T$). The first two resonances are caused by direct excitation, see the righthand side of (13.20). The 1/2 subharmonic resonance is a parametric resonance, since it is caused by parametric excitation (in the expression for the linear stiffness, a periodic time-dependent term is present), see (13.20). This parametric resonance is initiated at two period doubling bifurcations (indicated by two ◇ symbols) near $f = 2f_2$. All three resonances are of softening type due to the inertia nonlinearities and are qualitatively similar as found for the case of harmonic base-acceleration, which is numerically investigated in [6]. Cyclic fold bifurcations (indicated by o symbols) are found on all three resonance peaks. Experimentally, large frequency hysteresis intervals can be observed between 58–68 [Hz] and 92–142 [Hz]. The numerically obtained bifurcation points mark the boundaries for these frequency hysteresis intervals. At these bifurcation points, in the experiments large sudden jumps in the peak-to-peak values occur during the frequency sweep-up and the frequency sweep-down. Note that the frequency hysteresis interval for the second superharmonic resonance near $f \approx f_2/2$ is very small because the frequencies, at which the two cyclic fold bifurcations occur, are very close to each other. Therefore, sudden jumps are hardly noticeable. The boundaries of the frequency hysteresis interval associated with the harmonic resonance are marked by cyclic fold bifurcations at 58 [Hz] (a jump occurs to the low amplitude branch in the sweep-down) and 68 [Hz] (a jump occurs to the high amplitude branch in the sweep-up). The boundaries of the frequency hysteresis interval associated with the 1/2 subharmonic (parametric) resonance are marked by a cyclic fold bifurcation at 92 [Hz] (a jump occurs to the low amplitude branch in the sweep-down) and a subcritical period doubling bifurcation at 142 [Hz] (a jump occurs to the high amplitude branch in the sweep-up). Near the frequencies, where jumps occur, sometimes intermediate experimental dots are visible, e.g. near 58 and 92 [Hz]. These arise from transient effects, i.e. for these frequencies 50 periods are not enough to let the transient damp out related to the weakly damped first beam mode.

During the parameter identification process it appeared that inclusion of quadratic damping is essential to get good fit results, especially around the harmonic resonance and the 1/2 subharmonic resonance. The beneficial influence of the quadratic damping on the quality of the fit between numerical results and experimental results is also observed in [2, 18]. In general, the experimental steady-state results are in good correspondence with the semi-analytical results. However, some discrepancies can be noted. First of all, the experimental results show a somewhat larger amplitude in the peaks of the harmonic and the subharmonic resonance. Furthermore, in some (small) frequency regions, the experimentally obtained frequency-amplitude plots show small peaks and/or small jumps, which are not present in the semi-analytical results. Two of these peaks/jumps can be clearly observed near the top of the harmonic resonance ($f \approx 60$ [Hz]) and on the subharmonic resonance branch near $f = 120$ [Hz], see enlargements **A** and **B** in Fig. 13.3. Figure 13.4 shows projections of the experimental response on the phase plane spanned by $V_{L/4}(t)$ and $\dot{V}_{L/4}(t)$ and the corresponding Poincaré mappings, i.e. period T sampled values of $V_{L/4}$ plotted against period T sampled values of $\dot{V}_{L/4}$, close to these two jumps. At $f = 58$ [Hz], at the left side of the jump, see enlargement **A** in Fig. 13.3, the Poincaré map shows a single dot, indicating a harmonic response. However, at $f = 60$ [Hz], at the right side of the jump, see enlargement **A** in Fig. 13.3, the Poincaré map shows two dots, indicating a 1/2 subharmonic response. Similarly, for the scenario depicted in enlargement **B** in Fig. 13.3, for $f = 117$ [Hz] the response is 1/2 subharmonic and for $f = 119$ [Hz] the response has become 1/4 subharmonic.

13.4.2 *Steady-State Responses for the 2-MODE Model*

To examine if the experimentally observed period doubling behavior presented in Fig. 13.4 and the related small peaks/jumps in enlargements **A** and **B** of Fig. 13.3 are due to nonlinear interaction with the second beam mode v_2, a semi-analytical model with two beam modes (the 2-MODE model) is derived, see Sect. 13.2. For

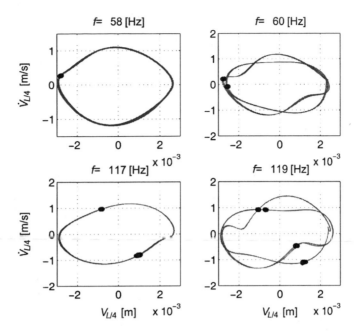

Fig. 13.4 Phase-plane projections and Poincaré mappings for four experimentally obtained responses for *beam 2* with $v_d = 0.03$ [V] and $m_t = 0.51$ [kg]

the 2-MODE model, the identified parameter values are listed in the last column of Table 13.3 and the eigenfrequencies and damping ratios linearized around the static equilibrium position are listed in the last column of Table 13.4. A very small geometric imperfection related to the second bending mode of the beam is identified: $e_2 = 0.04$ [-]. In the parameter identification for the second mode, no quadratic damping is found, i.e. $c_{q,2} = 0$ [kg/m]. In Fig. 13.5, the steady-state response predicted by the 2-MODE model is compared with the experimental results obtained for $v_d = 0.03$ [V] and $m_t = 0.51$ [kg] (same values as used in Fig. 13.3). The experimental results in Fig. 13.3 and Fig. 13.5 are obviously identical. As can be noted in Fig. 13.5, inclusion of the second beam mode in the model instigates a second harmonic resonance with softening around $f = 215$ [Hz]. This second harmonic resonance is observed at a slightly lower frequency in the experimental results. Furthermore, and in correspondence with the experimental results, in the semi-analytical results for the 2-MODE model at the 1/2 subharmonic branch near $f = 123$ [Hz], two period doubling bifurcations are observed (indicated by two ◊ symbols), resulting in a branch with 1/4 subharmonic responses, see enlargement **B** in Fig. 13.5. Note that for clarity no experimental results are shown in the enlargements. The 1/4 subharmonic branch exhibits three cyclic fold bifurcations (indicated by o symbols) resulting in two (separate) stable parts of the branch. Due to its complexity, this branch cannot be easily compared with the experimental results in this region, see the experimental results in enlargement **B** of Fig. 13.3 (here, only results for the frequency sweep-down are shown). Nevertheless, it can

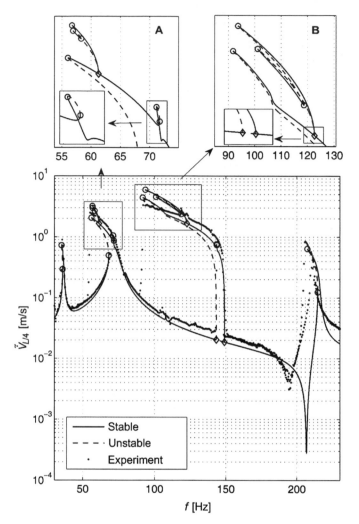

Fig. 13.5 Frequency-amplitude plot with $v_d = 0.03$ [V] and $m_t = 0.51$ [kg] (experimental versus semi-analytical results based on 2-MODE model)

be noted that for the semi-analytical results, the 1/4 subharmonic response continues to a lower excitation frequency as observed experimentally. A similar phenomenon occurs near the top of the first harmonic resonance. Again, in correspondence with the experimental results, near $f = 61$ [Hz] two nearly coinciding period doubling bifurcations occur (again indicated by two $\diamond$ symbols), from which now a 1/2 subharmonic branch bifurcates.

Furthermore, in the semi-analytical results near $f = 72 \approx f_3/3$ [Hz], see enlargement **A** in Fig. 13.5, a small 3rd superharmonic resonance can be distinguished, which is related to the second beam mode with eigenfrequency f_3, see the last column of Table 13.4. This superharmonic resonance cannot be seen in the experimental results. However, the frequency interval, in which this resonance occurs, is so small, that it is possibly missed in the frequency sweep due to a too coarse frequency step. Subsequently, a very small 2nd superharmonic resonance related to the second beam mode can be observed near $f = 108 \approx f_3/2$ [Hz], both on the high amplitude 1/2 subharmonic branch and on the low-amplitude harmonic branch. This resonance can be seen more clearly in the semi-analytical response than in the experimental response. Finally, a very small resonance can be seen on the 1/2 subharmonic branch near $f = 144 \approx 2f_3/3$ [Hz]. Two cyclic fold bifurcations associated with this resonance are indicated by o symbols. Also here, this resonance can be seen more clearly in the semi-analytical response than in the experimental response.

For further validation, power spectral densities (PSDs) of the experimentally and semi-analytically obtained transversal velocities, i.e. $\dot{V}_{L/4}(t)$ and $\dot{v}(t, L/4)$, are compared. First, this is carried out for the 1/2 subharmonic solution at an excitation frequency of $f = 60$ [Hz]. Both the experimentally and the semi-analytically obtained PSDs (obviously with a base frequency of 30 [Hz]) show dominating contributions of two frequency components, namely 60 [Hz] and 210 [Hz], indicating a two-to-seven combination resonance of the first and second beam mode. Subsequently, the comparison is carried out for the 1/4 subharmonic solution at an excitation frequency of $f = 119$ [Hz]. In this case, both the experimentally and the semi-analytically obtained PSDs (obviously with a base frequency of 29.75 [Hz]) again show dominating contributions of two frequency components, namely 59.5 [Hz] and 208.25 [Hz]. Again this indicates a two-to-seven combination resonance of the first and second beam mode. A two-to-seven internal resonance between the first two beam bending modes has also been observed experimentally in [12].

In conclusion, the 2-MODE model can qualitatively explain the experimentally observed small extra resonances/jumps and period doubling behavior near the top of the first harmonic resonance and on the 1/2 subharmonic branch. For a better quantitative match of these (and other) dynamic response details, the damping and imperfection parameters of the model may need to be further refined and, possibly, also more beam modes must be included in the model. Moreover, more detailed experiments based on smaller frequency sweep increments Δf may be necessary for (improved) identification of the (additional) damping and imperfection parameters. This, however, is considered to be out of the scope of this work. Nevertheless, it has been illustrated that, using a semi-analytical approach, the steady-state dynamic response can be studied in detail.

13.5 Conclusions

In this chapter, the dynamic stability of a slender beam carrying a top mass has been investigated. The weight of the top mass is well below the static buckling load. The beam is dynamically excited at its base by means of an amplifier-shaker system with a harmonic input voltage.

A semi-analytical model has been derived for a base-excited slender beam carrying a top mass. The beam has been assumed to be inextensible. The nonlinear inextensibility constraint as well as the nonlinear expression for the curvature of the beam have been approximated by third order Taylor series expansions. In the model, geometrical imperfections of the beam have been taken into account. Furthermore, the beam model includes linear as well as quadratic viscous damping forces. This structural model has been extended by coupling it to a linear model of the shaker (with amplifier). The resulting low-dimensional coupled model has been combined with numerical tools for efficient calculation of periodic solution branches of the system and their local stability, and for detection of bifurcation points.

An experimental set-up has been built and used to validate the numerically obtained steady-state responses. In the experiments, frequency sweep-up and sweep-down excitation has been performed via the harmonic input voltage. Shaker and amplifier parameters have been identified experimentally for the bare amplifier-shaker system. Damping and imperfection parameters of the semi-analytical model have been identified by using a least-squares method, which fits numerically obtained periodic solutions as good as possible to experimentally obtained periodic solutions.

Frequency-amplitude curves have been calculated for both one-mode and two-mode models of the beam. For the one-mode model, already a good match between numerical and experimental steady-state responses is obtained. The main resonances being a parametric 1/2 subharmonic resonance, a harmonic resonance, and a second superharmonic resonance (all related to the first beam mode) are predicted well. Also the frequencies, at which (period doubling and cyclic fold) bifurcations are calculated, correspond well with the frequencies, at which experimentally sudden jumps in the response amplitude are observed during the frequency sweep-up and sweep-down. The two-mode model shows some additional resonances. Next to the (expected) second harmonic resonance also some smaller resonances (and extra bifurcations associated with these resonances) occur in the low-frequency range. Some of these resonances can be identified as combination resonances of the two beam modes. These additional resonances are also found in the experiments.

The results presented in this chapter in principle only refer to the coupled shaker-beam system and thus depend on the particular shaker used. However, the nonlinear resonance phenomena presented in the current chapter still occur if the bottom of the beam is harmonically excited with constant amplitude. This is demonstrated by simulations in [6]. To some extent, this is not really a surprise, since the shaker dynamics are linear. Moreover, the resonance frequency (at 18.1 Hz) of the mode, where the shaker vibration dominates, is clearly below the interesting frequency range for the beam carrying the top mass (30–160 Hz). One could say that the shaker is "designed" in such a way, that its dynamics do not qualitatively alter the dynamics of the beam carrying the top mass with harmonic base excitation.

Acknowledgements This research is supported by the Dutch Technology Foundation STW, applied science division of NWO and the technology programme of the Ministry of Economic Affairs (STW project EWO.5792).

References

1. Amabili, M.: Nonlinear vibrations and stability of shells and plates. Cambridge University Press, Cambridge (2008)
2. Anderson, T., Nayfeh, A., Balachandran, B.:Experimental verification of the importance of the nonlinear curvature in the response of a cantilever beam. ASME J Vibr and Acoustics 118(1):21–27 (1996)
3. Doedel, E., Paffenroth, R., Champneys, A., Fairgrieve, T., Kuznetsov, Y., Oldeman, B., Sandstede, B., Wang, X.: AUTO97: Continuation and bifurcation software for ordinary differential equations (with HOMCONT). Technical Report, Concordia University (1998)
4. Gibson, C.: Elementary geometry of differentiable curves: An Undergraduate Introduction. Cambridge University Press, Cambridge (2001)
5. Mallon, N.: Dynamic stability of thin-walled structures: a semi-analytical and experimental approach. PhD thesis, Eindhoven University of Technology (2008)
6. Mallon, N., Fey, R., Nijmeijer, H.: Dynamic stability of a base-excited thin beam with top mass In: Proc. of the 2006 ASME IMECE, Nov. 5-10, Paper 13148, Chicago, USA, pp 1–10 (2006)
7. McConnell, K.: Vibration testing, Theory and Practice. Wiley, Chichester (1995)
8. Mettler, E.: Dynamic buckling. In: Handbook of engineering mechanics (Flügge, W., ed.). McGraw-Hill, London (1962)
9. Nayfeh, A., Pai, P.: Linear and nonlinear structural mechanics. Wiley-VCH, Weinheim (2004)
10. Noijen, S., Mallon, N., Fey, R., Nijmeijer, H., Zhang, G.: Periodic excitation of a buckled beam using a higher order semi-analytic approach. Nonlinear Dynamics 50(1-2):325–339 (2007)
11. Preumont, A.: Mechatronics, dynamics of electromechanical and piezoelectric systems. Springer, Dordrecht (2006)
12. Ribeiro, P., Carneiro, R.: Experimental detection of modal interactions in the non-linear vibration of a hinged-hinged beam. J of Sound and Vibr 277(4-5):943–954 (2004)
13. Son, I. S., Uchiyama, Y., Lacarbonara, W., Yabuno, H.: Simply supported elastic beams under parametric excitation. Nonlinear Dynamics 53:129–138 (2008)
14. Thomsen, J.: Vibrations and stability; advanced theory, analysis, and tools, second edition. Springer Verlag, Berlin (2003)
15. TueDACS: TUeDACS Advanced Quadrature Interface (2008)
16. Verbeek, G., de Kraker, A., van Campen, D.: Nonlinear parametric identification using periodic equilibrium states. Nonlinear Dynamics 7:499–515 (1995)
17. Virgin, L.: Vibration of axially loaded structures. Cambridge University Press, New York (2007)
18. Yabuno, H., Ide, Y., Aoshima, N.: Nonlinear analysis of a parametrically excited cantilever beam (effect of the tip mass on stationary response). JSME Int J 41(3):555–562 (1998)
19. Yabuno, H., Okhuma, M., Lacarbonara, W.: An experimental investigation of the parametric resonance in a buckled beam. In: Proceedings of the ASME DETC'03, Chicago, USA, pp 2565–2574 (2003)
20. Zavodney, L., Nayfeh, A.: The non-linear response of a slender beam carrying a lumped mass to a principal parametric excitation: theory and experiment. Int J Non-Linear Mech 24(2):105–125 (1989)

Chapter 14
A Study of the Onset and Stabilization of Parametric Roll by Using an Electro-Mechanical Device

Jonatan Peña-Ramírez and Henk Nijmeijer

14.1 Introduction

Maritime industry plays a major role in our life and in the economy of the world. Most of the merchandize and goods like electronics, cars, food, clothes, are transported from producers to end consumers by ship containers from one end of the world to the other end. Over the last decades, shipping industry has experienced a continuous growing both in their fleets and in the total trade volume. As a consequence of this growing, it has been necessary the design of new ships and vessels capable of transporting as much as possible of products. This is the reason why nowadays ships are designed using cutting edge technology in order to find an optimal design looking mainly at economic aspects. For instance, modern container ships hulls feature a bow flare and stern overhang in combination with a flow-optimized geometry below the water line. This design is twofold: at one hand it provides maximum space for container storage and at the other hand it provides a minimal water resistance. However, modern designs of vessels and ships seem to be prone to a phenomenon called *parametric roll*.

Parametric roll is an undesired phenomenon because it may produce cargo damage, delay or even suspension of the activities performed by the crew, seasickness in passengers and crew and in the limit case it can lead to the capsizing of the ship [13]. It has been suggested (c.f. [5]) that the onset of parametric roll is due to the occurrence of the following conditions: the ship is sailing in head seas, the natural period of roll is approximately twice the wave encounter period, the roll damping is low, the wave height exceeds a critical level and the wavelength is close to the ship length.

J. Peña-Ramírez (✉) • H. Nijmeijer
Department of Mechanical Engineering, Eindhoven University of Technology,
P.O. Box 513, 5600 MB Eindhoven, The Netherlands
e-mail: J.Pena@tue.nl; H.Nijmeijer@tue.nl

T.I. Fossen and H. Nijmeijer (eds.), *Parametric Resonance in Dynamical Systems*,
DOI 10.1007/978-1-4614-1043-0_14, © Springer Science+Business Media, LLC 2012

Probably, the earliest studies about this phenomenon are due to William Froude (1810–1879), who in 1857 started a serious research in order to find the causes and conditions that lead to parametric roll resonance. It was the time when the Great Eastern, a ship that was so big in comparison with the size of any other ship, was under construction. The designer, I. K. Brunel was concerned because he thought that the ship could behave in an unexpected manner. Brunel then asked Froude to start a theoretical study on parametric roll. One of Froude's early discoveries was that the roll angle can increase rapidly when the period of the ship is in resonance with the period of wave encounter. He also came to the conclusion that the roll motion is not produced by the waves hitting the side of the hull, but rather because of the pressure of the waves acting on the hull. Although Froude made several simplifications in his analysis because of the intractable mathematics, his research ended with a theory of rolling in waves and its stabilization by the introduction of bilge keels (c.f. [3, 13], and the references therein).

Because modern ships still experience dangerous roll motions, it continues being a hot topic not only in the research field but also in the maritime industry. We mention two incidents, where millions of dollars were lost. In 1998, a post-Panamax C11 class container was caught by a violent storm and experienced parametric roll with roll angles close to 40°. As a consequence one-third of the on-deck containers were lost overboard and a similar amount were severely damaged [5]. More recently, in January 2003, another Panamax container vessel encountered a storm in the North Atlantic. It was reported that the ship experienced violent rolling with angles close to 47°. As a result, 133 containers were lost overboard and other 50 presented severe damage [4].

So far, several models of different sophistication have been proposed by the researchers in order to analyze the dynamical behavior of a ship in a seaway. In particular, there are models that have been developed for the study of parametric roll, for instance, we mention the simplified nonlinear models presented in [8, 12], where three DOF are considered (heave, pitch, and roll). Furthermore, some authors have derived their models by using an analogy of the ship motions to a mechanical system [10, 16]. Besides the development of models there is the issue of finding stabilization techniques in order to cope with parametric roll phenomenon. Hence, in the literature we can find different stabilization techniques as for example, the use of bilge keels [10], passive and active U-tanks [1, 7], rudders [2] and fins stabilizers [6].

In this work, we present an experimental study of parametric roll occurring in a container ship. As a "towing tank" we use an electro-mechanical platform consisting of two (controllable) mass-spring-damper oscillators mounted on an elastically supported (controllable) beam. The stiffness and damping in the system have been identified experimentally and the state vector is reconstructed by using the position measurements. Then, via computer controlled feedback, the dynamical properties of the system are modified by canceling the inherent dynamics of the setup and enforcing the dynamics of a 3-DOF (heave, pitch, and roll) container ship. The heave and pitch motions are represented by the displacement of the oscillators and the supporting beam is used to mimic the roll motion. Additionally, the experimental setup is masked with the dynamics of a mechanical system consisting of two masses restrained by elastic springs and supporting two identical pendula. Such model was developed to simulate heave-pitch-roll motion of a ship in longitudinal waves.

At the end of the day, we want to show that our experimental setup is suitable for the experimental analysis of parametric roll and that can facilitate the understanding of ship dynamics. Actually, we want to show that the setup can be seen as a testbed for controllers ad hoc designed to stabilize the roll motion.

The rest of the manuscript is organized as follows. In Sect. 14.2 we describe in detail the experimental setup. Then, in Sect. 14.3 we present experimental results related to the onset and stabilization of parametric roll in a container ship. Next, in Sect. 14.4 we conduct an experimental analysis of the dynamics corresponding to a mechanical system developed to simulate the most general case of heave-pitch-roll motion in a vessel. Finally, in Sect. 14.5 we draw some conclusions.

14.2 The Experimental Setup

In this section, we describe the electro-mechanical device depicted in Fig. 14.1 which has been used for the experiments. It consists of two oscillators mounted on an elastically supported beam. The system has three DOF corresponding to the axial displacement of the oscillators and the beam. Moreover, each DOF is equipped with a voice coil actuator and with a linear variable differential transformer position

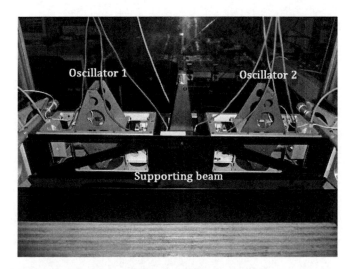

Fig. 14.1 Photo of the experimental setup at Eindhoven University of Technology

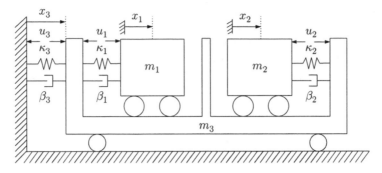

Fig. 14.2 Schematic model of the setup

sensor. The maximum stroke of the oscillators and the beam is approximately 6 mm. and the position sensors are calibrated such that 1 [V] = 5 [mm]. In the other hand, the actuators have a limited input of ± 0.42 [V].

The experimental setup is schematically depicted in Fig. 14.2. The masses corresponding to the oscillators are given by $m_i \in \mathbb{R}^+$ ($i = 1,2$) and the mass of the supporting beam is denoted by $m_3 \in \mathbb{R}^+$. This mass may be varied by a factor 10. The stiffness and damping characteristics present in the system are assumed to be linear with constants coefficients $\kappa_i, \beta_i \in \mathbb{R}^+$ respectively. However, the experimental setup allows modeling of different types of springs (for instance, linear or cubic) and any other desired effect within the physical limitations of the setup. The electric actuator force for subsystem i ($i = 1 \ldots 3$) is denoted as u_i. Finally, $x_i \in \mathbb{R}$ ($i = 1,2,3$) are the displacements of the oscillators and the supporting beam respectively.

Using Newton's 2nd law, it follows that the idealized – i.e. assuming that no friction is present – equations of motion of the system of Fig. 14.2 are

$$m_1 \ddot{x}_1 = -\kappa_1(x_1 - x_3) - \beta_1(\dot{x}_1 - \dot{x}_3) + u_1,$$

$$m_2 \ddot{x}_2 = -\kappa_2(x_2 - x_3) - \beta_2(\dot{x}_2 - \dot{x}_3) + u_2,$$

$$m_3 \ddot{x}_3 = \sum_{i=1}^{2} [\kappa_i(x_i - x_3) + \beta_i(\dot{x}_i - \dot{x}_3) - u_i] - \kappa_3 x_3 - \beta_3 \dot{x}_3 + u_3. \qquad (14.1)$$

For convenience, system (14.1) is written in the following manner

$$\ddot{x}_1 = -\omega_1^2(x_1 - x_3) - 2\zeta_1\omega_1(\dot{x}_1 - \dot{x}_3) + \frac{1}{m_1}u_1,$$

$$\ddot{x}_2 = -\omega_2^2(x_2 - x_3) - 2\zeta_2\omega_2(\dot{x}_2 - \dot{x}_3) + \frac{1}{m_2}u_2,$$

$$\ddot{x}_3 = \sum_{i=1}^{2} \mu_i \left[\omega_i^2(x_i - x_3) + 2\zeta_i\omega_i(\dot{x}_i - \dot{x}_3) - \frac{1}{m_i}u_i \right] - \omega_3^2 x_3 - 2\zeta_3\omega_3\dot{x}_3 + \frac{1}{m_3}u_3,$$

$$(14.2)$$

Table 14.1 Parameters values for the experimental setup according to model (14.2)

	Oscillator 1	Oscillator 2	Supporting platform
ω_i [rad s^{-1}]	12.5521	14.0337	9.7369
ζ_i [−]	0.3362	0.4226	0.0409
m_i [kg]	0.198	0.210	4.1

where $\omega_i = \sqrt{\frac{\kappa_i}{m_i}}$ [rad s^{-1}], $\zeta_i = \frac{\beta_i}{2\omega_i m_i}$ [−] are the angular eigenfrequency and dimensionless damping coefficient present in subsystem i ($i = 1, 2, 3$), the coupling strength is denoted by $\mu_i = \frac{m_i}{m_3}$ ($i = 1, 2$). After a suitable parametric identification of model (14.2) the parameters presented in Table 14.1 were obtained (see [14]).

The potential of this experimental setup to perform experiments on several dynamical systems relies on the fact that the properties of the system can be adjusted or modified by a suitable design of the control inputs u_i. These inputs are generated as follows: a data acquisition system reads data from the sensors and forwards the converted data to a computer. In the computer the state vector is reconstructed by an observer [15] and the reconstructed state vector is used to construct the new desired dynamics. With this data, the output u_i is generated, it consists of a feed forward part (to cancel the original dynamics) plus compensation terms plus the desired dynamics. Then, it is clear that by means of state feedback, the original dynamics are masked with the dynamics that we want. For example, in [14] the setup is used for experimentally testing synchronization of coupled oscillators.

By means of two experiments we show the capabilities of the experimental setup to conduct experiments on parametric roll. As a first example we implement the dynamics of a 3-DOF nonlinear container ship model navigating in head seas and as a second example the dynamics of a mechanical model for simulating heave-pitch-roll motion of a ship in longitudinal waves.

In both cases, a controller is implemented in order to stabilize the parametric roll resonance condition. Since the experimental setup is fully actuated, the choice of the controller is arbitrary and therefore many controllers can be implemented and tested in the setup.

14.3   Case 1: A High-Fidelity 3-DOF Nonlinear Container Ship Model

A ship can be seen as a rigid body that can be modeled as a 6-DOF system. Three of these DOF, named surge, sway, and yaw, correspond to unrestored motions in the horizontal plane while the other three DOF named heave, pitch, and roll, correspond to oscillating motions in the vertical plane. However, some simplifications can be done in the model, depending on the desired analysis. For instance, for the study of parametric roll, some authors [8, 9, 12, 17] agree in the fact that a 3rd order model (considering heave, pitch, and roll) is enough to study the phenomenon, since the restoring forces that may produce resonance on the ship roll motion only act when the ship is subjected to motions in the vertical plane.

In this section, the dynamical behavior of the experimental setup presented in the previous section is modified in order to mimic the dynamics of a nonlinear container ship model developed in [8]. Additionally, we implement a control strategy for the stabilization of the roll motion. Such control law is presented in [6] and is based in a combined speed and fin stabilizer control.

14.3.1 The Model and Its Implementation in the Setup

Consider the following nonlinear container ship model

$$\ddot{s} = (M+A)^{-1}\left(c_{\text{ext}}(\zeta,\dot{\zeta},\ddot{\zeta}) - B(\dot{\phi})\dot{s} - c_{\text{res}}(s,\zeta)\right), \tag{14.3}$$

where

$$s(t) = [z(t) \quad \phi(t) \quad \theta(t)]^{\text{T}} \tag{14.4}$$

is the generalized vector which contains the three restoring degrees of freedom, heave, roll, and pitch, respectively. $M \in \mathbb{R}^{3\times3}$ is the generalized mass matrix, $A \in \mathbb{R}^{3\times3}$ describes the hydrodynamic added mass matrix and $B \in \mathbb{R}^{3\times3}$ represents the hydrodynamic damping matrix. $c_{\text{res}} \in \mathbb{R}^{3\times1}$ contains the nonlinear restoring forces and moments dependent on the relative motions between ship hull and wave elevation $\zeta(t)$. Finally, the generalized vector $c_{\text{ext}} \in \mathbb{R}^{3\times1}$ contains the external forces exerted by the waves. These forces are depending on weave heading, encounter frequency, wave amplitude, and time. For the derivation of the model, as well as for definitions and expressions for M, A, B, c_{res}, and c_{ext}, the reader is referred to [8, 12].

Indeed, system (14.3) has the following structure

$$\begin{bmatrix} \ddot{z} \\ \ddot{\phi} \\ \ddot{\theta} \end{bmatrix} = \underbrace{\begin{bmatrix} a_{11} & a_{12} & a_{13} \\ a_{21} & a_{22} & a_{23} \\ a_{31} & a_{32} & a_{33} \end{bmatrix}}_{(M+A)^{-1}} \left(\underbrace{\begin{bmatrix} c_{\text{ext}z} \\ c_{\text{ext}\phi} \\ c_{\text{ext}\theta} \end{bmatrix}}_{c_{\text{ext}}(\zeta,\dot{\zeta},\ddot{\zeta})} - \underbrace{\begin{bmatrix} c_{\text{res}z} \\ c_{\text{res}\phi} \\ c_{\text{res}\theta} \end{bmatrix}}_{c_{\text{res}}(z,\phi,\theta,\zeta)} \right) - \underbrace{\begin{bmatrix} c_{11} & c_{12} & c_{13} \\ c_{21} & c_{22} & c_{23} \\ c_{31} & c_{32} & c_{33} \end{bmatrix} \begin{bmatrix} \dot{z} \\ \dot{\phi} \\ \dot{\theta} \end{bmatrix}}_{(M+A)^{-1}B}. \tag{14.5}$$

For convenience, we rewrite (14.5) as

$$\ddot{z} = F_z(z,\phi,\theta,\zeta,\dot{\zeta},\ddot{\zeta}) - c_{11}\dot{z} - c_{12}\dot{\phi} - c_{13}\dot{\theta}, \tag{14.6}$$

$$\ddot{\phi} = F_\phi(z,\phi,\theta,\zeta,\dot{\zeta},\ddot{\zeta}) - c_{21}\dot{z} - c_{22}\dot{\phi} - c_{23}\dot{\theta}, \tag{14.7}$$

$$\ddot{\theta} = F_\theta(z,\phi,\theta,\zeta,\dot{\zeta},\ddot{\zeta}) - c_{31}\dot{z} - c_{32}\dot{\phi} - c_{33}\dot{\theta}, \tag{14.8}$$

where $F_z(\cdot) = a_{11}c_{\text{ext}z} + a_{12}c_{\text{ext}\phi} + a_{13}c_{\text{ext}\theta} - a_{11}c_{\text{res}z} - a_{12}c_{\text{res}\phi} - a_{13}c_{\text{res}\theta}$. Expressions for F_ϕ and F_θ can be derived from (14.5).

As we mention before, the experimental setup not only allows for the implementation of other dynamics (like the dynamics of a ship) but also a wide variety of controllers can be implemented and validated within the physical limits of the setup. Hence, a controller is incorporated in order to stabilize the parametric roll resonance condition occurring in system (14.6)–(14.8).

For the time being, we use the controller presented in [6]. This controller has two objectives: to avoid that the encounter frequency approaches twice the roll natural frequency ω_ϕ, and to increase the damping in roll.

Since in deep water, the encounter frequency (c.f. [11]) is given by

$$\omega_e = \omega - \frac{\omega^2}{g} U \cos(\mu), \tag{14.9}$$

where ω is the wave frequency, U is the forward velocity of the ship and μ is the heading angle. Then, it is clear that the first objective is achieved by varying the forward velocity of the vessel. Different to [6], we do not generate the velocity in a dynamical way; rather we consider that the velocity is given by a setpoint that can be increased/decreased with a prescribed acceleration/deceleration rate. The setpoint is changed whenever the roll angle achieves a certain threshold. In this way, we do not need to increase the model to a 4th order model.

The second objective is achieved by including fin stabilizers. The hydraulic machinery, that generates the fin-induced roll moment τ_ϕ is modeled as follows (see [6])

$$\dot{\tau}_\phi = \frac{1}{t_r} \tau_{\max} \text{sat}\left(\frac{\tau_c}{\tau_{\max}}\right) - \frac{1}{t_r} \tau_\phi, \tag{14.10}$$

where $\tau_{\max}$ is the maximum moment that can be provided by the fins, τ_c is the moment generated by the controller and is given in [6]. The time constant t_r corresponds to the time constant of the hydraulic machinery.

Consequently, it follows from (14.6)–(14.8) and (14.10) that the simplified nonlinear ship container model with fin stabilizer control is given by

$$\ddot{z} = F_z(z,\phi,\theta,\zeta,\dot{\zeta},\ddot{\zeta}) - c_{11}\dot{z} - c_{12}\dot{\phi} - c_{13}\dot{\theta}, \tag{14.11}$$

$$\ddot{\phi} = F_\phi(z,\phi,\theta,\zeta,\dot{\zeta},\ddot{\zeta}) + \tau_\phi - c_{21}\dot{z} - c_{22}\dot{\phi} - c_{23}\dot{\theta}, \tag{14.12}$$

$$\ddot{\theta} = F_\theta(z,\phi,\theta,\zeta,\dot{\zeta},\ddot{\zeta}) - c_{31}\dot{z} - c_{32}\dot{\phi} - c_{33}\dot{\theta}. \tag{14.13}$$

The experimental setup depicted in Fig. 14.1 can be adjusted to mimic the container ship dynamics (14.11)–(14.13). First, we make an analogy between the electro-mechanical experimental setup and the heave, roll, and pitch motion of the ship. In other experiments (related with synchronization), we have found that under some circumstances, the oscillators can be in an oscillating state while the beam is at rest. Therefore, the following choice seems to be logic: displacement of oscillator 1 will correspond to heave displacement, displacement of oscillator 2 will represent the rotation angle in pitch and the supporting beam will denote the rotation angle in roll.

Next, the virtual coordinate system $s = [z \; \phi \; \theta]^{\mathrm{T}}$ is obtained by choosing the appropriate coordinate transformation. Since in the experimental setup all the displacements are translational, the coordinate transformation should be chosen such that translational displacements are mapped to rotation angles. Then, we write:

$$
\begin{bmatrix} z \\ \phi \\ \theta \end{bmatrix} = \begin{bmatrix} \frac{1}{x_1^*} & 0 & 0 \\ 0 & 0 & \frac{\gamma_1}{x_3^*} \\ 0 & \frac{\gamma_2}{x_2^*} & 0 \end{bmatrix} \begin{bmatrix} x_1 \\ x_2 \\ x_3 \end{bmatrix},
\tag{14.14}
$$

where x_i^*, $i = 1, 2$, are the maximal displacements of the oscillators and x_3^* is the maximal displacement of the supporting beam. In the sequel, these values are taken to be $x_1^* = x_2^* = x_3^* = 5$ [mm]. This mapping assures rotation angles of $\pm \gamma_i$ [rad].

In order to complete the adjustment of the experimental setup we choose the actuator forces as follows

$$
u_1 = m_1 \left(\omega_1^2 \Delta x_1 + 2\zeta_1 \omega_1 \dot{\Delta} x_1 + F_z(\cdot) - c_{11}\dot{z} - c_{12}\dot{\phi} - c_{13}\dot{\theta} \right),
\tag{14.15}
$$

$$
u_2 = m_2 \left(\omega_2^2 \Delta x_2 + 2\zeta_2 \omega_2 \dot{\Delta} x_2 + F_\theta(\cdot) - c_{31}\dot{z} - c_{32}\dot{\phi} - c_{33}\dot{\theta} \right),
\tag{14.16}
$$

$$
\begin{aligned}
u_3 = m_3 \big(&\omega_3^2 x_3 - 2\zeta_3 \omega_3 \dot{x}_3 + \mu_1 \left(F_z(\cdot) - c_{11}\dot{z} - c_{12}\dot{\phi} - c_{13}\dot{\theta} \right) \\
&+ \mu_2 \left(F_\theta(\cdot) - c_{31}\dot{z} - c_{32}\dot{\phi} - c_{33}\dot{\theta} \right) + F_\phi(\cdot) \\
&+ \tau_\phi - c_{21}\dot{z} - c_{22}\dot{\phi} - c_{23}\dot{\theta} \big),
\end{aligned}
\tag{14.17}
$$

where $\Delta x_i = (x_i - x_3)$, $\dot{\Delta} x_i = (\dot{x}_i - \dot{x}_3)$.

In closed loop, the dynamics of system (14.2) with controllers (14.15)–(14.17) coincides with dynamics (14.11)–(14.13). Therefore, the electro-mechanical experimental setup has been "converted" into a container ship.

14.3.2 Experimental and Numerical Analysis

In order to demonstrate that the experimental setup can actually mimic the dynamical behavior of a ship, in particular the onset and stabilization of parametric roll, we present two experiments: one corresponding to the uncontrolled case, where oscillations in roll appear and the other one corresponding to the controlled case, where parametric roll is stabilized. Indeed, the second experiment shows the capability of the setup to test and validate controllers that have been designed to cope with the problem of parametric roll.

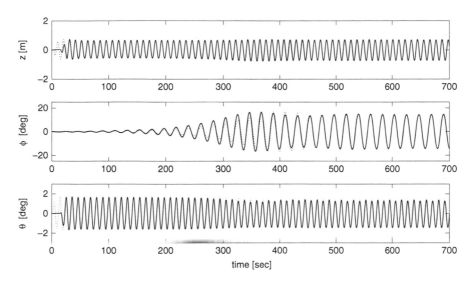

Fig. 14.3 Onset of parametric roll. *Solid line*: experiment, *dotted line*: simulation

For the first experiment, we consider system (14.11)–(14.13) with parameters listed in [8]. Such parameters correspond with a 1 : 45 scale model ship. Furthermore, the following experimental conditions are assumed: wave amplitude $A_\omega = 2.5$ [m], wave frequency $\omega = 0.4640$ [rad/s], encounter angle $\mu = 180°$, and encounter frequency, $w_e = 0.5842$ [rad/s]. For this experiment the forward velocity of the ship is taken to be 5.4806 [m/s]. Note that the forward velocity of the ship is directly related with the surge motion of the ship and therefore it cannot be related to the velocity of the experimental setup, where heave, pitch, and roll motions are being reproduced by the oscillators and the beam respectively. In this analysis, the forward velocity is considered as a control parameter given by a setpoint implemented in software and its changes are reflected in adjustments in the value of the encounter frequency (see (14.9)). Ultimately, this is reflected in changes in the dynamical behavior of system (14.11)–(14.13).

The initial conditions for the oscillators and the beam are as follows: $x_1(0) = 125$ [μm], $x_2(0) = 0$, $x_3(0) = 58.178$ [μm], $\dot{x}_1 = \dot{x}_2 = \dot{x}_3 = 0$. These initial conditions are related to the initial conditions of system (14.11)–(14.13) by means of (14.14). In this experiment, we want to investigate the onset of parametric roll, therefore the control input τ_ϕ in (14.12) is taken to be zero.

Figure 14.3 shows the experimental (solid line) and numerical (dotted line) results corresponding to heave, roll, and pitch. The onset of parametric roll becomes immediately clear from the graph in the middle of Fig. 14.3 and after 400 s it stabilizes with an amplitude of $\pm 15°$. The figure also reveals that the experimental results are in fair agreement with the simulation results. Indeed, in steady-state it is hard to distinguish the difference.

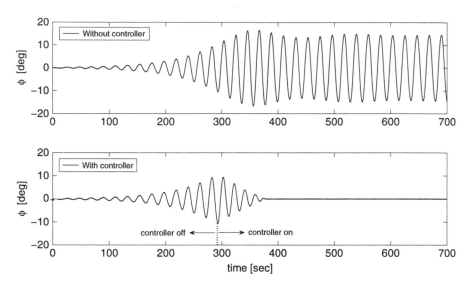

Fig. 14.4 Experimental result. Parametric roll is stabilized

In a second experiment, we implement controller (14.10) in order to stabilize parametric roll. All the initial conditions and parameters are as in experiment one. The controller is activated when the rotational angle in roll achieves a threshold angle of $10°$. Furthermore, when controller (14.10) is activated, the forward velocity of the vessel is increased 35% with an acceleration rate of 0.04 $[m/s^2]$. This increment in the forward velocity is also reflected in the value of the encounter frequency, which is also increased from 0.5842 [rad/s] to 0.6263 [rad/s]. In the same way, the value of the external wave forces and the values of the entries of the added mass matrix and hydrodynamic damping matrix are updated. In our experiment, the hydraulic machinery has been implemented in software and we have considered a time constant $t_r = 1$ [msec] since the data acquisition system of the setup has a maximum sampling period of 1 [msec].

Initially, the controller is switched-off, but when the rotation angle in roll reaches the threshold value $\phi = 10°$, the controller is switched-on and after the transient, the oscillations in roll are "quenched" as depicted in Fig. 14.4.

For the mapping (14.14) we have used $\gamma_1 = 0.035$ [rad], which assures a maximal rotation in pitch of $\theta = \pm2°$ and $\gamma_2 = 0.3$ [rad], which yields a maximal rotation in roll of $\phi = \pm17°$. This mapping not only allows to convert translational displacements to rotational angles but also yields the signals in a range that is suitable for the experimental setup as can be seen in Fig. 14.5, where the inputs u_i (see equations (14.15)–(14.17)) are depicted. From this figure it is evident that the inputs of the actuators are far from saturation, since the maximum voltage input allowed by the actuators is ±0.42 [V].

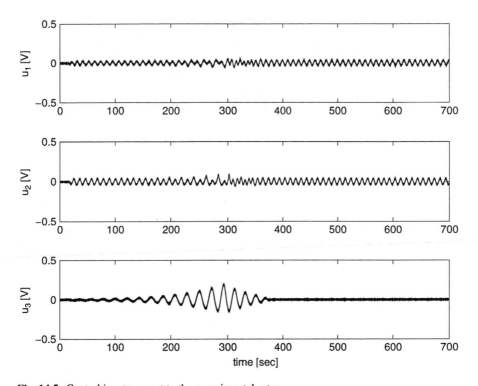

Fig. 14.5 Control inputs u_i sent to the experimental setup

14.4   Case 2: Mechanical Model for Simulating Heave-Pitch-Roll Motions

In this section, we investigate the onset of parametric roll by using the mechanical model depicted in Fig. 14.6, which has been developed to simulate heave-pitch-roll motions of a ship in longitudinal waves and it has been presented in [17]. As in the previous case, we present experiments related to the uncontrolled and controlled situations.

14.4.1   The Model and Its Implementation in the Setup

Consider the mechanical system depicted in Fig. 14.6. It consists of two masses restrained by elastic springs and supporting two equal pendulums rigidly connected by means of a weightless rod. Each mass is externally excited by a harmonic force. These external forces have the same amplitude and frequency but there is a phase lag between them. This phase shift is to include the delayed effects of the wave

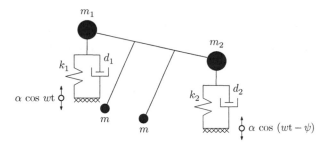

Fig. 14.6 Mechanical model for simulating heave-pitch-roll motion

propagating along the ship. This model was developed to simulate the heave-pitch-roll motion of a ship in longitudinal waves. For more details about the model, the reader is referred to [17].

The equations of motion for the system of Fig. 14.6 are:

$$(m_1 + m)\left(\ddot{z}_1 - \alpha\omega^2\cos\omega t\right) + d_1\dot{z}_1 + k_1 z_1 + ml\left(\ddot{\varphi}\sin\varphi + \dot{\varphi}^2\cos\varphi\right) = 0, \quad (14.18)$$

$$(m_2 + m)\left(\ddot{z}_2 - \alpha\omega^2\cos(\omega t - \psi)\right) + d_2\dot{z}_2 + k_2 z_2 + ml\left(\ddot{\varphi}\sin\varphi + \dot{\varphi}^2\cos\varphi\right) = 0, \quad (14.19)$$

$$\frac{1}{2}ml\left(\ddot{z}_1 - \alpha\omega^2\cos\omega t + \ddot{z}_2 - \alpha\omega^2\cos(\omega t - \psi)\right)\sin\varphi$$
$$+ml^2\ddot{\varphi} + c\dot{\varphi} + mgl\sin\varphi = \tau_{\mathrm{p}}, \quad (14.20)$$

where τ_{p} is an external torque for the pendula. By defining the new time variable $\tau = \sqrt{\frac{g}{l}}t$, system (14.18)–(14.20) is rewritten in the following dimensionless form (see [17])

$$\omega_1'' + \kappa_1\omega_1' + q_1^2\omega_1 + \mu_{m1}\left(\varphi''\sin\varphi + \varphi'^2\cos\varphi\right) = a\eta^2\cos\eta\tau, \quad (14.21)$$

$$\omega_2'' + \kappa_2\omega_2' + q_2^2\omega_2 + \mu_{m2}\left(\varphi''\sin\varphi + \varphi'^2\cos\varphi\right) = a\eta^2\cos(\eta\tau - \psi), \quad (14.22)$$

$$\frac{1}{2}\left[\omega_1'' - a\eta^2\cos\eta\tau + \omega_2'' - a\eta^2\cos(\eta\tau - \psi)\right]\sin\varphi + \varphi'' + \kappa_0\varphi' + \sin\varphi = \tau_\varphi, \quad (14.23)$$

where $\omega_i = \frac{z_i}{l}$, $\kappa_i = \frac{d_i}{\omega_0(m_i + m)}$, $q_i^2 = \frac{k_i}{\omega_0^2(m_i + m)}$, $\mu_{mi} = \frac{m}{(m_i + m)}$ for $i = 1, 2$ and $\omega_0 = \sqrt{\frac{g}{l}}$, $\kappa_0 = \frac{c}{\omega_0 ml^2}$, $\eta = \frac{\omega}{\omega_0}$, $a = \frac{\alpha}{l}$ and $\tau_\varphi = \frac{\tau_{\mathrm{p}}}{ml^2\omega_0^2}$.

It can be shown that this system (with $\tau_\varphi = 0$) has a steady-state solution which has the property $\sum_{i=1}^{2}[\omega_i^2 + \dot{\omega}_i^2] \neq 0$, $\varphi = \dot{\varphi} = 0$. This solution can become unstable in certain intervals of the frequency of excitation, denoted as ω in (14.18)–(14.20). When the solution becomes unstable, parametric roll resonance will appear ($\varphi \neq 0$).

For the case where parametric roll resonance appears, it is necessary to stabilize it. Then, we derive a controller by following the derivations presented in [6]. This controller is designed by using backstepping. Note however that, at this stage, we can adopt any controller for the experimental setup to be tested.

For the design of the control, we use the uncoupled equation for roll ((14.23) with $\omega_1'' = \omega_2'' = 0$) and it follows that the control model verifies:

$$\varphi'' + \kappa_0 \varphi' + \sin\varphi + \frac{1}{2}\left[-a\eta^2\cos\eta\tau - a\eta^2\cos(\eta\tau - \psi)\right]\sin\varphi = \tau_\varphi, \quad (14.24)$$

$$\dot{\tau}_\psi + \frac{1}{t_r}\tau_\psi = \frac{1}{t_r}\tau_{max}\,\text{sat}\left(\frac{\tau_c}{\tau_{max}}\right), \quad (14.25)$$

where t_r is a time constant that coincides with the sampling period of the data acquisition system of the experimental setup (1 msec), τ_{max} is the maximum input that can be delivered to the system and τ_c verifies

$$\tau_c = -Q_3 z_2 - Q_2 z_1 - \kappa_0 Q_1 \varphi + \sin\varphi + Q_1^2\varphi - Q_2\dot{z}_1 t_r, \quad (14.26)$$

where $z_1 = \dot{\varphi} + Q_1\varphi$, $z_2 = \tau_\varphi + Q_2 z_1 + \kappa_0 Q_1 \varphi - \sin\varphi - Q_1^2\varphi$, $Q_1 > 0$, $Q_2 > (Q_1 + 2\gamma - \kappa_0)$, $\gamma = \frac{a\eta^2}{2}$, $Q_3 > 0$.

After the derivation of the controller, the system (14.21)–(14.23) with controller (14.25) is implemented in the experimental setup of Fig. 14.1. The analogy between system of Fig. 14.6 and the setup of Fig. 14.1 is as follows: the vertical displacement corresponding to mass 1 is represented by oscillator 1, the vertical displacement of mass 2 is represented by oscillator 2 and the rotation angle of pendula is represented by the supporting beam.

The next step is to obtain the virtual coordinate system $s := \begin{bmatrix} \omega_1 & \omega_2 & \varphi \end{bmatrix}^T$. Then, we use the transformation

$$\begin{bmatrix} \omega_1 \\ \omega_2 \\ \varphi \end{bmatrix} = \begin{bmatrix} \frac{\varepsilon_1}{x_1^*} & 0 & 0 \\ 0 & \frac{\varepsilon_2}{x_2^*} & 0 \\ 0 & 0 & \frac{\alpha}{x_3^*} \end{bmatrix} \begin{bmatrix} x_1 \\ x_2 \\ x_3 \end{bmatrix}, \quad (14.27)$$

where x_i^* has the same meaning as in (14.14). With this transformation, the translational displacement of the supporting beam is mapped to rotation angle and assures angles in roll of $\pm\alpha$ [rad]. The constants $\varepsilon_i > 0$ are scaling factors used in order to leave the signal corresponding to the vertical displacement of mass i between suitable ranges for the setup.

The adjustment continues by defining the actuator forces of the setup as follows:

$$u_1 = m_1 \left(\omega_1^2 \Delta x_1 + 2\zeta_1 \omega_1 \dot{\Delta} x_1 + \omega_1'' \right), \tag{14.28}$$

$$u_2 = m_2 \left(\omega_2^2 \Delta x_2 + 2\zeta_2 \omega_2 \dot{\Delta} x_2 + \omega_2'' \right), \tag{14.29}$$

$$u_3 = m_3 \left(\omega_3^2 x_3 + 2\zeta_3 \omega_3 \dot{x}_3 + \mu_1 \omega_1'' + \mu_2 \omega_2'' + \varphi'' \right) \tag{14.30}$$

with ω_i'', $i = 1,2$ and φ'' as given in (14.21)–(14.23). It is clear that the closed-loop dynamics of the experimental setup coincides with the dimensionless dynamics of the mechanical system depicted in Fig. 14.6.

14.4.2 Experimental and Simulation Results

Some experimental results are provided in order to show the capability of the experimental setup to mimic the dynamics of the mechanical model of Fig. 14.6 used to simulate the heave-pitch-roll motion of a ship. The onset of parametric roll is analyzed for the controlled and uncontrolled situations. We also analyze the effect in the roll motion when the mass of the pendula (corresponding to the mass of the supporting beam) is varied.

For the experiments, we consider model (14.21)–(14.23) with the following parameters: $m = 8.1$ [kg], $m_1 = m_2 = 0.210$ [kg], $l = 9.81$ [m], $\alpha = 0.5689$ [-], $\psi = \frac{\pi}{8}$ [rad], $g = 9.81$ [m/s^2], $k_1 = k_2 = 8.0698$ [N/s] $d_1 = d_2 = 67.06$ [Ns/m], $c = 60$ [Nms/rad], $\omega = 2$ [rad/s].

In the first experiment, we investigate the occurrence of parametric roll resonance. The initial conditions for the oscillators and the beam are as follows: $x_1(0) = 0.0011$ [m], $x_2(0) = 0.001$ [m], $x_3(0) = 0.00001$ [m], $\dot{x}_1(0) = \dot{x}_2(0) = \dot{x}_3(0) = 0$. These initial conditions are related with the initial conditions of system (14.21)–(14.23) by means of (14.27). In this experiment, parametric roll is not stabilized, hence we consider $\tau_\varphi = 0$ in (14.23).

Figure 14.7 shows the time series for heave, pitch, and roll. The oscillations in roll are slowly increasing until certain steady-state value (approximately 30°) as becomes evident from the graph at the bottom of the figure. The behavior in heave and pitch motions is as expected, since the masses are excited with the same amplitude and frequency but with a phase lag of $\frac{\pi}{8}$ [rad]. However, the phase difference is a bit lower, in part, due to the "disturbance" produced by the oscillations in roll, since we have verified in other experiments that when parametric roll does not appear, the phase difference in heave and pitch is precisely $\frac{\pi}{8}$ [rad]. The experimental and numerical results are fairly comparable and in steady-state (around 350 s) the differences are negligible. The small differences between experimental and numerical results observed in the transient are rather quantitative than qualitative and the most probable cause is the slightly different initial conditions and the natural damping present in the setup which is not perfectly canceled.

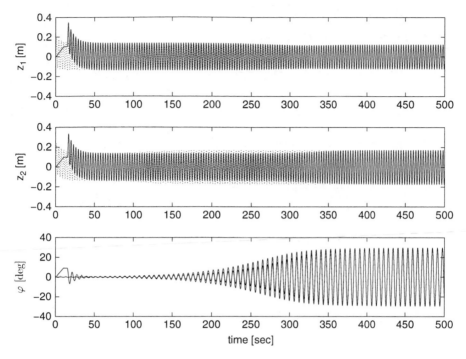

Fig. 14.7 Heave, pitch, and roll motions. *Solid line*: experimental results. *Dotted line*: simulation results

In a second experiment, controller (14.25) is included in order to stabilize the parametric roll resonance condition. The parameters of the controller are $\tau_{max} = 0.01$ [Nm], $t_r = 1$ [msec], with gains $Q_1 = 5$, $Q_2 = 10.155$, and $Q_3 = 7$. The parameters for the model and the initial conditions are as in experiment one except for $x_3(0) = 0.0002$ and $m = 6$ [kg]. The controller is activated when the roll angle achieves the threshold angle $\varphi = 10°$. The experimental results are presented in Fig. 14.8. From this figure it is possible to realize that parametric roll has been stabilized. The control law τ_φ has been implemented in software and is sent to the setup by means of the data acquisition system. Finally, we present an experiment in which one of the parameters of the system is varied during the experiment. In [17] it has been shown that the stability threshold of the semi-trivial solution $\sum_{i=1}^{2}[\omega_i^2 + \dot{\omega}_i^2] \neq 0$, $\varphi, \dot{\varphi} = 0$ is dependent on the parameters of the system. Indeed, in experiments one and two, we have chosen the parameters such that the semi-trivial solution is unstable (parametric roll occurs). However, we also find that by considering $m = 4.1$ [kg] and with the same parameters as in experiment one, the stability threshold is not violated and therefore no parametric roll appears. This is illustrated experimentally. First, experiment one is repeated but with $m = 4.1$ [kg], therefore no parametric roll occurs.

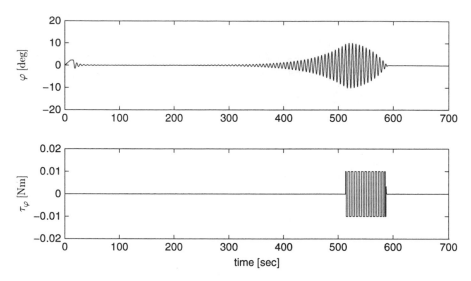

Fig. 14.8 Parametric roll is stabilized

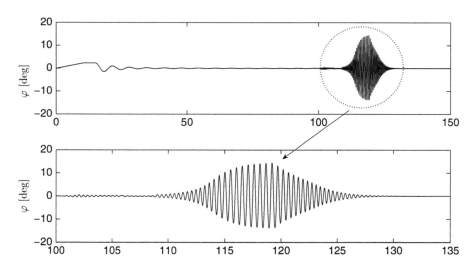

Fig. 14.9 In this experiment parametric roll is triggered by varying the mass of the beam

At t = 100 s, we add extra mass in the supporting beam. As a consequence, we can observe resonance in the roll motion. At $t \approx 120$ s, we remove the extra mass and the resonance in roll disappears, as depicted in Fig. 14.9. Clearly, we can see that by varying the mass of the supporting beam we can trigger the onset of parametric roll.

14.5   Conclusions

We have presented an electro-mechanical setup which is capable of conducting parametric roll experiments. The dynamical behavior of the system has been modified such that we were able to mimic the dynamics of a simplified 3-DOF nonlinear container ship model and the dynamics of a mechanical model which simulates the heave-pitch-roll motion of a ship in a longitudinal sea. In both cases, the oscillators are used to represent the heave and pitch motions and the supporting beam is used to reproduce the roll motion. The experiments have been supported by numerical results and are quite comparable with results that already have been presented in the literature.

One of the advantages of this approach is the low implementation cost since for the experiments we do not require additional equipment. The only requirement is a computer and a data acquisition system. The rest is implemented in software. This brings another advantage: in this setup it is possible to implement any external excitation and hence, it is possible to create, for example, an excitation corresponding to the waves of a rough sea or the situation in which the ship is navigating in random seas. In the same way, it is possible to study the influence of specific parameters in the onset of parametric roll, since we are able to change/update any parameter at any time. Ultimately, this experimental testbed can be seen as an alternative for the validation of models and mainly, for testing controllers ad hoc designed for the stabilization of parametric roll, with the final aim of improving its performance in real applications.

Acknowledgements The first author acknowledges the support of the Mexican Council for Science and Technology (CONACYT).

References

1. Abdel Gawad, A. F., Ragab, S. A., Nayfeh, A. H., Mook, D. T.: Roll stabilization by anti-roll passive tanks, Ocean Engineering, **28**:457–469 (2001)
2. Amerongen van, J., Klugt, van der P. G. M., Nauta Lemke, van H. R.: Rudder roll stabilization for ships. Automatica. **26**(4):679–690 (1990)
3. Brown, D. K.: The way of a ship in the midst of the sea: the life and work of William Froude. Periscope Publishing Ltd. ISBN-10: 1904381405 (2006)
4. Carmel, S. M.: Study of parametric rolling event on a panamax container vessel. Journal of the Transportation Research Board, **1963**:56–63 (2006)
5. France, W. N., Levadou, M., Trakle, T. W., Paulling, J. R., Michel, R.K., Moore, C.: An investigation of head-sea parametric rolling and its influence on container lashing systems, Marine Technology **40**(1):1-19 (2003)
6. Galeazzi, R., Holden, C., Blanke, M., Fossen, T. I.: Stabilization of parametric roll resonance by combined speed and fin stabilizer control. Proceedings of the 10th European Control Conference, Budapest, Hungary, pp. 4895–4900 (2009)

7. Holden, C., Galeazzi, R., Fossen, T. I., Perez, T.: Stabilization of parametric roll resonance with active u-tanks via Lyapunov control design. Proceedings of the 10th European Control Conference, Budapest, Hungary, pp. 4889–4894 (2009)
8. Holden, C., Galeazzi, R., Rodríguez, C. A., Perez, T., Fossen, T. I., Blanke, M., Neves, M. A. S.: Nonlinear container ship model for the study of parametric roll resonance. Modeling, identification and control **28**(4):87–103 (2007)
9. Ibrahim, R. A., Grace, I. M.: Modeling of ship roll dynamics and its coupling with heave and pitch. Mathematical Problems in Engineering, vol. 2010, 32 pages, doi:10.1155/2010/934714 (2010)
10. Baniela, I. S.: Roll motion of a ship and the roll stabilising effect of bilge keels. The Journal of Navigation, **61**:667–686, doi:10.1017/S0373463308004931 (2008)
11. Lloyd, A. R. J. M.: Seakeeping: ship behaviour in rough weather, Ellis Horwood Limited, Chichester (1989)
12. Neves, M. A. S., Rodrguez, C. A.: A coupled non-linear mathematical model of parametric resonance of ships in head seas. Applied Mathematical Modelling, **33**(6):2630–2645, doi: 10.1016/j.apm.2008.08.002 (2009)
13. Perez, T.: Ship motion control: course keeping and roll stabilisation using rudder and fins. Advances in Industrial Control. London, United Kingdom (2005)
14. Pogromsky, A., Rijlaarsdam, D., Nijmeijer, H.: Experimental Huygens synchronization of oscillators. Nonlinear Dynamics and Chaos: Advances and Perspectives. Springer series Understanding Complex Systems. Editors: Marco Thiel, Jurgen Kurths, M. Carmen Romano, Gyorgy Karolyi and Alessandro Moura. pp 195–210 (2010)
15. Rosas Almeida, D. I., Alvarez, J., Fridman, L.: Robust observation and identification of n-DOF Lagrangian systems. International Journal of Robust and Nonlinear control. **17**:842–861 (2007)
16. Tondl, A., Nabergoj, R.: Simulation of parametric ship roll and hull twist oscillations. Nonlinear Dynamics, **3**(1):41-56, doi: 10.1007/BF00045470 (1992)
17. Tondl, A., Ruijgrok, T., Verhulst, F., Nabergoj, R.: Autoparametric resonance in mechanical systems. Cambridge, United Kingdom (2000)

Chapter 15
Controlling Parametric Resonance: Induction and Stabilization of Unstable Motions

Roberto Galeazzi and Kristin Y. Pettersen

15.1 Parametric Resonance: Threat or Advantage?

Parametric resonance is a well-known resonant phenomenon, which can determine the instability of a system in response to small periodic variations of one of its parameters. In the light of this simple description, the common sense suggests that parametric resonance is a threat for any system where it can potentially onset. As a matter of fact, if we restrain the analysis to marine structures and automotive systems the former answer perfectly fits. For the last 12 years parametric roll resonance on ships has been in focus of the maritime community as one of the top stability related issues, and still it is. Several control strategies have been proposed, which try to stabilize the large roll motion: backstepping controllers have been designed to damp the resonant oscillations using, for example, active U-tanks [5] or fin stabilizers [3]; an extremum seeking controller has been proposed to detune the frequency synchronization by altering ship's speed and/or course [2]. A considerable effort has also been produced by the automotive research community, in particular focusing on how periodic variations of the road profile can induce unstable steering oscillations in motorcycles [8, 13].

However, if we look at a completely different class of systems it is possible to find several applications where the onset of parametric resonance is an advantage.

R. Galeazzi (✉)
Department of Electrical Engineering, Technical University of Denmark, DTU Electrical
Engineering, DK 2800 Kgs. Lyngby, Denmark

Center for Ships and Ocean Structures, Norwegian University of Science and Technology,
NO-7491 Trondheim, Norway
e-mail: rg@elektro.dtu.dk

K.Y. Pettersen
Department of Engineering Cybernetics, Norwegian University of Science and Technology,
NO-7491 Trondheim, Norway
e-mail: Kristin.Ytterstad.Pettersen@itk.ntnu.no

T.I. Fossen and H. Nijmeijer (eds.), *Parametric Resonance in Dynamical Systems*,
DOI 10.1007/978-1-4614-1043-0_15, © Springer Science+Business Media, LLC 2012

In micro-electromechanical systems parametric resonance phenomena are induced in order to, for example, reduce the parasitic signal in capacitive sensing [16], or to increase the sensitivity of mass sensors at the pico scale $(10^{-12}$ g) [17], or to increase robustness against parameter variations in micro-gyroscopes [12]. Analogous interests in capitalizing the large energy released by parametric resonant oscillations to boost specific system features is also raising in the field of wave energy exploitation. Here the idea is to induce parametric resonance in order to increase the amount of energy producible by the converter [10, 11].

Therefore, by looking at the large variety of systems and possible applications where parametric resonance may naturally occur or can be artificially induced, it appears natural to investigate how active control strategies can be used in order to either trigger the resonant phenomenon or to stabilize it. Starting from this consideration, in this chapter the authors revisit some of the theory of parametric resonance through the use of a mechanical equivalent, which can represent many of the systems aforementioned, and they cast both the induction and the stabilization of resonant oscillations as a tracking problem. An input–output feedback linearizing controller is then designed and shown to be capable both of triggering parametric resonance and stabilizing the unstable motion.

In particular, Sect. 15.2 introduces the mechanical system used in the analysis, namely the pendulum with moving support, as a member of the class of autoparametric systems. Lagrangian description of the system's dynamics is provided, and the stability analysis of the open loop system is carried out. Section 15.3 formulates the control problems, and it presents the design of the controller based on feedback linearization theory. Section 15.4 illustrates the performance of the closed loop system through simulation results. Section 15.5 draws some conclusions, and it highlights possible future research paths.

15.2 Autoparametric Systems

Autoparametric systems consist of two or more vibrating components, which interact in a nonlinear fashion [14]. The components are divided into the *primary system*, which is usually in a vibrating state, and the *secondary system*, which is usually at rest while the primary system is oscillating. This state is called semi-trivial solution of the autoparametric system.

The excitation acts on the primary system under the form of external forcing, self-excitation, parametric excitation, or a combination of those. Within certain intervals of the excitation frequency the semi-trivial solution can become unstable, and the system enters in autoparametric resonance. The vibrations of the primary system act as parametric excitation on the secondary one, which will be no longer at rest.

Autoparametric systems in resonance condition can display different behaviors including periodic, quasi-periodic, non-periodic, and also chaotic behaviors. Moreover, the occurrence of the resonance often goes along with saturation phenomena. In particular, when the secondary system enters into parametric resonance it

Fig. 15.1 Pendulum with moving support. The reference frame is right-handed, therefore positive rotations are counterclockwise

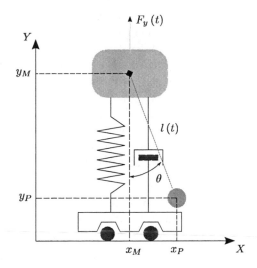

functions as an energy absorber by draining energy from the external excitation through the primary system. This entails a large increase of the amplitude of the displacements of the secondary system whereas the oscillation's amplitude of the primary system is maintained as almost constant.

The energy absorbency of the secondary system can determine both undesirable and desirable results. When parametric roll onsets on a ship, the roll motion becomes the sink of the wave energy exciting the vessel, and the springing of violent roll oscillations is definitely a troublesome outcome. Conversely if the resonance could be induced in a wave energy converter, the capability of draining more energy out of the wave motion will clearly be beneficial.

The difference between autoparametric resonance and parametric resonance resides in the presence of a primary system driving the onset of the resonant condition. For instance parametric roll on ships can be seen as either a parametric resonance or an autoparametric resonance phenomenon depending on whether we consider only the roll subsystem or we include the heave and/or pitch dynamics. In the former case the parametric resonance origins as a results of the quasi-periodic variations of the ship's metacentric height, which is a parameter within the roll dynamics; in the latter case are the oscillations in heave and/or pitch (primary system), directly excited by the wave motion, which determine the onset of the resonant behavior in roll.

15.2.1   Pendulum with Moving Support

A well-known autoparametric system is the pendulum with moving support, as that represented in Fig. 15.1. The system consists of a pendulum, whose pivot point is connected to a mass-spring-damper, which in turn is placed atop a cart. The position of the pivot point can change both along the X and Y directions due to

the vertical oscillation of the mass-spring-damper or to the horizontal displacement
of the cart. The mass-spring-damper together with the cart represent the primary
system, whereas the pendulum is the secondary system. The system is assumed
to be underactuated since no direct control action can be performed either on
the vertical motion of the mass-spring-damper or on the swinging motion of the
pendulum.

The mass-spring-damper is in a vibration state due to the action of an external
sinusoidal force; conversely the pendulum's bob is not directly subject to any
force/moment apart from gravity, hence the secondary system is initially at rest.
In order to induce parametric resonance into the secondary system two conditions
must be fulfilled:

- The equilibrium position of the pendulum's bob must be altered.
- The natural frequency of the pendulum must be approximately equal to half the
  frequency of the external excitation.

Since the swinging of the pendulum cannot be directly actuated due to the absence
of a torque acting on the pivot point, the first requirement can be met by changing
the position of the cart, which will produce an inertia effect about the pivot point.
Considering that the natural frequency of a pendulum is a function of the length of
the pendulum's rod, it is evident that in order to achieve the second requirement the
pendulum must have a variable length rod.

A four degrees-of-freedom model is first derived by applying Lagrange's theory.
Then a stability analysis under the assumption of external sinusoidal excitation is
carried out to determine the stability conditions to be infringed in order to trigger
the resonant phenomenon.

15.2.1.1 Lagrangian Model

Consider a mass-spring-damper system of mass m_1 oscillating under the action of
an external sinusoidal force $F_y(t)$. A second mass m_2 is attached to the bottom end
of a massless rod of variable length, whose pivot point is joint to the first mass m_1.
The two masses are placed on top of a massless cart that is free to move along the
horizontal direction.

Said l_0 the nominal length of the pendulum, the variable rod's length is given by

$$l(t) = l_0 + \delta_l(t), \qquad \forall t \; \delta_l(t) > -l_0,$$

where δ_l is the deviation from the nominal value. Then the vector of generalized
coordinates is defined as $\mathbf{q} \triangleq [x_M, y_M, \theta, \delta_l]^T$, where (x_M, y_M) is the position of the
mass m_1, and θ is the oscillation angle of the pendulum.

The equations of motion can be derived from *Lagrange's equations*

$$\frac{d}{dt}\left(\frac{\partial \mathcal{L}(\mathbf{q},\dot{\mathbf{q}})}{\partial \dot{\mathbf{q}}}\right) - \frac{\partial \mathcal{L}(\mathbf{q},\dot{\mathbf{q}})}{\partial \mathbf{q}} = \tau, \tag{15.1}$$

where

$$\mathscr{L}(\mathbf{q},\dot{\mathbf{q}}) = \mathscr{T}(\mathbf{q},\dot{\mathbf{q}}) - \mathscr{V}(\mathbf{q}) \tag{15.2}$$

is the Lagrangian given by the difference between the kinetic energy $\mathscr{T}$ and the potential energy $\mathscr{V}$; τ is the vector of the generalized forces that accounts for unknown external forces (disturbances) τ^e and for control inputs τ^c

$$\tau = \tau^e + \tau^c$$

$$= \begin{bmatrix} 0 \\ F_y^e \\ 0 \\ 0 \end{bmatrix} + \begin{bmatrix} F_x^c \\ 0 \\ 0 \\ F_\delta^c \end{bmatrix}.$$

Given the position (x_M, y_M) of the mass m_1, the position of the mass m_2 at any given instant in time is given by the vector (in the following the notation s_θ and c_θ stands for $\sin\theta$ and $\cos\theta$, respectively)

$$\mathbf{r}_P: \begin{cases} x_P = x_M + (l_0 + \delta_l)s_\theta \\ y_P = y_M - (l_0 + \delta_l)c_\theta \end{cases}, \tag{15.3}$$

and its velocity by the vector

$$\mathbf{v}_P: \begin{cases} \dot{x}_P = \dot{x}_M + (l_0 + \delta_l)\dot{\theta}c_\theta + \dot{\delta}_l s_\theta \\ \dot{y}_P = \dot{y}_M + (l_0 + \delta_l)\dot{\theta}s_\theta - \dot{\delta}_l c_\theta \end{cases}. \tag{15.4}$$

The kinetic energy of the (m_1, m_2)-system is then given by

$$\mathscr{T}(\mathbf{q},\dot{\mathbf{q}}) = \frac{1}{2}\dot{\mathbf{q}}^T \mathbf{M}(\mathbf{q})\dot{\mathbf{q}}$$

$$= \frac{1}{2}m_1\left(\dot{x}_M^2 + \dot{y}_M^2\right) + \frac{1}{2}m_2\left(\dot{x}_P^2 + \dot{y}_P^2\right)$$

$$= \frac{1}{2}(m_1+m_2)\left(\dot{x}_M^2 + \dot{y}_M^2\right) + \frac{1}{2}m_2\left[(l_0+\delta_l)^2\dot{\theta}^2 + \dot{\delta}_l^2\right.$$

$$\left. + 2(l_0+\delta_l)(\dot{x}_M c_\theta + \dot{y}_M s_\theta)\dot{\theta} + 2\dot{\delta}_l(\dot{x}_M s_\theta - \dot{y}_M c_\theta)\right], \tag{15.5}$$

where $\mathbf{M}(\mathbf{q})$ is the mass-inertia matrix

$$\mathbf{M}(\mathbf{q}) = \begin{bmatrix} m_1+m_2 & 0 & m_2(l_0+\delta_l)c_\theta & m_2 s_\theta \\ 0 & m_1+m_2 & m_2(l_0+\delta_l)s_\theta & -m_2 c_\theta \\ m_2(l_0+\delta_l)c_\theta & m_2(l_0+\delta_l)s_\theta & m_2(l_0+\delta_l)^2 & 0 \\ m_2 s_\theta & -m_2 c_\theta & 0 & m_2 \end{bmatrix}$$

$$\mathbf{M}(\mathbf{q}) = \mathbf{M}^T(\mathbf{q});$$

whereas the potential energy reads

$$\mathscr{V}(\mathbf{q}) = \frac{1}{2}ky_M^2 + m_2 g(l_0 + \delta_l)(1 - c_\theta) \tag{15.6}$$

with k being the spring constant of the mass-spring-damper. The Lagrangian is hence given by

$$\mathscr{L}(\mathbf{q}, \dot{\mathbf{q}}) = \frac{1}{2}(m_1 + m_2)(\dot{x}_M^2 + \dot{y}_M^2) + \frac{1}{2}m_2\left[(l_0 + \delta_l)^2\dot{\theta}^2 + \dot{\delta}_l^2\right.$$
$$\left. + 2(l_0 + \delta_l)(\dot{x}_M c_\theta + \dot{y}_M s_\theta)\dot{\theta} + 2\dot{\delta}_l(\dot{x}_M s_\theta - \dot{y}_M c_\theta)\right] - \frac{1}{2}ky_M^2$$
$$- m_2 g(l_0 + \delta_l)(1 - c_\theta), \tag{15.7}$$

which gives rise to the following equations of motion

$$(m_1 + m_2)\ddot{x}_M + d_1\dot{x}_M$$
$$+ m_2\left[(l_0 + \delta_l)\ddot{\theta}c_\theta - (l_0 + \delta_l)\dot{\theta}^2 s_\theta + 2\dot{\delta}_l\dot{\theta}c_\theta + \ddot{\delta}_l s_\theta\right] = F_x^c, \tag{15.8}$$

$$(m_1 + m_2)\ddot{y}_M + d_2\dot{y}_M + ky$$
$$+ m_2\left[(l_0 + \delta_l)\ddot{\theta}s_\theta + (l_0 + \delta_l)\dot{\theta}^2 c_\theta + 2\dot{\delta}_l\dot{\theta}s_\theta - \ddot{\delta}_l c_\theta\right] = F_y^e, \tag{15.9}$$

$$m_2(l_0 + \delta_l)^2\ddot{\theta} + d_3\dot{\theta} + m_2(l_0 + \delta_l)gs_\theta$$
$$+ m_2(l_0 + \delta_l)\left[\ddot{x}_M c_\theta + \ddot{y}_M s_\theta + 2\dot{\delta}_l\dot{\theta}\right] = 0 \tag{15.10}$$

$$m_2\ddot{\delta}_l + d_4\dot{\delta}_l$$
$$- m_2\left[(l_0 + \delta_l)\dot{\theta}^2 + \ddot{x}_M s_\theta - \ddot{y}_M c_\theta + g(1 - c_\theta)\right] = F_\delta^c \tag{15.11}$$

where linear damping terms $d_i\dot{q}_i$ have been introduced. System (15.8)–(15.11) can be rewritten in dimensionless form as

$$\ddot{x} + \mu_1\dot{x} + \alpha\left[(1 + \delta)(\ddot{\theta}c_\theta - \dot{\theta}^2 s_\theta) + 2\dot{\delta}\dot{\theta}c_\theta + \ddot{\delta}s_\theta\right] = \Phi_x^c, \tag{15.12}$$

$$\ddot{y} + \mu_2\dot{y} + \omega_y^2 y + \alpha\left[(1 + \delta)(\ddot{\theta}s_\theta + \dot{\theta}^2 c_\theta) + 2\dot{\delta}\dot{\theta}s_\theta - \ddot{\delta}c_\theta\right] = \Phi_y^e, \tag{15.13}$$

$$\ddot{\theta} + \frac{\mu_3}{(1 + \delta)^2}\dot{\theta} + \frac{2}{1 + \delta}\dot{\delta}\dot{\theta} + \frac{1}{1 + \delta}(\omega_\theta^2 + \ddot{y})s_\theta + \frac{1}{1 + \delta}\ddot{x}c_\theta = 0, \tag{15.14}$$

$$\ddot{\delta} + \mu_4\dot{\delta} - (1 + \delta)\dot{\theta}^2 - \ddot{y}c_\theta + \ddot{x}s_\theta + \omega_\theta^2(1 - c_\theta) = \Phi_\delta^c, \tag{15.15}$$

where

$$x = \frac{x_M}{l_0}, \qquad y = \frac{y_M}{l_0}, \qquad \delta = \frac{\delta_l}{l_0}$$

and the parameters are given by

$$\alpha = \frac{m_2}{m_1 + m_2}, \qquad \omega_\theta^2 = \frac{g}{l_0}, \qquad \omega_y^2 = \frac{k}{m_1 + m_2},$$

$$\mu_1 = \frac{d_1}{m_1 + m_2}, \qquad \mu_2 = \frac{d_2}{m_1 + m_2}, \qquad \mu_3 = \frac{d_3}{m_2 l_0^2}, \qquad \mu_4 = \frac{d_4}{m_2 l_0},$$

$$\Phi_x^c = \frac{F_x^c}{(m_1 + m_2) l_0}, \qquad \Phi_y^e = \frac{F_y^e}{(m_1 + m_2) l_0}, \qquad \Phi_\delta^c = \frac{F_\delta^c}{m_2 l_0}.$$

By means of matrix notation the following compact form can be achieved (the symbol ˜ denotes non-dimensional quantities)

$$\widetilde{\mathbf{M}}(\tilde{\mathbf{q}})\, \ddot{\tilde{\mathbf{q}}} + \widetilde{\mathbf{D}}\dot{\tilde{\mathbf{q}}} + \widetilde{\mathbf{C}}(\tilde{\mathbf{q}},\dot{\tilde{\mathbf{q}}})\, \dot{\tilde{\mathbf{q}}} + \tilde{\mathbf{g}}(\tilde{\mathbf{q}}) = \tilde{\tau}, \qquad (15.16)$$

where $\tilde{\mathbf{q}} = [x, y, \theta, \delta]^{\mathrm{T}}$, and $\tilde{\tau} = \left[\Phi_x^c, \Phi_y^e, 0, \Phi_\delta^c \right]^{\mathrm{T}}$. $\widetilde{\mathbf{M}}(\tilde{\mathbf{q}})$ is the scaled mass-inertia matrix

$$\widetilde{\mathbf{M}}(\tilde{\mathbf{q}}) = \begin{bmatrix} 1 & 0 & \alpha(1+\delta)c_\theta & \alpha s_\theta \\ 0 & 1 & \alpha(1+\delta)s_\theta & -\alpha c_\theta \\ (1+\delta)c_\theta & (1+\delta)s_\theta & (1+\delta)^2 & 0 \\ s_\theta & -c_\theta & 0 & 1 \end{bmatrix},$$

$\widetilde{\mathbf{D}}$ is the viscous damping matrix

$$\widetilde{\mathbf{D}} = \begin{bmatrix} \mu_1 & 0 & 0 & 0 \\ 0 & \mu_2 & 0 & 0 \\ 0 & 0 & \mu_3 & 0 \\ 0 & 0 & 0 & \mu_4 \end{bmatrix}, \qquad \widetilde{\mathbf{D}} > 0,$$

$\widetilde{\mathbf{C}}(\tilde{\mathbf{q}},\dot{\tilde{\mathbf{q}}})$ is the Coriolis-centripetal matrix

$$\widetilde{\mathbf{C}}(\tilde{\mathbf{q}},\dot{\tilde{\mathbf{q}}}) = \begin{bmatrix} 0 & 0 & -\alpha\left((1+\delta)\dot\theta s_\theta - \dot\delta c_\theta\right) & \alpha\dot\theta c_\theta \\ 0 & 0 & \alpha\left((1+\delta)\dot\theta c_\theta + \dot\delta s_\theta\right) & \alpha\dot\theta s_\theta \\ 0 & 0 & (1+\delta)\dot\delta & (1+\delta)\dot\theta \\ 0 & 0 & -(1+\delta)\dot\theta & 0 \end{bmatrix},$$

and $\tilde{\mathbf{g}}(\tilde{\mathbf{q}})$ is the vector of gravitational-restoring forces and moments

$$\tilde{\mathbf{g}}(\tilde{\mathbf{q}}) = \begin{bmatrix} 0 \\ \omega_y^2 y \\ (1+\delta)\,\omega_\theta^2 s_\theta \\ \omega_\theta^2(1-c_\theta) \end{bmatrix}.$$

15.2.2 Stability Analysis

In this section, a stability analysis under the assumption of external sinusoidal excitation is carried out, to determine the conditions to be infringed in order to trigger the resonant phenomenon.

Consider the mass-spring-damper (15.13) driven by $\Phi_y^e = \Phi_0 \cos \omega_e t$, and no control action is performed, that is $\Phi_x^c = \Phi_\delta^c = 0$. Then the semi-trivial solution of the system (15.12)–(15.15) is given by

$$x(t) = 0, \tag{15.17}$$
$$y(t) = Y_0 \cos(\omega_e t + \psi_y), \tag{15.18}$$
$$\theta(t) = 0, \tag{15.19}$$
$$\delta(t) = \Delta_0 \cos(\omega_e t + \psi_\delta), \tag{15.20}$$

where the pairs of parameters (Y_0, ψ_y) and (Δ_0, ψ_δ) can be found by substituting (15.18) and (15.20) into the linear system

$$\ddot{y} + \mu_2 \dot{y} + \omega_y^2 y - \alpha \ddot{\delta} = \Phi_0 \cos \omega_e t, \tag{15.21}$$
$$\ddot{\delta} + \mu_4 \dot{\delta} - \ddot{y} = 0. \tag{15.22}$$

The stability of the semi-trivial solution is investigated by looking at its behavior in a neighborhood defined as

$$x(t) = 0 + u_x(t), \tag{15.23}$$
$$y(t) = Y_0 \cos \omega_e t + u_y(t), \tag{15.24}$$
$$\theta(t) = 0 + u_\theta(t), \tag{15.25}$$
$$\delta(t) = \Delta_0 \cos \omega_e t + u_\delta(t), \tag{15.26}$$

where $u_x(t)$, $u_y(t)$, $u_\theta(t)$, and $u_\delta(t)$ are small perturbations, and the phase shifts ψ_y and ψ_δ have been arbitrarily set to zero. Substituting (15.23)–(15.26) into the system (15.12)–(15.15) and linearizing around the semi-trivial solution the following variational system in nondimensional time $\zeta = \frac{1}{2}\omega_e t$ is obtained

$$u_x'' + \tilde{\mu}_1 u_x' - \alpha\left(\sigma - 2\tilde{\mu}_4 \Delta_0 \sin(2\zeta)\right) u_\theta - \frac{\alpha \tilde{\mu}_3}{1 + \Delta_0 \cos(2\zeta)} u_\theta' = 0, \quad (15.27)$$

$$u_y'' + \tilde{\mu}_2 u_y' + \frac{4\omega_y^2}{\omega_e^2} u_y + \tilde{\mu}_4 u_\delta' = 0, \quad (15.28)$$

$$u_\theta'' + \frac{1}{1 + \Delta_0 \cos(2\zeta)}\left[\left(\frac{\tilde{\mu}_3}{1 + \Delta_0 \cos(2\zeta)} + 4(1-\alpha)\Delta_0 \sin(2\zeta)\right) u_\theta' \right.$$

$$\left. + (\sigma + \varepsilon \cos(2\zeta)) u_\theta - \tilde{\mu}_1 u_x'\right] = 0, \quad (15.29)$$

$$u_\delta'' + \tilde{\mu}_4 u_\delta' + \tilde{\mu}_2 u_y' + \frac{4\omega_y^2}{\omega_e^2} u_y = 0. \quad (15.30)$$

where $\sigma = \frac{4\omega_\theta^2}{\omega_e^2}$, $\varepsilon = \frac{4Y_0}{\omega_e^2}$, and $\tilde{\mu}_i = \frac{2\mu_i}{\omega_e}$. Equations (15.28) and (15.30) form a marginally stable linear system whose solution (u_y, u_δ) converges to $(0, \bar{u}_\delta)$ for ζ going to infinity. Therefore the stability of the overall system is solely determined by the (u_x, u_θ)-subsystem.

The (u_x, u_θ)-subsystem (15.27) and (15.29) is a linear periodic system of the form

$$\dot{z} = A(\zeta) z, \quad A(\zeta + T) = A(\zeta), \quad (15.31)$$

where $z = [u_x, u_x', u_\theta, u_\theta']^T$, and the time-varying dynamical matrix $A(t)$ is

$$A(\zeta) = \begin{bmatrix} 0 & 1 & 0 & 0 \\ 0 & -\tilde{\mu}_1 & \alpha(\sigma - 2\tilde{\mu}_4 \Delta_0 \sin(2\zeta)) & \frac{\alpha\tilde{\mu}_3}{1+\Delta_0\cos(2\zeta)} \\ 0 & 0 & 1 & 0 \\ 0 & \frac{\tilde{\mu}_1}{1+\Delta_0\cos(2\zeta)} & -\frac{\sigma+\varepsilon\cos 2\zeta}{1+\Delta_0\cos(2\zeta)} & -\frac{\tilde{\mu}_3}{(1+\Delta_0\cos(2\zeta))^2} - \frac{4(1-\alpha)\Delta_0\sin(2\zeta)}{1+\Delta_0\cos(2\zeta)} \end{bmatrix}, \quad (15.32)$$

whose entries are periodic functions of period $T = \pi$. According to Floquet theory [4] the system (15.31) admits three different kinds of solutions, that is stable, unstable, or periodic, depending on the characteristic multipliers associated to the system. Further, the overall stability of the (u_x, u_θ)-subsystem relies on the stability of the u_θ dynamics as shown by (15.27), which admits a solution $u_x \neq 0$ only if $u_\theta \neq 0$.

If we assume that $\Delta_0 \ll 1$ then (15.29) reduces to the linear damped Mathieu equation [9] linked to the cart dynamics through a velocity coupling. Therefore, it is plausible that the stability properties of the u_θ dynamics are quite similar to those of the Mathieu equation with damping. In order to confirm this assumption the Fourier spectral method [1, 15] is applied to numerically derive the stability chart of the system (15.31).

Figure 15.2 shows the stability diagram around the first region of instability derived for the following values of the system's parameters: $\tilde{\mu}_1 = 0.5$, $\tilde{\mu}_3 = 0.1$, $\tilde{\mu}_4 = 0.8$, $\Delta_0 = 0.02$, $\alpha \approx 0.09$. In particular, the dash-dotted lines are the transition

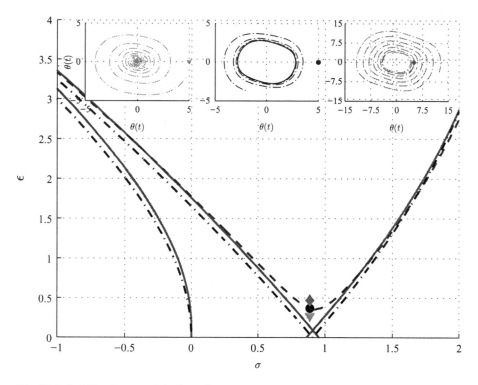

Fig. 15.2 Stability diagram of the (u_x, u_θ)-subsystem: stable, periodic, and unstable solutions can alternate depending upon the value of the parameter set (σ, ε)

curves of the linear Mathieu equation coupled with the cart dynamics with $\bar{\mu}_i = 0$; the solid lines are the transition curves of system (15.31) with $\bar{\mu}_i = 0$; the dashed lines are the transition curves of system (15.31) with damping coefficients equal to the values above mentioned.

By comparing the transition curves of the linear Mathieu equation (dash-dotted lines) with those of system (15.31) when damping is not included (solid lines) we can see that the effect of the time-varying rod length is to slightly push up the origin of the unstable tongue detaching it from the σ-axis. Moreover, as for the standard damped Mathieu equation, the effect of the damping is to increase the size of the stable regions. As expected, system (15.31) shows three different behaviors: stable if the (σ, ε) pair is below the transition curve; periodic if the (σ, ε) pair lies on the transition curve; unstable if the (σ, ε) pair is above the transition curves. The three different scenarios are illustrated in the inserts within Fig. 15.2.

Concluding, the stability of the (u_x, u_θ)-subsystem is determined by three parameters: the frequency tuning σ, which has to be close to 1 (i.e. $\omega_\theta \approx \frac{1}{2}\omega_e$) in the first region of instability; the system damping, which defines the smallest amplitude of the parametric excitation needed in order to trigger the resonance; the amplitude of the parametric excitation ε, which determines the magnitude of the system response.

15.3 Parametric Resonance Control

As the stability analysis pointed out, parametric resonance is an instability phenomenon, whose onset is due to the concurrent fulfillment of two conditions:

- The frequency of the parametric excitation is approximately equal to twice the natural frequency of the secondary system (principal parametric resonance condition).
- The amplitude of the parametric excitation is larger than the damping of the secondary system.

Moreover, the secondary system must be perturbed from its stable equilibrium point in order to trigger the resonance.

Control strategies aiming at stabilizing or inducing parametric resonance into the system have to act on the primary or secondary system such that the aforementioned conditions are failed or met, respectively. It is worth to note that the stabilization of parametric resonance can be achieved by not satisfying only one of the requirements, for example increasing the damping of the secondary system; however by failing both of them a faster convergence to a stable mode is obtained. Conversely, the induction of parametric resonance requires that both prerequisites are attained.

The authors decided to focus on the frequency coupling condition in order to both induce and stabilize the resonant oscillations, assuming that the damping condition is implicitly satisfied. In particular, the induction of parametric resonance is achieved by bringing the system into the principal parametric resonance region, that is, where $\omega_\theta = \frac{1}{2}\omega_e$.

15.3.1 Parametric Resonance Induction

Assuming that the frequency of the external excitation ω_e acting on the mass-spring-damper is retrievable by means of low-level signal processing, the induction of parametric resonance into the system (15.16) can be set up as an *output tracking problem*.

Problem 15.1. Let $\omega_I(t) = \frac{1}{2}\omega_e$ be the induction reference frequency at time t. Find a control law $\Phi_\delta^c = \Phi_\delta^c(\tilde{q}, \dot{\tilde{q}}, \omega_I(t))$ such that ω_θ converges asymptotically to the prescribed reference frequency trajectory $\omega_I(t)$.

The solution of the output tracking problem results in designing the control law Φ_δ^c such that the length of the pendulum rod δ converges to δ^* for $t \to \infty$, where $\delta^* = (4l_e - l_0)/l_0$ with l_e being the length of the rod of a virtual pendulum oscillating at the natural frequency ω_e. However this control action alone is not sufficient to trigger parametric resonance; in fact a small perturbation is necessary to bring the pendulum away from its stable equilibrium point $\theta = 0$. Problem 15.1 can then be reformulated as

Parametric Resonance Tracking. Given the system (15.16) and the induction reference frequency $\omega_l(t)$, find a control law $\tilde{\tau}^c = [\Phi_x^c(\tilde{\mathbf{q}}, \dot{\tilde{\mathbf{q}}}, x^*), \Phi_\delta^c(\tilde{\mathbf{q}}, \dot{\tilde{\mathbf{q}}}, \omega_l(t))]^\mathsf{T}$ such that ω_θ converges asymptotically to the prescribed frequency $\omega_l(t)$, and $\theta(t) \neq 0$ for all $t > t_c$, where t_c is the time instant where the control action starts.

The second control goal is achieved by arbitrarily changing the position of the cart along the X axis, with no specific preference about the value of the new set point or the direction of the motion. Therefore the controller goal in this case is limited to stabilize the cart around the new chosen position x^*.

Consider the multivariable nonlinear system described in state space form as

$$\dot{\mathbf{x}} = \mathbf{f}(\mathbf{x}) + \mathbf{b}(\mathbf{x})\mathbf{u} + \mathbf{p}(\mathbf{x})\mathbf{w}, \qquad (15.33)$$

$$\mathbf{y} = \mathbf{h}(\mathbf{x}) \qquad (15.34)$$

in which $\mathbf{x}$ is the state vector split into the position $\mathbf{x}_1 = [x_1, x_2, x_3, x_4]^\mathsf{T} \triangleq [x, y, \theta, \delta]^\mathsf{T}$ and the velocity $\mathbf{x}_2 = [x_5, x_6, x_7, x_8]^\mathsf{T} \triangleq [\dot{x}, \dot{y}, \dot{\theta}, \dot{\delta}]^\mathsf{T}$, $\mathbf{u} = [\Phi_x^c, \Phi_\delta^c]^\mathsf{T}$ is the vector of control inputs, $\mathbf{w} = \Phi_y^e$ is the disturbance. The following smooth vector fields defined in an open set of $\mathbb{R}^8$

$$\mathbf{f}(\mathbf{x}) \triangleq \begin{bmatrix} \mathbf{x}_2 \\ -\widetilde{\mathbf{M}}(\mathbf{x}_1)^{-1}(\widetilde{\mathbf{D}}\mathbf{x}_2 + \widetilde{\mathbf{C}}(\mathbf{x}_1, \mathbf{x}_2)\mathbf{x}_2 + \tilde{\mathbf{g}}(\mathbf{x}_1)) \end{bmatrix},$$

$$\mathbf{b}(\mathbf{x}) \triangleq \frac{1}{1-\alpha} \begin{bmatrix} 0\,0\,0\,0 & 1 & 0 & -\frac{\cos x_3}{1+x_4} & -\sin x_3 \\ 0\,0\,0\,0 & -\alpha \sin x_3 & \alpha \cos x_3 & 0 & 1 \end{bmatrix}^\mathsf{T},$$

$$\mathbf{p}(\mathbf{x}) \triangleq \frac{1}{1-\alpha} \begin{bmatrix} 0, & 0, & 0, & 0, & 0, & 1, & -\frac{\sin x_3}{1+x_4}, & \cos x_3 \end{bmatrix}^\mathsf{T}$$

describe the state dynamics, whereas the smooth functions

$$\mathbf{h}(\mathbf{x}) \triangleq [x_1, x_4]^\mathsf{T}$$

describe the output evolution.

Proposition 15.1. *The transformation of variables*

$$\mathbf{z} = \begin{bmatrix} \xi \\ \eta \end{bmatrix} = \mathbf{T}(\mathbf{x})$$

$$= \begin{bmatrix} x_1 \\ x_5 \\ x_4 \\ x_8 \\ x_2 \\ x_3 \\ \frac{\alpha x_8 \cos x_3 - \alpha(1+x_4)x_7 \sin x_3 - x_6}{\alpha \cos x_3} \\ x_7 + \frac{x_5 \cos x_3 + x_6 \sin x_3}{1+x_4} \end{bmatrix} \qquad (15.35)$$

is a local diffeomorphism on the domain $D_x = \left\{ \mathbf{x} \in \mathbb{R}^8 | x_3 \neq \frac{\pi}{2} + n\pi \wedge x_3 \neq \right.$
$\arcsin\left(\alpha^{-\frac{1}{2}}\right) \wedge x_4 \neq -1 \right\}$, *which brings the system* (15.33)–(15.34) *into the normal form*

$$\dot{\eta} = \mathbf{f}_0(\eta, \xi), \tag{15.36}$$

$$\dot{\xi} = \mathbf{A}_c \xi + \mathbf{B}_c \Gamma(\mathbf{x})[\mathbf{u} - v(\mathbf{x}) - \upsilon(\mathbf{x})], \tag{15.37}$$

$$\mathbf{y} = \mathbf{C}_c \xi, \tag{15.38}$$

where $\xi \in \mathbb{R}^4$, $\eta \in \mathbb{R}^4$, *and* $(\mathbf{A}_c, \mathbf{B}_c, \mathbf{C}_c)$ *is a canonical form representation of a chain of integrators.*

Proof The transformation of variables $\mathbf{T}(\mathbf{x})$ is obtained by exploiting the notion of vector relative degree and applying Proposition 5.1.2 in [6]. System (15.33)–(15.34) has vector relative degree $\{\rho_1, \rho_2\} = \{2, 2\}$ on the domain $D_1 = \left\{ \mathbf{x} \in \mathbb{R}^8 | x_3 \neq \right.$ $\arcsin\left(\alpha^{-\frac{1}{2}}\right) \right\}$; in fact by using the Lie derivative we obtain

$$L_{b_j} h_i = 0, \qquad \text{for } 1 \leq j \leq 2, \, 1 \leq i \leq 2$$

and

$$L_{b_1} L_f h_1(\mathbf{x}) = \frac{1}{1-\alpha}, \tag{15.39}$$

$$L_{b_1} L_f h_2(\mathbf{x}) = -\frac{\sin x_3}{1-\alpha}, \tag{15.40}$$

$$L_{b_2} L_f h_1(\mathbf{x}) = -\frac{\alpha \sin x_3}{1-\alpha}, \tag{15.41}$$

$$L_{b_2} L_f h_2(\mathbf{x}) = \frac{1}{1-\alpha}. \tag{15.42}$$

Moreover the matrix

$$\Gamma(\mathbf{x}) = \begin{bmatrix} L_{b_1} L_f h_1(\mathbf{x}) & L_{b_2} L_f h_1(\mathbf{x}) \\ L_{b_1} L_f h_2(\mathbf{x}) & L_{b_2} L_f h_2(\mathbf{x}) \end{bmatrix}$$
$$= \begin{bmatrix} \frac{1}{1-\alpha} & -\frac{\alpha \sin x_3}{1-\alpha} \\ -\frac{\sin x_3}{1-\alpha} & \frac{1}{1-\alpha} \end{bmatrix} \tag{15.43}$$

is nonsingular on D_1, since its determinant

$$\det(\Gamma(\mathbf{x})) = \frac{1 - \alpha \sin^2 x_3}{(1-\alpha)^2}$$

is zero if $x_3 = \arcsin\left(\alpha^{-\frac{1}{2}}\right)$. Therefore we can define the first four new variables as

$$\xi_1 \triangleq h_1(\mathbf{x}) = x_1, \tag{15.44}$$

$$\xi_2 \triangleq L_f h_1(\mathbf{x}) = x_5, \tag{15.45}$$

$$\xi_3 \triangleq h_2(\mathbf{x}) = x_4, \tag{15.46}$$

$$\xi_4 \triangleq L_f h_2(\mathbf{x}) = x_8 . \tag{15.47}$$

Since $\rho = \rho_1 + \rho_2 = 4$ and the state space is eight dimensional, it is possible to find four other functions $\eta_i(\mathbf{x})$ such that the mapping

$$\mathbf{T}(\mathbf{x}) = [\xi_1, \ldots, \xi_4, \eta_1, \ldots, \eta_4]^{\mathsf{T}}$$

is a local diffeomorphism. By noting that the distribution $B = \mathrm{span}\{b_1, b_2\}$ is involutive on D_1, we can determine the functions η_i by solving the set of linear partial differential equations

$$L_{b_j}\eta_i(\mathbf{x}) = 0$$

$$\Rightarrow \begin{cases} \frac{\partial \eta_i}{\partial x_5} - \frac{\cos x_3}{1+x_4}\frac{\partial \eta_i}{\partial x_7} - \sin x_3 \frac{\partial \eta_i}{\partial x_8} = 0 \\ -\alpha \sin x_3 \frac{\partial \eta_i}{\partial x_5} + \alpha \cos x_3 \frac{\partial \eta_i}{\partial x_6} + \frac{\partial \eta_i}{\partial x_8} = 0 \end{cases}. \tag{15.48}$$

A set of functions satisfying system (15.48) is given by

$$\eta_1 \triangleq x_2, \tag{15.49}$$

$$\eta_2 \triangleq x_3, \tag{15.50}$$

$$\eta_3 \triangleq \frac{\alpha x_8 \cos x_3 - \alpha(1+x_4)x_7 \sin x_3 - x_6}{\alpha \cos x_3}, \tag{15.51}$$

$$\eta_4 \triangleq x_7 + \frac{x_5 \cos x_3 + x_6 \sin x_3}{1+x_4} . \tag{15.52}$$

which is defined on the domain $D_2 = \left\{\mathbf{x} \in \mathbb{R}^8 | x_3 \neq \frac{\pi}{2} + n\pi \wedge x_4 \neq -1\right\}$. Therefore, the change of variables (15.35) qualifies as a diffeomorphism since its jacobian matrix is nonsingular on the domain $D_x = D_1 \cap D_2$.

Finally by applying the transformation $\mathbf{T}(\mathbf{x})$ to the system (15.33)–(15.34) we obtain the normal form (15.36)–(15.38) where $\Gamma(\mathbf{x})$ is given by (15.43) and

$$v(\mathbf{x}) = \begin{bmatrix} L_f^2 h_1(\mathbf{x}) \\ L_f^2 h_2(\mathbf{x}) \end{bmatrix}$$

$$= \begin{bmatrix} \frac{1}{1-\alpha}\left(-\mu_1 x_5 + \alpha\mu_4 x_8 \sin x_3 + \frac{\alpha\mu_3}{1+x_4} x_7 \cos x_3 + \alpha\omega_\theta^2 \sin x_3\right) \\ \frac{1}{1-\alpha}\left(\begin{array}{l} -\mu_4 x_8 + \mu_1 x_5 \sin x_3 - \mu_2 x_6 \cos x_3 \\ +(1-\alpha)(1+x_4)x_7^2 - \omega_\theta^2(1-\cos x_3) - \omega_y^2 x_2 \cos x_3 \end{array}\right) \end{bmatrix},$$

$$(15.53)$$

$$\upsilon(\mathbf{x}) = \begin{bmatrix} L_p L_f h_1(\mathbf{x}) \\ L_p L_f h_2(\mathbf{x}) \end{bmatrix}$$

$$= \begin{bmatrix} 0 \\ \frac{\cos x_3}{1-\alpha} \end{bmatrix}. \qquad (15.54)$$

□

Equation (15.54) shows that the disturbance Φ_y^e affects the output y_2, whereas the output y_1 is insensitive to it. Hence the control design should also address the problem of disturbance decoupling together with the tracking of parametric resonance.

Proposition 15.2. *Consider the system in normal form* (15.36)–(15.38), *and the reference vector* $\mathbf{y}_R = [y_{1R}(t), \dot{y}_{1R}(t), y_{2R}(t), \dot{y}_{2R}(t)]^T = [x_1^*(t), x_5^*(t), x_4^*(t), x_8^*(t)]^T$. *The input–output feedback linearizing control law*

$$\mathbf{u} = \Gamma(\mathbf{x})^{-1}(v(\mathbf{x}) + \upsilon(\mathbf{x})\mathbf{w} + \phi(\mathbf{x})),$$

where

$$\phi(\mathbf{x}) = \mathbf{v} + \ddot{\mathbf{y}}_R(t)$$

solves the Parametric Resonance Tracking *problem and decouples the disturbance from the output* y_2. *Moreover, the internal dynamics* $\dot{\eta} = \mathbf{f}_0(\eta, \xi)$ *is bounded for all* $t \geq 0$.

Proof. Let

$$\mathbf{e} = \begin{bmatrix} \xi_1 - y_{1R}(t) \\ \xi_2 - \dot{y}_{1R}(t) \\ \xi_3 - y_{2R}(t) \\ \xi_4 - \dot{y}_{2R}(t) \end{bmatrix} = \xi - \mathbf{y}_R$$

be the tracking error vector. Introducing the tracking error into the normal form (15.36) and (15.37) yields

$$\dot{\eta} = \mathbf{f}_0(\eta, \mathbf{e} + \mathbf{y}_R), \qquad (15.55)$$

$$\dot{\mathbf{e}} = \mathbf{A}_c \mathbf{e} + \mathbf{B}_c \Gamma(\mathbf{x}) \left\{ [\mathbf{u} - v(x) - \upsilon(\mathbf{x})] - \begin{bmatrix} \ddot{y}_{1R} \\ \ddot{y}_{2R} \end{bmatrix} \right\}. \qquad (15.56)$$

Hence the state feedback control

$$\mathbf{u} = \boldsymbol{\Gamma}(\mathbf{x})^{-1}(\nu(\mathbf{x}) + \upsilon(\mathbf{x})\mathbf{w} + \boldsymbol{\phi}(\mathbf{x})) \tag{15.57}$$

reduces the system (15.36)–(15.37) to the cascade system

$$\dot{\boldsymbol{\eta}} = \mathbf{f}_0(\boldsymbol{\eta}, \mathbf{e} + \mathbf{y}_R), \tag{15.58}$$

$$\dot{\mathbf{e}} = \mathbf{A}_c \mathbf{e} + \mathbf{B}_c \boldsymbol{\phi}(\mathbf{x}), \tag{15.59}$$

where

$$\boldsymbol{\phi}(\mathbf{x}) = -\mathbf{K}_I \mathbf{e} + \begin{bmatrix} \ddot{y}_{1R} \\ \ddot{y}_{2R} \end{bmatrix}. \tag{15.60}$$

By selecting the gain matrix $\mathbf{K}_I$ such that the matrix $\mathbf{A}_c - \mathbf{B}_c \mathbf{K}_I$ is Hurwitz, then the *Parametric Resonance Tracking* problem is solved.

The normal form (15.36) and (15.37) has an equilibrium point at $(\boldsymbol{\eta}, \boldsymbol{\xi}) = (0,0)$. In particular, the zero dynamics $\dot{\boldsymbol{\eta}} = \mathbf{f}_0(\boldsymbol{\eta}, 0)$ given by

$$\dot{\eta}_1(\eta, 0) = -\frac{\alpha \eta_3 \cos \eta_2 + \alpha \eta_4 \sin \eta_2}{1 - \alpha \sin^2 \eta_2}, \tag{15.61}$$

$$\dot{\eta}_2(\eta, 0) = \frac{\eta_4 + \frac{1}{2}\alpha \eta_3 \sin(2\eta_2)}{1 - \alpha \sin^2 \eta_2}, \tag{15.62}$$

$$\dot{\eta}_3(\eta, 0) = \frac{\mu_2(\eta_3 \cos \eta_2 + \eta_4 \sin \eta_2)}{(\alpha \sin^2 \eta_2 - 1)\cos \eta_2} - \frac{(\eta_4 + \frac{1}{2}\alpha \eta_3 \sin(2\eta_2))\eta_3 \sin(2\eta_2)}{2(\alpha \sin^2 \eta_2 - 1)\cos \eta_2}$$
$$+ \frac{\omega_y^2 \eta_1}{\alpha \cos \eta_2}, \tag{15.63}$$

$$\dot{\eta}_4(\eta, 0) = \frac{\eta_4 + \frac{1}{2}\alpha \eta_3 \sin(2\eta_2)}{\alpha \sin^2 \eta_2 - 1}\left(\mu_3 - \frac{(\eta_3 \cos \eta_2 + \eta_4 \sin \eta_2)\alpha \cos \eta_2}{\alpha \sin^2 \eta_2 - 1}\right)$$
$$- \omega_\theta^2 \sin \eta_2 \tag{15.64}$$

is locally asymptotically stable in $\eta = 0$. In fact by linearizing the zero dynamics around $\eta = 0$ we obtain the following matrix

$$\mathbf{A}_0 = \begin{bmatrix} 0 & 0 & -\alpha & 0 \\ 0 & 0 & 0 & 1 \\ \frac{\omega_y^2}{\alpha} & 0 & -\mu_2 & 0 \\ 0 & -\omega_\theta^2 & 0 & -\mu_3 \end{bmatrix}, \tag{15.65}$$

whose eigenvalues

$$\lambda_{1,2} = -\frac{1}{2}\mu_2 \pm \sqrt{\mu_2^2 - 4\omega_y^2}, \qquad (15.66)$$

$$\lambda_{3,4} = -\frac{1}{2}\mu_3 \pm \sqrt{\mu_3^2 - 4\omega_\theta^2} \qquad (15.67)$$

have always negative real part. Hence, applying Theorems 4.16 and 4.18 in [7] it follows that for sufficiently small initial conditions of the error and internal dynamics $\mathbf{e}(0)$, $\eta(0)$, and for a small reference trajectory $\omega_1(t)$ with small derivative, the state $\eta(t)$ will be bounded for all $t \geq 0$. $\qquad \square$

Remark 15.1. Note that the proof about the local boundedness of the internal dynamics is valid only if the reference trajectory is small. If this is not the case the time-varying nonlinear system

$$\dot{\eta} = \mathbf{f}_0(\eta, \mathbf{y}_R(t)) \qquad (15.68)$$

should be considered instead.

15.3.2 *Parametric Resonance Stabilization*

The choice of focusing on the frequency coupling condition allows to exploit the control law (15.57) also for stabilizing the system once parametric resonance has fully developed. This can be attained by defining a stabilizing trajectory $\omega_S(t) \neq \frac{1}{2}\omega_e$ to be tracked by the closed-loop system. The control problem can be formulated as

Parametric Resonance Stabilization. Assume that $\omega_e = 2\omega_\theta$ and that the system (15.16) is in parametric resonance. Given the stabilizing reference frequency $\omega_S(t) = \bar{\omega} \neq \frac{1}{2}\omega_e(t)$, find a control law $\Phi_\delta^c = \Phi_\delta^c(\tilde{\mathbf{q}}, \dot{\tilde{\mathbf{q}}}, \omega_S(t))$ such that ω_θ converges asymptotically to $\omega_S(t)$.

The solution of the *Parametric Resonance Stabilization* problem results in designing the control law Φ_δ^c such that the length of the pendulum rod δ converges to δ^* for $t \to \infty$, where $\delta^* = (\bar{l} - l_0)/l_0$ with $\bar{l}$ being the length of the rod of a virtual pendulum oscillating at the natural frequency $\bar{\omega}$.

Note that since the stabilization is achieved by detuning the frequency coupling condition $\omega_\theta = \frac{1}{2}\omega_e$ there is no need for controlling the horizontal position of the cart. Therefore in the following analysis it is assumed that the controller does not actuate the cart, which will remain at the position $x = x^*$ where it was originally placed.

Proposition 15.3. *Assume that* $\omega_e = 2\omega_\theta$ *and that the system in normal form* (15.36)–(15.37) *is in parametric resonance. Consider the reference vector* $\mathbf{y}_R = [y_{2R}(t), \dot{y}_{2R}(t)]^T = [x_4^*(t), x_8^*(t)]^T$. *The input–output feedback linearizing control law*

$$\mathbf{u} = \Gamma(\mathbf{x})^{-1}(\nu(\mathbf{x}) + \upsilon(\mathbf{x})\mathbf{w} + \phi(\mathbf{x})),$$

where

$$\phi(\mathbf{x}) = \mathbf{v} + \ddot{\mathbf{y}}_R(t)$$

solves the Parametric Resonance Stabilization *problem and decouples the disturbance from the output* y_2. *Moreover the internal dynamics* $\dot{\eta} = \mathbf{f}_0(\eta, \xi)$ *is bounded for all* $t \geq 0$.

Proof. Analogously to the proof of Proposition 15.2 we define the tracking error as

$$\mathbf{e} = \begin{bmatrix} 0 \\ 0 \\ \xi_3 - y_{2R}(t) \\ \xi_4 - \dot{y}_{2R}(t) \end{bmatrix},$$

where the first two entries are equal to zero because the cart is assumed to maintain its position. The normal form (15.36) and (15.37) then reads

$$\dot{\eta} = \mathbf{f}_0(\eta, \mathbf{e} + \mathbf{y}_R), \tag{15.69}$$

$$\dot{\mathbf{e}} = \mathbf{A}_c\mathbf{e} + \mathbf{B}_c\Gamma(\mathbf{x})\left\{[\mathbf{u} - \nu(x) - \upsilon(\mathbf{x})] - \begin{bmatrix} \ddot{y}_{1R} \\ \ddot{y}_{2R} \end{bmatrix}\right\}. \tag{15.70}$$

Therefore the state feedback control law

$$\mathbf{u} = \Gamma(\mathbf{x})^{-1}(\nu(\mathbf{x}) + \upsilon(\mathbf{x})\mathbf{w} + \phi(\mathbf{x})) \tag{15.71}$$

with $\phi(\mathbf{x}) = -\mathbf{K}_S\mathbf{e} + [0, \ddot{y}_{2R}]^T$ reduces the normal form to the cascade

$$\dot{\eta} = \mathbf{f}_0(\eta, \mathbf{e} + \mathbf{y}_R), \tag{15.72}$$

$$\dot{\mathbf{e}} = (\mathbf{A}_c - \mathbf{B}_c\mathbf{K}_S)\mathbf{e}, \tag{15.73}$$

where $\mathbf{A}_c - \mathbf{B}_c\mathbf{K}_S$ is Hurwitz.

The boundedness of the internal dynamics $\eta(t) = \mathbf{f}_0(\eta, \xi)$ can be demonstrated analogously as in Proposition 15.2. □

15.4   Simulation Results

The efficacy of the proposed control strategies for inducing and stabilizing parametric resonance has been tested in simulation.

Figure 15.3 shows an example of induction and tracking of parametric resonance. For $0 < t < 100$ s the mass-spring-damper oscillates under the action of the external sinusoidal disturbance Φ_y^e while the pendulum is at rest. At $t = 100$ s a new reference trajectory $\delta^*(t)$, which ensures the tuning of the frequency coupling condition is provided. As a consequence, the controller (15.57) enforces that the output $y_2(t)$ follows the reference trajectory and that the frequency condition for the onset of parametric resonance is fulfilled. At the same time the controller destabilizes the pendulum by driving the cart to its new set point x^*. This produces the sparkle for the onset of parametric resonance into the pendulum, which for $100 \leq t < 400$ s develops oscillations of increasing amplitude at a frequency $\omega_\theta = \frac{1}{2}\omega_{e,1}$. At $t = 400$ s the frequency of the external excitation decreases determining a temporary frequency ratio $\omega_\theta / \omega_{e,2} = 1$ as shown in Fig. 15.4. Hence the controller increases the rod's length δ up to the new reference trajectory maintaining the parametric resonance alive. The amplitude of the pendulum oscillations further increases due

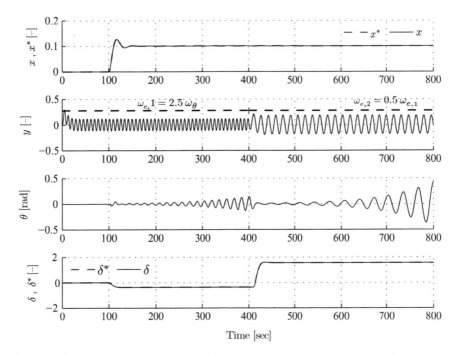

Fig. 15.3 Parametric resonance tracking: multiple variations of external excitation frequency ω_e are tracked by the controller

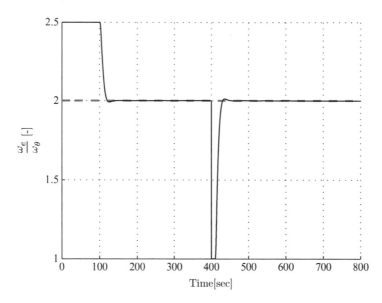

Fig. 15.4 Parametric resonance tracking: the controller enforces that the frequency ratio ω_e/ω_θ is kept equal to 2

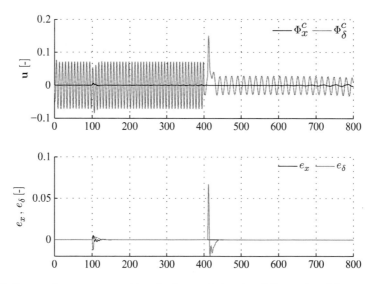

Fig. 15.5 Parametric resonance tracking: (*top*) control signals and (*bottom*) position errors of the controlled variables

to the larger amplitude of the parametric excitation provided by y. Figure 15.5 illustrates the commanded control signals and the evolution of the position error.

Figure 15.6 shows an example of stabilization of parametric resonance after it has been triggered according to the former description. At $t = 600$ s a stabilizing

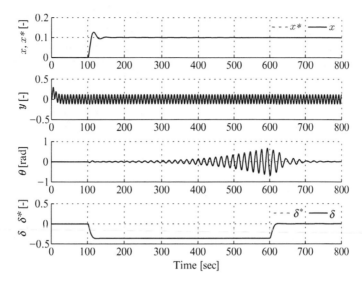

Fig. 15.6 Parametric resonance stabilization: the controller stabilizes the pendulum about $\theta = 0$ by detuning the frequency coupling condition

trajectory that detunes the frequency coupling condition is provided. Consequently, the controller (15.71) enforces the output y_2 to track the new reference signal, and it brings the pendulum out of the principal parametric resonance condition within 200 s. The decay rate of the pendulum oscillations could be increased if the proposed control strategy is coupled with a damping injection into the secondary system. This could be done by increasing the moment due to dissipative forces, for example, by applying a direct torque on the pendulum's pivot point or by moving the cart in counter phase with respect to the pendulum oscillations.

15.5 Conclusions

Parametric resonance is a widespread phenomenon that may be threatening or beneficial according to the particular system where it takes place. A four degrees-of-freedom Lagrangian model of a pendulum with moving support has been derived in order to, first, revisit some of the stability theory of autoparametric resonant systems by applying Floquet theory, and, second, to design control strategies to induce and stabilize the unstable oscillations. Two control problems, namely the *Parametric Resonance Tracking* and the *Parametric Resonance Stabilization*, have been set up as output tracking problems where induction and stabilization of parametrically resonant behaviors are achieved by tracking a reference frequency, which enforces or not the frequency coupling condition $\omega_\theta = \frac{1}{2}\omega_e$. An input–output feedback linearizing controller has been designed and analytically proven to solve both control problems. The efficacy of the proposed control strategy has been also verified in simulation, where both induction and stabilization of parametric resonance into the pendulum with moving support have been successfully obtained.

The authors believe that control strategies aiming at inducing and tracking parametric resonance will be of particular interest for future application as energy conversion systems where the obvious goal is to increase the energy throughput while maintaining constant or even reducing the effort to produce such energy. In this respect parametric resonance can become a very useful phenomenon since very large oscillations can be generated by a rather small parametric excitation. However the same feature that makes parametric resonance appealing should create awareness of the potential danger hidden in such phenomenon. That is why a sound and profound knowledge of the dynamics of the system where parametric resonance is wished to be induced is needed in order to capture the energy flow between the different system modes when the resonance takes places. Therefore models and control methods which rely on the concept of energy exchange in the system and through interacting systems seems to be particularly suited for these kinds of applications.

References

1. Boyd, J.P.: Chebyshev and Fourier Spectral Methods. DOVER Publications (2000)
2. Breu, D., Fossen, T.I.: Extremum seeking speed and heading control applied to parametric roll resonance. In: Proceedings of the 8th IFAC Conference on Control Applications in Marine Systems. IFAC, Germany (2010)
3. Galeazzi, R., Holden, C., Blanke, M., Fossen, T.I.: Stabilization of parametric roll resonance by combined speed and fin stabilizer control. In: Proceedings of the 10th European Control Conference. EUCA, Hungary (2009)
4. Grimshaw, R.: Nonlinear Ordinary Differential Equations. CRC Press, United States of America (1993)
5. Holden, C., Galeazzi, R., Perez, T., Fossen, T.I.: Stabilization of parametric roll resonance with active U-tanks via Lyapunov control design. In: Proceedings of the 10th European Control Conference. EUCA, Hungary (2009)
6. Isidori, A.: Nonlinear Control Systems – Third Edition. Springer, Great Britain (1995)
7. Khalil, H.K.: Nonlinear Systems – third edition. Prentice Hall, United States of America (2002)
8. Limebeer, D.J.N., Sharp, R.S., Evangelou, S.: Motorcycle steering oscillations due to road profiling. Transacations of the ASME **69**, 724–739 (2002)
9. Mathieu, E.: Mémoire sur le movement vibratoire d'une membrane de forme elliptique. Journal de Mathématiques Pures et Appliquées **13**, 137–203 (1868)
10. Olvera, A., Prado, E., Czitrom, S.: Performance improvement of OWC systems by parametric resonance. In: Proceedings of the 4th European Wave Energy Conference (2001)
11. Olvera, A., Prado, E., Czitrom, S.: Parametric resonance in an oscillating water column. Journal of Engineering Mathematics **57(1)**, 1–21 (2007)
12. Oropeza-Ramos, L.A., Burgner, C.B., Turner, K.L.: Robust micro-rate sensor actuated by parametric resonance. Sensors and Actuators A: Physical **152**, 80–87 (2009)
13. Shaeri, A., Limebeer, D.J.N., Sharp, R.S.: Nonlinear steering oscillations of motorcycles. In: Proceedings of the 43rd IEEE Conference on Decision and Control. IEEE, United States of America (2004)

14. Tondl, A., Ruijgrok, T., Verhulst, F., Nabergoj, R.: Autoparametric Resonance in Mechanical Systems. Cambridge University Press, United States of America (2000)
15. Trefethen, L.N.: Spectral methods in MATLAB. SIAM (2000)
16. Turner, K.L., Miller, S.A., Hartwell, P.G., MacDonald, N.C., Strogatz, S.H., Adams, S.G.: Five paramrametric resonances in a microelectromechanical system. Nature **396**, 149–152 (1998)
17. Zhang, W., Turner, K.L.: Application of parametric resonance amplification in a single-crystal silicon micro-oscillator nased mass sensor. Sensors and Actuators A: Physical **122**, 23–30 (2005)

Index

T.I. Fossen and H. Nijmeijer (eds.), *Parametric Resonance in Dynamical Systems*,
DOI 10.1007/978-1-4614-1043-0, © Springer Science+Business Media, LLC 2012

CPSIA information can be obtained at www.ICGtesting.com
Printed in the USA
LVOW100647170712

290371LV00007B/15/P